Rajeev Arora, PhD
Editor

Adaptations and Responses of Woody Plants to Environmental Stresses

Pre-publication REVIEWS, COMMENTARIES, EVALUATIONS . . .

"VERY INFORMATIVE. . . . Provides both an historical overview and recent developments in the understanding of both cold acclimation and dormancy and the factors that affect these processes. There is a good balance between cold acclimation and dormancy plus two chapters on nitrogen and one on pathogen-related proteins."

L. V. Gusta, PhD
Professor
Department of Plant Sciences
University of Saskatchewan

More pre-publication
REVIEWS, COMMENTARIES, EVALUATIONS . . .

"Presents STATE-OF-THE-ART RESEARCH and understanding of plant hardiness in terms of molecular biology, biochemistry, and physiology. . . . Provides readers with a view of recent advances in perennial woody plant hardiness research. Five chapters review freezing stress. . . . Three chapters review how the plant initiates its endodormancy/cold acclimation. . . . Other chapters deal with the role of bark storage proteins, nitrogen fixation in non-legume woody perennials, and pathogenesis-related proteins."

Paul H. Li, PhD
Professor
Horticultural Science
& Plant Biological Sciences
University of Minnesota

Food Products Press
An Imprint of The Haworth Press, Inc.

Adaptations and Responses of Woody Plants to Environmental Stresses

Adaptations and Responses of Woody Plants to Environmental Stresses

Editor
Rajeev Arora, PhD

International Book Distributing Co.
(Publishing Division)

Published by

INTERNATIONAL BOOK DISTRIBUTING CO.
(Publishing Division)
Chaman Studio Building, 2nd Floor,
Charbagh, Lucknow 226 004 U.P. (INDIA)
Tel. : Off. : 2450004, 2450007, 2459058 Fax : 0522-2458629
E-Mail : ibdco@sancharnet.in

First Indian Reprint 2006

Price: Rs 1155

ISBN 81-8189-107-4

Published under arrangement with Food Products Press®, an imprint of The Haworth Press, Inc., 10 Alice Street, Binghamton, NY 13904-1580 USA

Printed at:
Efficient Offset Printers
215, Shahzada Bagh Ind. Complex,
Phase-II, Delhi - 110 035
Phone : 25933736, 25933762

Adaptations and Responses of Woody Plants to Environmental Stresses

CONTENTS

Introduction
Rajeev Arora

Ice Nucleation, Propagation, and Deep Supercooling in Woody Plants
M. Wisniewski
M. Fuller
J. Palta
J. Carter
R. Arora

Characterization of a Novel Y_nSK_n Class of Dehydrin-Like cDNAs from Cold Acclimated Red-Osier Dogwood (*Cornus sericea* L.) Xylem
E. Sarnighausen
D. T. Karlson
Y. Zeng
P. B. Goldsbrough
K. G. Raghothama
E. N. Ashworth

Inheritance of Cold Hardiness and Dehydrin Genes in Diploid Mapping Populations of Blueberry 37
Ganesh R. Panta
Lisa J. Rowland
Rajeev Arora
Elizabeth L. Ogden
Chon-Chong Lim

Molecular Genetic and Physiological Analysis of the Cold-Responsive Dehydrins of Blueberry 53
L. J. Rowland
G. R. Panta
S. Mehra
C. Parmentier-Line

Photoperiodic Regulation of Apical Growth Cessation in Northern Tree Species: The Role of Phytochrome and Gibberellin 77
Jorunn E. Olsen
John B. Jensen
Jörgen A. Mölmann
Arild Ernstsen
Olavi Junttila

Cell-Cell Communication as a Key Factor in Dormancy Cycling 113
Päivi L. H. Rinne
Christiaan van der Schoot

Hormones and Endodormancy Induction in Woody Plants 157
Karen K. Tanino

Freezing Tolerance and Injury in Grapevines 201
Anne Fennell

Physiology and Regulation of Seasonal Nitrogen Cycling in Woody Plants 237
Gary D. Coleman

Physiology and Molecular Biology of a Family of Pathogenesis-Related PR-10 Proteins in Conifers 261
Abul K. M. Ekramoddoullah

Nitrogen Fixation as a Stress-Avoidance Strategy Among Actinorhizal (Non-Legume) Trees and Shrubs 281
Heidi A. Kratsch
William R. Graves

Index 305

ABOUT THE EDITOR

Rajeev Arora, PhD, is Associate Professor in the Department of Horticulture at Iowa State University. Dr. Arora is the author or co-author of over 50 contributions to scientific journals and is Associate Editor (Developmental Physiology) of *Journal of American Society for Horticultural Science*. He is also on the editorial board of *Tree Physiology*. Primary focus of Dr. Arora's research program is to understand the cellular physiology, molecular biology and genetics of plant's response to low temperature stress. His research on the specific membrane perturbations during freeze-thaw injury in herbaceous plants and physiological and genetic analysis of cold acclimation and endodormancy in woody perennials has been published in the leading plant biology journals and is widely cited. Dr. Arora is a member of American Society for Horticultural Science (ASHS), American Society of Plant Biologists (ASPB) and American Rhododendron Society, and has presented several invited seminars at national and international conferences on Plant Cold Hardiness research. He has received Outstanding Paper/Research Presentation Awards from USDA-ARS and ASHS and is a recipient of Gamma Sigma Delta Faculty Merit Award.

Introduction

Perennial woody plants are subjected to a variety of stresses (both biotic and abiotic) throughout the year. Cold winters, hot and dry summers, insect pests, and diseases–all take their toll on fruit trees, forest plantings, and especially those landscape trees and shrubs that grow in poor edaphic conditions of urban situations and/or are exposed to stressful city microclimates. Woody plants are enormously important and beneficial to our well-being (as source of food, fiber, lumber, pulp, recreation and visual aesthetics, just to name few). Therefore, we must gain fundamental understanding (at molecular level) of how these plants respond to various stresses, and strive to develop tougher plants with improved resistance to environmental stress and/or the strategies to protect them from stressful conditions.

Lack of cold tolerance is the single most limiting factor for survival of woody plants in various parts of the United States and in other northern climates of the world. Whereas significant breakthroughs progress has been made in the molecular understanding of plant responses to low temperature stress in herbaceous plants, particularly *Arabidopsis* model system, progress on this front with woody perennials has been relatively slow and limited. Molecular and genetic studies with trees have been hampered due to many limitations that are unique to working with woody perennial systems such as lack of characterized mutants, long generation times, heterozygosity of various traits in parental populations, ability to tolerate little inbreeding, and complexities associated with the study of cold acclimation in woody perennials; the latter arising from the simultaneous seasonal transitions in dormancy and cold hardi-

[Haworth co-indexing entry note]: "Introduction." Arora, Rajeev. Co-published simultaneously in *Journal of Crop Improvement* (Food Products Press, an imprint of The Haworth Press, Inc.) Vol. 10, No. 1/2 (#19/20), 2004, pp. 1-4; and: *Adaptations and Responses of Woody Plants to Environmental Stresses* (ed: Rajeev Arora) Food Products Press, an imprint of The Haworth Press, Inc., 2004, pp. 1-4. Single or multiple copies of this article are available for a fee from The Haworth Document Delivery Service [1-800-HAWORTH, 9:00 a.m. - 5:00 p.m. (EST). E-mail address: docdelivery@haworthpress.com].

http://www.haworthpress.com/web/JCRIP

Digital Object Identifer: 10.1300/J411v10n01_01

ness status during the annual growth cycle making it difficult to study the two events separately.

The goal of this special thematic issue is to cover recent advancements in the fundamental understanding of the various aspects of woody plants response to environmental stresses. Authors of the manuscripts in this issue are world-renowned researchers who have made significant contributions to the knowledge of woody plant stress physiology and molecular biology. It is my sincere hope that this publication will provide readers with some of the most significant advances in woody plant stress research.

Michael Wisniewski (USDA-ARS, West Virginia, USA), thorough his manuscript, introduces us to the use of infrared thermography to reveal the ways ice nucleation and its propagation occurs in plant tissues and how structural features of tissues can affect these processes. Direct visualization of the freezing process in intact plant tissues both in field and lab can provide some valuable insights into devising freeze-protection strategies. A series of papers (three in total) document the importance of a unique class of stress proteins called 'dehydrins' in cold hardening of blueberry and red-osier dogwood. Panta et al. (USDA, Maryland, USA) describe the use of mapping populations of diploid blueberry to study co-segregation of *dehydrin* genes and cold hardiness in segregating progenies whereas Rowland et al. (USDA, Maryland, USA) present their work on the cloning and expression studies of unique *dehydrin* genes in blueberry germplasm differing in cold acclimation abilities. Sarnighausen et al. (Purdue University, USA) implicate a novel *dehydrin* gene in cold hardening (acclimation) of xylem tissues of extremely cold-hardy shrub red-osier dogwood. Studies with *Arabidopsis* model system have recently revealed that dehydrins are part of the cold-regulon (a suite of genes under control of cold responsive CBF regulatory genes) whose induction is sufficient to confer cold tolerance under non-inductive conditions. These findings have put dehydrins on the center stage of cold acclimation research. Work presented in this issue, by above mentioned authors, further extends the dehydrin research in the realm of woody perennials where cold acclimation is much more complex process than in herbaceous/model plant systems.

Three articles of this issue focus on recent advances made in our understanding of the biology of growth cessation and dormancy cycling (during an annual cycle) in woody perennials vis-à-vis the seasonality of cold acclimation and deacclimation. Olsen et al. (University of Finland, Helsinki, Finland), besides reviewing recent literature in this field, discuss recent findings of their work with transgenic hybrid aspen (over expressing oat *PHYA* gene) and suggest that growth cessation and initiation of cold hardening in trees in response to short photoperiods in fall can be controlled through phytochrome and GA mediated systems. Rinne and van der Shoot (Agricultural University of Norway, Norway) describe how cell-cell communication in the apical shoot meristems may regulate photoperiodically induced dormancy induction, and its maintenance or release in woody plants. Their data suggest that reversible blockage of cell-cell communication (due to closing of plasmodesmatal entrances) might be central to dormancy cycling in birch and poplar. This newly acquired knowledge may facilitate future efforts to diagnose and manipulate endodormancy in woody plants and, in turn, transitions in low temperature stress tolerance. A comprehensive review by Karen Tanino (University of Saskatchewan, Canada) provides an up-to-date analysis and synthesis of literature on the hormonal aspects of dormancy control in woody plants and should provide the reader with the current status of this field. Anne Fennell focuses her research on the physiological and molecular understanding of low temperature stress tolerance and dormancy cycling in grapes, the most important fruit crop in the world. In her manuscript she reviews the mechanisms of freezing tolerance in grapevines and how they are influenced by genotypic, phonological, and environmental factors.

The article by Gary Coleman (University of Maryland, USA) describes molecular biology of seasonal nitrogen cycling. Past research indicates that majority of leaf nitrogen of over wintering woody perennials is translocated from senescing leaves during autumn to the root and shoot bark and stored as bark storage proteins (BSPs). Gary Coleman describes here that *bspA* expression is a photoperiod response mediated via phytochrome and involves Ca^{2+} and protein kinases as part of signaling pathways. Abul Ekramoddoullah (Canadian Forest Service, Canada) studies physiology and molecular biology of PR-10 proteins in conifers and reviews his work on the seasonal and pathogen-responsive expression patterns of members of PR-10 gene family and ultra structural localization of these proteins in response to blister rust pathogen. He also discusses the possibility that since PR-10 genes are induced by diverse stimuli such as cold and pathogens, their expression could serve as mechanistic link, at the molecular level, between biotic and abiotic stress tolerance in pine. Finally, Kratsch and Graves (Iowa State University, USA) focus on the nitrogen-fixation as a strategy used by certain woody plants to better tolerate, or altogether avoid, various biotic and abiotic stresses. Their paper emphasizes biological aspects of actinorhizal symbioses (in contrast with leguminous symbioses) and the horticultural potential of temperate woody species that form this symbioses vis-à-vis their use in urban landscapes and other stressful environments.

It was a great pleasure working with the authors who contributed to this issue. They all submitted their contributions on schedule and responded promptly and positively to reviewers' comments. I wish to thank all of my colleagues, who served as reviewers, for their generosity and for providing useful comments. This book would not have been possible without the outstanding cooperation and hard work by the editorial and administrative staff of Haworth Press/Food Products Press. I sincerely appreciate the opportunity to bring out this thematic issue.

Finally, I owe special thanks to my wife Usha and daughters, Uttara and Shivgami, for their love, support, and patience that were vital for the success of this volume.

Rajeev Arora
Department of Horticulture
Iowa State University
Ames, IA 50014

Ice Nucleation, Propagation, and Deep Supercooling in Woody Plants

M. Wisniewski
M. Fuller
J. Palta
J. Carter
R. Arora

SUMMARY. The response of woody plants to freezing temperatures is complex. Species vary greatly in their ability to survive freezing temperatures and the resulting dehydrative and mechanical stresses that occur as a result from the presence of ice. Initially, this is presented by the ability to inhibit the formation of ice (ice nucleation) by supercooling. Significant questions exist about the role of internal and external ice nucleating

M. Wisniewski is Plant Physiologist, U.S. Department of Agriculture, Agricultural Research Service, Appalachian Fruit Research Station, Kearneysville, WV 25430 (E-mail: mwisniew@afrs.ars.usda.gov).

M. Fuller is affiliated with Seale-Hayne Faculty of Agriculture, University of Plymouth, Newton Abbot, UK.

J. Palta is affiliated with the Department of Horticulture, University of Wisconsin, Madison, WI 53706.

J. Carter is affiliated with the Department of Horticulture, University of Minnesota, St. Paul, MN 55108.

R. Arora is affiliated with the Department of Horticulture, Iowa State University, Ames, IA 50011.

Journal paper No. J-19880 of the Iowa Agriculture and Home Economics Experiment Station, Ames, Iowa, Project No. 3601, and, in part, by Hatch Act and State of Iowa funds.

[Haworth co-indexing entry note]: "Ice Nucleation, Propagation, and Deep Supercooling in Woody Plants." Wisniewski, M. et al. Co-published simultaneously in *Journal of Crop Improvement* (Food Products Press, an imprint of The Haworth Press, Inc.) Vol. 10, No. 1/2 (#19/20), 2004, pp. 5-16; and: *Adaptations and Responses of Woody Plants to Environmental Stresses* (ed: Rajeev Arora) Food Products Press, an imprint of The Haworth Press, Inc., 2004, pp. 5-16. Single or multiple copies of this article are available for a fee from The Haworth Document Delivery Service [1-800-HAWORTH, 9:00 a.m. - 5:00 p.m. (EST). E-mail address: docdelivery@haworthpress.com].

http://www.haworthpress.com/web/JCRIP

Digital Object Identifer: 10.1300/J411v10n01_02

agents in determining the extent to which any particular plant can supercool. Additionally, little is known about how plant structure can affect ice nucleation and propagation. In this review, the ability of high-resolution infrared thermography to reveal significant details about the freezing process is demonstrated. In general, the presence of effective, intrinsic nucleators appear to be common in woody plants. The nucleators appear to be as effective as external ice nucleators and induce stems to freeze at warm, subzero temperatures. Barriers appear to exist, however, that prevent ice propagation into lateral appendages such as buds, or newly extended primary tissues. Deep supercooling represents a unique adaptation of woody plants to avoid freezing injury by dramatically suppressing ice formation in specific tissues. The extent of suppression is limited by the homogeneous nucleation temperature of water (−38°C) and therefore deep supercooling is characteristic of moderately hardy woody plants. In contrast, it has been proposed that the most cold-hardy woody plants have the ability to form glassed solutions. These solutions are very stable as long as the cell remains below the melting temperature of the glass and so allows tissues to become relatively impervious to the stresses associated with extremely low temperatures.

KEYWORDS. Infrared thermography, frost protection, cold acclimation, antifreeze proteins

INTRODUCTION

In the course of a year, perennial plants in temperate climates are exposed to several types of freezing stress including low temperature extremes and unseasonable episodes of frost. In response to short daylength and low temperature, woody plants cold acclimate (Wisniewski and Arora, 1993; Sakai and Larcher, 1987; Artlip and Wisniewski, 2001). This is a dynamic process that not only exhibits a distinct seasonality but the mechanisms that confer cold hardiness in midwinter may differ from those in late winter or early spring. This is further complicated by the fact that in many perennial woody plants can respond very differently to freezing temperatures.

Freezing injury can occur in plants only after the formation of ice within their tissues. Knowledge of the patterns of ice nucleation and propagation is important to understanding freeze-stress-resistance mechanisms in plants (Workmaster et al., 1999). While ice formation in plants has been studied by thermal analyses using thermocouples, this technique does not yield direct data about either the location of ice initiation or the temperature at that specific time point. In contrast, infrared thermography has recently been used to visualize ice nucleation and propagation in plants (Wisniewski et al., 1997; Wisniewski and Fuller, 1999; Wisniewski et al., 2001; Fuller and Wisniewski, 1998; Pearce and Fuller, 2001). This technique allows for the direct visualization of the freezing process in intact plants under both laboratory and field conditions so one can obtain data on the number and location of initial freezing events, whether or not they occurred on the surface or internally, the temperature at which ice formed, and the direction and rate of ice propagation. Data from these studies relevant to ice formation in woody plants will be discussed.

Also unique to the buds and xylem tissues of several species of temperate woody plants is the ability to avoid freezing stress by deep supercooling. A comprehensive list of woody species that exhibit this trait has been presented by George et al. (1982). During a deep supercooling event, cellular water remains liquid within cells at very low temperatures by remaining isolated from heterogeneous ice nuclei and the nucleating effect of extracellular ice. Supercooled water is in a metastable condition and will form intracellular ice in response to a heterogeneous nucleation event or when the homogeneous nucleation temperature (−38°C) of water is reached. A general review of this topic will also be presented.

ICE NUCLEATION AND PROPAGATION

The subject of ice nucleation in plants has been comprehensively reviewed by Ashworth and Kieft (1995). Ice formation can be induced by either intrinsic or extrinsic nucleating sources. Once ice has been initiated, barriers and avenues to ice propagation, both within the plant and at the plant surface, determine the subsequent pattern of ice formation and are a factor in the survival mechanism exhibited by a given plant tissue or organ. Barriers to ice propagation within a plant can be permanent or temporary (Workmaster et al., 1999), i.e., they can prevent or just delay the growth of ice crystals. Some barriers may also only be present at particular times of the year, or upon completion of particular stages of development. The inability of ice to propagate into buds, due to a lack of differentiated vascular tissue (Ashworth, 1982; Quamme et al., 1995) is an example of such a temporal barrier. In wheat (*Triticum aestivum* L.) ice propagation was impeded for several hours at stem and rachis nodes (Single, 1964).

Wisniewski et al. (1997) demonstrated in a comprehensive manner that infrared thermography could reveal details about the freezing process that could not be easily obtained with the use of thermocouples. Using this technology, they were able to demonstrate the freezing of water droplets on the surface of plants occurring as independent events, and that droplets containing populations of the ice-nucleating-active bacterium, *Pseudomonas syringae*, would freeze at warmer temperatures than plain water alone. The presence of these water droplets could induce open flowers of peach (*Prunus persica* L. Batsch) and apple (*Malus domestica* L.) to freeze. However, the presence of efficient intrinsic nucleators, that were active at warm sub-zero temperatures, in these plants was also documented. In most cases, ice would be initiated in the outer bark of stem tissues and then propagate into expanded primary tissues such as flower buds or young shoots, despite the fact that these tissues were supercooled to lower temperatures (due to radiative cooling) than the stem tissues themselves. An example of ice propagation in a stem of peach is illustrated in Figure 1. Ice was initiated in the bark tissues and progressed more rapidly around the stem than across it. Importantly, this study also demonstrated that thick leaf cuticles could also act as a barrier to ice propagation. Water droplets frozen on the surface of leaves of rhododendron (*Rhododendron* sp.) could not induce the leaves to freeze. Rather, ice formation was initiated inside the stem and propagated into the lateral appendages (Figure 2). Preliminary studies have indicated that infrared thermography could also be used to study the freezing process in conifers (Sutinen et al., 2001).

In a detailed study of the freezing of cranberry (*Vaccinium macrocarpon* Ait.) uprights and fruit, Workmaster et al. (1999) used infrared thermography to reveal unique details about the freezing process. Their results indicated that leaves were nucleated by ice penetration via stomata located on the abaxial surface and that the thick cuticle present on the adaxial surface was an effective barrier to extrinsic nucleation. Furthermore, cranberry fruit have the ability to supercool due to the presence of barriers present at both the pedicel and fruit surface. Only frozen moisture present at the calyx end of the fruit in the remnant nectary could induce the fruit to freeze, most likely through stomata. Leaves only froze when ice was present on the abaxial surface. Once initiated, ice propagated to the stem and then readily to other leaves. In both unripe and ripe fruit, ice propagation from the stem to the fruit was never observed

FIGURE 1. Infrared video images of ice propagation across a stem section of peach. Ice formation, visible as light area, was initiated in the outer bark tissues (A) and then slowly propagated across the stem (B and C).

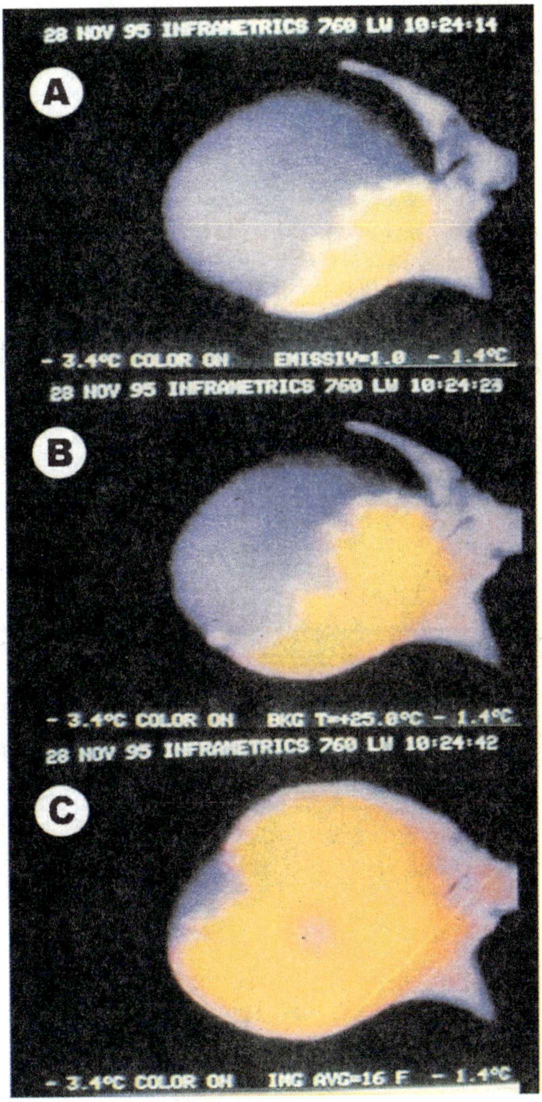

FIGURE 2. Infrared video images of ice nucleation and propagation in a detached shoot section of rhododendron cv. Olga. A frozen droplet of water containing ice-nucleating active bacteria can be seen as a bright spot on the surface of an upper leaf. Despite the presence of the frozen droplet, ice formation was initiated internally in a lower portion of the stem. In (A), the upper leaflets have already begun to freeze while the two lowest visible leaves have not begun to freeze. The freezing event can be seen as a warming of the leaf tissue due to the heat of fusion that is released when water undergoes a phase transition from a liquid to a solid. The camera temperature range is presented at the bottom of each picture. Brighter colors represent warmer temperatures while the darker colors represent colder temperatures within the set temperature range. In B, ice can be seen propagating in the mid vein of the lower leaf (arrow). In C and D, all leaves have frozen and the heat released during the freezing process is beginning to dissipate and the leaf temperature is beginning to approach the ambient, background temperature.

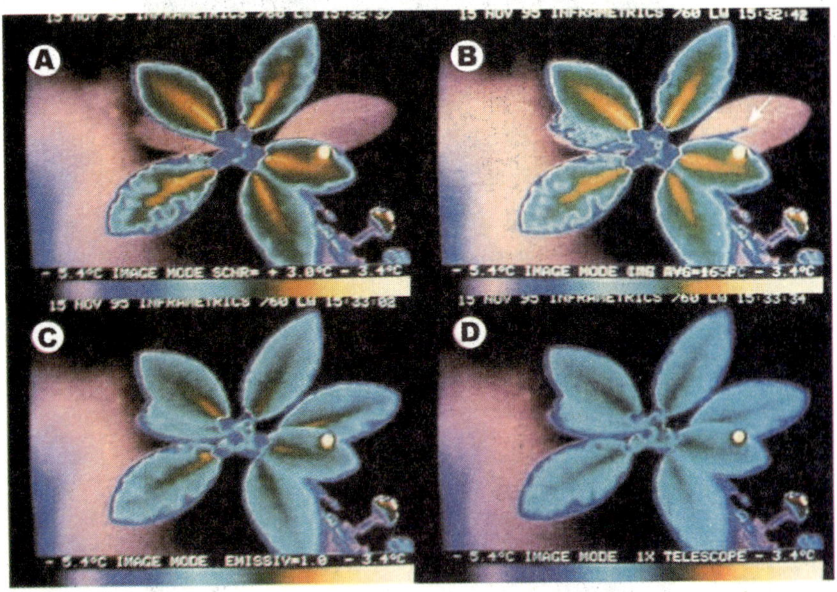

(Figure 3). Fruit could remain supercooled for up to one hour (the duration of the experiment). Ripe berries supercooled to colder temperatures and for longer durations than unripe berries.

Carter et al. (1999, 2001) have also used infrared thermography to study the freezing process in flowering shoots of blackcurrant (*Ribes nigrum* L.). In the initial study (Carter et al., 1999), it was observed that differences in field ratings of flower cold hardiness of cultivars could

FIGURE 3. Ice propagation in fruit-bearing, cranberry uprights. Ice can be seen propagating in the stem of the upright on the right. Despite propagation of ice throughout the stem and leaves, ice did not propagate through the pedicel into the stem. This provides evidence that some type of barrier exists that prevents ice propagation through the pedicel and thus allows the fruit to supercool.

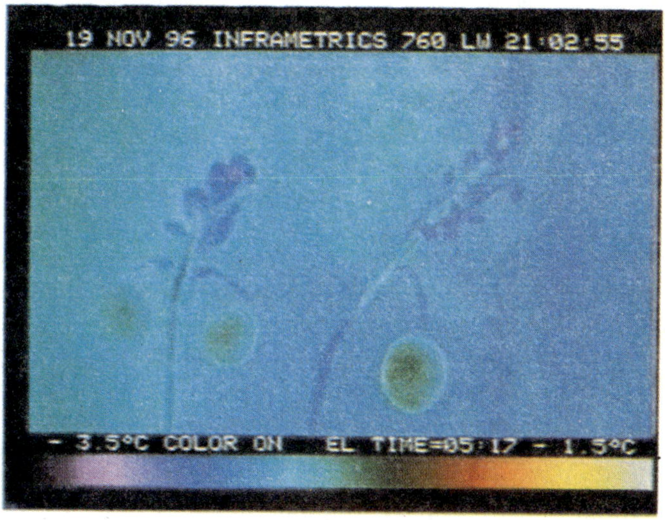

not be correlated with differences in flower bud hardiness obtained in laboratory freezing studies. This indicated that perhaps the flowering strigs of blackcurrant were supercooling. This was supported by the observation that ice nucleation in stem tissues occurred at around −2°C while detached flowers could supercool to −9°C. The patterns of freezing injury in shoots with flowering strigs frozen to −5°C also indicated the possibility that barriers to ice propagation from stem tissue to the raceme may exist and play an important role in the mechanism of frost resistance. In the subsequent study (Carter et al., 2001), patterns of ice formation and movement were studied. The pattern was discerned to be quite complex. The stem tissues of blackcurrant were always the first to freeze, however, it was common to see several independent movements of ice into the peduncle of a raceme and that individual flowers attached to the same peduncle could also freeze independently. It was concluded that barriers to movement of ice exist at specific anatomical junctions within the plant, notably where the peduncle of an inflorescence attaches to a stem and where a flower pedicel joins a peduncle. The time required for ice to pass through these barriers was inversely related to the degree of supercooling that had occurred prior to freezing.

The above mentioned studies clearly indicate that barriers to ice nucleation and propagation exist in plants and can play an important role in frost resistance. Furthermore, they indicate that the creation and/or enhancement of barriers could be used to improve frost protection. Such an approach could involve both selection for specific anatomical or structural traits or the physical application of materials that provide a barrier to ice nucleation. The latter approach has been recently explored in herbaceous plants (Wisniewski et al., 2002). In this study, an application of a hydrophobic particle film was administered to tomato plants (*Lycopersicon esculentum* L.) prior to the application of water containing ice-nucleation-active bacteria. In studies conducted in an environmental chamber, noncoated plants froze at temperatures around −2.5°C, while coated plants exhibited the ability to supercool to temperatures as low as −6.0°C. Studies conducted in an environmental chamber specifically designed to simulate radiative frost conditions also resulted in significant differences between coated and noncoated plants. When water was sprayed on the plants to simulate dew, it collected on noncoated plants but did not collect on the plants coated with the hydrophobic particle film (Figure 4). The water present on the surface of the noncoated plants resulted in leaf temperatures lower than chamber temperatures due to evaporative cooling. This is seen as large black areas on the noncoated plants illustrated in Figure 4.

DEEP SUPERCOOLING AND THE FORMATION OF GLASSES

Deep supercooling of flower buds and xylem tissues, in which water in cells or tissues remain isolated from the nucleating and dehydrative effects of extracellular ice, has been comprehensively reviewed (Ashworth, 1992; Wisniewski and Arora, 1993; Wisniewski, 1995; Wisniewski and Fuller, 1999; Wisniewski and Arora, 2000). Supercooling of floral buds stands in contrast to extraorgan freezing where water migrates to sites of extracellular ice away from the bud primordia (Sakai and Larcher, 1987) and equilibrium freezing in bark and xylem tissues where cells lose water to ice crystals present in the surrounding extracellular space (Chen et al., 1995). In both cases, the amount of water that is lost from a cell is directly dependent on the vapor pressure (and hence temperature) of the ice in the surrounding tissue. The lower the temperature the more cellular water is displaced. As a result, the concentration of solutes present in

FIGURE 4. Tomato plants either coated with a hydrophobic particle film (left) or non-coated (right). Both plants were sprayed with water and then placed in an environmental chamber and cooled. Water on the non-coated plant can be seen as large dark patches that represent temperatures outside of the set temperature range (−0.2 to 1.7°C). The presence of water resulted in lower leaf temperatures due to evaporative cooling. Water did not collect on the coated plants.

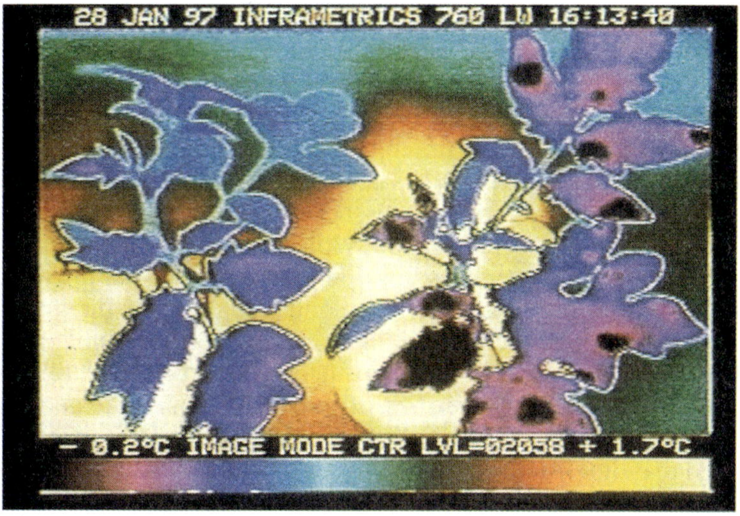

the cell increases and the freezing point decreases. Generally speaking, the major stress experienced by frozen tissues is believed to be severe dehydration and the cellular changes (membrane folding, protein denaturation, increased level of toxic solutes) that result (Artlip and Wisniewski, 2001; Chen et al., 1995; Wisniewski and Arora, 2000). In the present review, only deep supercooling of xylem tissues will be discussed and the reader is referred to other reviews for details on deep supercooling of floral buds vs. extraorgan freezing.

Of the many aspects of plant cold hardiness, deep supercooling is perhaps the most enigmatic (Wisniewski and Arora, 2000). The ability of some plants to maintain symplastic water in an unfrozen condition and without movement of the water into the apoplast is a remarkable adaptation that has not failed to impress both biophysicists and plant physiologists. Although the ability of woody plant tissues to avoid freezing by deep supercooling was first suggested in the 1960s (see review by

Wisniewski, 1995), the mechanism that allows small domains of water to avoid freezing, despite the presence of extracellular ice, remains little understood. This partly is due to the fact that the properties which allow deep supercooling to occur apparently rely on the structural organization of the tissue or organ. This feature has made it very difficult to manipulate plant material in a way to discover the fundamental mechanism and/or properties that allow deep supercooling to occur.

Deep supercooling of xylem tissues is a common characteristic of many temperate species of woody plants. Using differential thermal analysis (DTA), one can observe freezing events occurring in xylem tissues at very low temperatures (−25 to −40°C) whose appearance is correlated with death of the tissue (Wisniewski and Arora, 1993). Because of this association, DTA has been used extensively to evaluate the degree of cold hardiness of stem tissues of many important species of fruit and landscape species of woody plants. In order for deep supercooling to occur, a barrier must exist in xylem tissues that prevents the rapid loss of water to extracellular ice and also prevents the growth of ice crystals into living cells. In this regard, the porosity and/or permeability of the cell wall appears to play an essential role in the regulation of deep supercooling. In a series of papers (Wisniewski and Davis, 1989; Wisniewski et al., 1991a; Wisniewski et al., 1991b), it was demonstrated that the structure and composition of the pit membrane of xylem parenchyma cells appear to play an important role in regulating the extent of deep supercooling of xylem tissues. The pit membrane is a thin portion of the cell wall that allows for the passage of solutes and other materials, including plasmodesmatal connections, between cells. It is composed mainly of cellulose and pectic materials and unlike secondary wall material is unglignified for at least two years of development. By altering (chemically or enzymatically) the compostion of this layer of the cell wall, it was demonstrated that the extent of deep supercooling could be reduced or eliminated. In particular, the pectic component of the pit membrane appeared to play a key role in regulating its porosity, although a unique arabinogalactan-rich glycoprotein was also identified in the amorphous layer subtending the pit membrane. The role of this protein is unknown.

Another unique response of cells in woody plants is the ability to form glassed cell solutions (Chen et al., 1995; Hirsh et al., 1985). This believed to be a natural adaptation that occurs in extremely hardy plant species that can survive cooling to −196°C in liquid nitrogen. The formation of glasses is a unique metastable condition that cannot be terminated below the glass transition temperature. Aqueous glasses are extremely viscous brought about by high solute (sugar) concentrations at a sufficiently low temperature. In poplar (*Populus tremuloides*), glasses can form below −20°C (Hirsh et al., 1985). Although glassed solutions are extremely metastable and under a high degree of supercooling and hydrostatic tension, they are not subject to ice nucleation, solute crystallization, or water vapor cavitation so long as the solution remains below the melting temperature of the glass. Thus, the cytoplasm and its contents are extremely stable and relatively unaffected by the stresses associated with low temperature and the presence of ice. The glassed state is frequently relied upon for the survival of plant tissues during cryopreservation (Steponkus et al., 1992).

REFERENCES

Artlip, A. and M. Wisniewski. (2001). Induction of proteins in response to biotic and abiotic stress. Pp. 657-679. In: Handbook of Plant Physiology, 2nd ed., M. Pessarakli. Ed. Marcel Dekker, NY.

Ashworth, E.N. (1992). Formation and spread of ice in plant tissues. Hort. Rev. 13:215-255.

Ashworth, E.N. (1982). Properties of peach flower buds which facilitate supercooling. Plant Physiol. 70:1475-1479.

Ashworth, E.N. and T.L. Kieft. (1995). Ice nucleation activity associated with plants and fungi. Pp. 137-162. In: Biological Ice Nucleation and Its Applications, eds. R.E. Lee, Jr., G.J. Warren and L.V. Gusta. APS Press: Minneapolis, MN.

Carter, J., R. Brennan and M. Wisniewski. (1999). Low-temperature tolerance of blackcurrant flowers. HortScience 34:855-859.

Carter, J., R. Brennan and M. Wisniewski. (2001). Patterns of ice formation and movement in blackcurrant. HortScience 36:1027-1032.

Chen, T.H.H., M.J. Burke and L.V. Gusta. (1995). Freezing tolerance in plants: An overview. Pp. 115-136. In: Biological Ice Nucleation and Its Applications, eds. R.E. Lee, Jr., G.J. Warren and L.V. Gusta. APS Press: Minneapolis, MN.

Fuller, M. and M. Wisniewski. (1998). The use of infrared thermal imaging in the study of ice nucleation and freezing of plants. J. Thermal Biol. 23:81-89.

George, M., R. Becwar and M.J. Burke. (1982). Freezing avoidance by deep undercooling of tissue water in winter-hardy plants. Cryobiology 19:638-639.

Hirsh, A.G., R.J. Williams and H.T. Merryman. (1985). A novel method of natural cryopreservation: Intracellular glass formation in deeply frozen *Populus*. Plant Physiol. 79:41-56.

Pearce, R.S. and M.P. Fuller. (2001). Freezing of barley studied by infrared video thermography. Plant Physiol. 125:227-240.

Quamme, H.A., W.A. Su and L.J. Veto. (1995). Anatomical features facilitating supercooling of the flower within the dormant peach bud. J. Amer. Soc. Hort. Sci. 12:814-822.

Sakai, A. and W. Larcher. (1987). Frost Survival of Plants. Springer, Berlin.

Sutinen, M-J., Arora, R., M. Wisniewski, E. Ashworth, R. Strimbeck and J. Palta. (2001). Mechanisms of frost survival and freeze-damage in nature. Pp. 89-120. In: Conifer Cold Hardiness. F.J. Bigras and S.J. Colombo. Eds. Kluwer Academic Pub. Dordrecht. The Netherlands.

Single, W.V. (1964). Studies on frost injury in wheat. II. Ice formation within the plant. Austral. J. Agr. Res. 15:869-875.

Steponkus, P.L., R. Lang and S. Fujikawa. (1992). Cryopreservation of plant tissues by vitrifications. Pp. 1-61. In: Advances in Low-Temperature Biology. P.L. Steponkus, ed., JAI Press, Ltd. London.

Wisniewski, M. (1995). Deep supercooling in woody plants and the role of cell wall structure. Pp. 163-182. In: Biological Ice Nucleation and Its Applications, eds. R.E. Lee, Jr., G.J. Warren and L.V. Gusta. APS Press: Minneapolis, MN.

Wisniewski, M. and R. Arora. (2000). Structural and biochemical aspects of cold hardiness in woody plants. Pp. 419-438. In: Molecular Biology of Woody Plants, vol. 2. S.M. Jain and S.C. Minocha, eds. Kluwer Academic Pub. Dordrecht, The Netherlands.

Wisniewski, M. and R. Arora. (1993). Adaptation and response of fruit trees to freezing temperatures. Pp. 299-320. In: Cytology, Histology, and Histochemistry of Fruit Tree Diseases. A.R. Biggs, ed., CRC Press, Boca Raton.

Wisniewski, M. and G. Davis. (1989). Evidence for the involvement of a specific cell wall layer in the regulation of deep supercooling of xlem parenchyma. Plant Phsyiol. 139:105-116.

Wisniewski, M. and M. Fuller. (1999). Ice nucleation and deep supercooling plants: New insights using infrared thermography. Pp. 105-118. In: Cold Adapted Organisms: Ecology, Physiology, Enzymology, and Molecular Biology. R. Margesin and F. Schinner, eds. Springer-Verlag. Berlin.

Wisniewski, M., G. Davis and R. Arora. (1991a). Effect of macerase, oxalic acid, and EGTA on deep supercooling and pit membrane structure of xylem parenchyma of peach. Plant Physiol. 139:1354-1359.

Wisniewski, M., G. Davis, and K. Schaffer. (1991b). Mediation of deep supercooling of peach and dogwood by enzymatic modifications in cell-wall structure. Planta 184:254-260.

Wisniewski, M., D.M. Glenn and M. Fuller. (2002). Use of a hydrophobic particle film as a barrier to extrinsic ice nucleation in tomato plants. J. Amer. Soc. Hort. Sci. 127:358-364.

Wisniewski, M., S.E. Lindow and E.N. Ashworth. (1997). Observations of ice nucleation and propagation in plants using infrared thermography. Plant Physiol. 113: 327-334.

Wisniewski, M., M. Fuller, D.M. Glenn, J. Palta, J. Carter., L. Gusta, M. Griffith and J. Duman. (2001). Factors involved in ice nucleation and propagation in plants: An overview based on new insights gained from the use of infrared thermography. Buvisindi Icel. Agr. Sci. 14: 41-47.

Workmaster, B.A., J.P. Palta and M. Wisniewski. (1999). Ice nucleation and propagation in cranberry uprights and fruit using infrared video thermography. J. Amer. Soc. Hort. Sci. 124:619-625.

Characterization of a Novel Y_nSK_n Class of Dehydrin-Like cDNAs from Cold Acclimated Red-Osier Dogwood (*Cornus sericea* L.) Xylem

E. Sarnighausen
D. T. Karlson
Y. Zeng
P. B. Goldsbrough
K. G. Raghothama
E. N. Ashworth

SUMMARY. Red-osier dogwood (*Cornus sericea* L.) is among the most freeze-tolerant plant species. Four cDNAs were isolated from cold

---

E. Sarnighausen, D. T. Karlson, Y. Zeng, P. B. Goldsbrough, K. G. Raghothama, and E. N. Ashworth are affiliated with the Department of Horticulture and Landscape Architecture, Purdue University, West Lafayette, IN 47907-1165.

Present addresses: E. Sarnighausen, Institut für Biologie II, Albert-Ludwigs Universität Freiburg, Freiburg, Germany; D. T. Karlson, Department of Low Temperature Sciences, National Agricultural Research Center, Hitsujigaoka, Sapporo 0628555, Japan; Y. Zeng, VA Medical Center, Indianapolis, IN 46202.

Address correspondence to: E. N. Ashworth, Department of Horticulture and Landscape Architecture, Purdue University, West Lafayette, IN 47907-1165 USA (E-mail: ashworth@hort.purdue.edu).

This research was partially supported by the United States Department of Agriculture through the National Research Initiative Competitive Grants Program and through the Horticultural Research Institute. E. Sarnighausen and D. T. Karlson contributed equally to this manuscript and should be both viewed as senior authors.

[Haworth co-indexing entry note]: "Characterization of a Novel Y_nSK_n Class of Dehydrin-Like cDNAs from Cold Acclimated Red-Osier Dogwood (*Cornus sericea* L.) Xylem." Sarnighausen, E. et al. Co-published simultaneously in *Journal of Crop Improvement* (Food Products Press, an imprint of The Haworth Press, Inc.) Vol. 10, No. 1/2 (#19/20), 2004, pp. 17-35; and: *Adaptations and Responses of Woody Plants to Environmental Stresses* (ed: Rajeev Arora) Food Products Press, an imprint of The Haworth Press, Inc., 2004, pp. 17-35. Single or multiple copies of this article are available for a fee from The Haworth Document Delivery Service [1-800-HAWORTH, 9:00 a.m. - 5:00 p.m. (EST). E-mail address: docdelivery@haworthpress.com].

http://www.haworthpress.com/web/JCRIP

Digital Object Identifer: 10.1300/J411v10n01_03

acclimated red-osier dogwood xylem and found to encode novel dehydrin-like proteins. The four deduced amino acid sequences were extremely hydrophilic and contained sequence motifs characteristic of the dehydrin family. ROD (red-osier dehydrin) 44, 48 and 60 sequences each contained the Y, S, and K motifs characteristic of dehydrins. Whereas ROD 25 contained the Y and S motifs but lacked the lysine-rich repeat (K segment). A unique aspect of these sequences was that ROD 25, 44, 48 and 60 included 14, 33, 33 and 35 copies, respectively, of a sequence similar to DEYGNP. This is quite different from other dehydrin-like proteins, which typically contained only one to three Y motifs. The function of these highly repeated sequences is not known. Alignment of the deduced amino acid sequences of ROD 44 and ROD 48 revealed 69% identity. Alignment of overlapping regions of ROD 25 and ROD 60 revealed 89% identity. Transcript expression of Rod genes changed seasonally in red-osier dogwood and was correlated to periods of cold acclimation.

KEYWORDS. Cold acclimation, freeze-stress, desiccation stress, *Cornus sericea*

INTRODUCTION

Many plants respond to environmental cues, such as prolonged exposure to low temperature, and acquire increased tolerance to sub-zero temperatures (Weiser, 1970; Guy, 1990). This process, which is termed cold acclimation, is accompanied by changes in gene expression and the accumulation of stress induced proteins (Guy, 1990; Thomashow, 1990; 1999; Pearce, 1999). Many of the genes expressed during cold acclimation are also induced in response to drought stress. This is not surprising, as both drought and freeze-stress lead to cellular desiccation (Burke et al., 1976; Levitt, 1980; Guy, 1990). The first desiccation related proteins were identified in studies on cotton seed development. Several proteins were found to be induced during late stages of seed development and were termed Late Embryogenesis Abundant (LEA) proteins (Dure et al., 1981; Dure, 1993). LEA proteins have been divided into 6 unique groups based upon sequence characteristics (Baker et al., 1988; Wood, 1994). These classes of proteins, with the exception of group 6, share common characteristics such as boiling stability, hydrophilicity, deficiency in cysteine and tryptophan residues and the presence of putative amphipathic α-helices or random coils (Wood, 1994). Group 2 (D-11) dehydrin proteins have been identified in numerous plants exposed to desiccation stresses such as salinity, drought and freezing (reviewed by Close et al., 1993a; 1993b; Close,1996; 1997; Campbell and Close, 1997). Dehydrin-like proteins have also been shown to accumulate in several woody plant species during cold acclimation (Muthalif and Rowland, 1994; Arora and Wisniewski, 1994; 1996; Arora et al., 1996; Salzman et al., 1996; Wisniewski et al., 1996; Welling et al., 1997; Rinne et al., 1998; 1999; Lim et al., 1999).

Comparisons of dehydrin sequences within higher plants have identified three conserved motifs: DEYGNP, a poly-serine stretch, and a lysine-rich consensus sequence EKKGIMDKIKEKLPG, which have been designated as the Y, S and K segments, respectively (Close et al., 1993b; Close, 1996; 1997, Campbell and Close, 1997). The presence and number of these motifs have been used to subdivide dehydrins into 5 distinct categories using "YSK shorthand" (Y_nSK_2, K_n, K_nS, SK_n and Y_2K_n) (Close, 1996). Interspersed between the Y, S and K domains are less conserved regions of polar amino acids, which are termed "ϕ segments" (Campbell and Close, 1997). The majority of the dehydrins belong to the Y_nSK_2 class (Close, 1996).

We previously identified a 24 kDa protein related to cold acclimation in an extremely freeze tolerant woody shrub, red-osier dogwood (*Cornus sericea* L.) (Sarnighausen et al., 2002). This 24 kDa protein accumulated to greatest quantities during mid-winter when freeze tolerance was maximal, and was not detected during late spring and early summer when freeze tolerance was minimal. We were interested in further characterizing this protein and determining its role in cold acclimation.

IDENTIFICATION AND ISOLATION OF cDNAs

Samples of the current season's growth were harvested from a clonally propagated stand of *C. sericea* growing in the vicinity of Purdue University, West Lafayette, IN. Bark tissue was removed and the remaining xylem tissue was immediately plunged into liquid nitrogen. Samples were subsequently lyophilized and stored at −80°C until analyzed. RNA was isolated from lyophilized cold acclimated xylem as described by Chang et al. (1993) with minor modifications. Oligo(dT)-

cellulose was used to isolate polyadenylated mRNA as described by Aviv and Leder (1972). RNA blot analysis was performed using an adaptation of a procedure described by Alwine et al. (1977).

Polyadenylated mRNA from cold acclimated xylem was used as a template for the generation of a cDNA library using a Lambda ZAP cDNA kit according to manufacturer's instructions (Stratagene).

The cDNA library was screened with a polyclonal antibody with specifiity for a cold acclimation inducible 24 kDa protein from *Cornus sericea*. This antibody was raised in a chicken as previously described (Sarnighausen et al., 2002). Nitrocellulose filters (0.45 μm) were autoclaved and treated with IPTG. Approximately 600,000 plaques were screened in the first round. Plates containing approximately 50,000 plaques were overlaid with nitrocellulose impregnated with IPTG and maintained at 37°C for 5 hours. Plates were chilled for 30 minutes, the membranes removed and washed for 30 minutes each with TBS (Tris buffered saline, 50 mM Tris-HCl, pH 7.5, 200 mM NaCl), TBST (TBS + 0.05% w:v Tween-20), and TBS. Filters were blocked for 3 hours with 3% BSA in TBS. After blocking, filters were washed for 30 minutes each with TBS, TBST, and TBS. Primary antibody incubation (1:5000 anti-24 kDa polyclonal antibody) was performed overnight at 4°C in TBST with 0.05% sodium azide. Filters were washed for 30 minutes in the following solutions: TBS, TBST and TBS. Rabbit anti-chicken antibody (Sigma, St. Louis, MO) was diluted 1:100,000 in TBST and incubated with the filters for 2 hours at room temperature. Post-incubation, filters were washed as previously described, with an additional 30 minute TBS rinse. Color development was done utilizing 100 ml alkaline phosphatase buffer (100 mM Tris-HCl, pH 9.5, 100 mM NaCl, and 5 mM $MgCl_2$) with 667 μl 5% (w:v) NBT (nitroblue tetrazolium) in 70% (v:v) dimethylformamide and 333 μl 5% (w:v) BCIP (5-bromo-4-chloro-3-indoyl phosphate) in 100% dimethylformamide. Several positive plaques were selected for phage isolation. Phages were isolated overnight with gentle shaking at 4°C and subjected to further rounds of screening. *In vivo* excision of the pBluescript phagemids (containing the cDNA clones) was performed per supplier's instructions using the Exassist/SOLR system (Stratagene).

NUCLEIC ACID SEQUENCE ANALYSIS

The cDNA clones in pBluescript SK+ obtained from immunoscreening the cDNA library served as the template for double stranded DNA sequencing to obtain partial sequence at the 3′ and 5′ ends. These partial sequences contained highly repetitive segments, thus no suitable priming site could be found for primer-based sequencing. As an alternative, the cDNAs were subcloned into smaller fragments for complete bi-directional sequencing. Three fragments (474, 571, 591 base pairs) from Rod 48 were subcloned utilizing internal ClaI and KpnI restriction sites. Four fragments from Rod 60 (228, 504, 523, 666 base pairs) were subcloned by using internal EcoRI and KpnI restriction sites. These fragments were subcloned into pBluescript® SK+ plasmid that was previously cut with the appropriate combination of restriction enzymes and ligated with T4 DNA ligase (Promega, Inc., Madison, WI). Rod 44 was subcloned into smaller fragments by making sets of overlapping deletions for both strands using exonuclease III, followed by S1 nuclease and Klenow DNA polymerase I treatments, according to the method outlined in the Erase-a-base system (Promega, Inc., Madison, WI). Rod 25 was completely sequenced without subcloning into smaller fragments. Complete bi-directional sequencing was obtained using T3 and T7 primers from the 5′ and 3′ ends, respectively. DNA sequences were determined by the dideoxy chain termination method (Sanger et al., 1971) using Big Dye Kit (Perkin Elmer Corporation) at the Iowa State University DNA Sequencing Facility. Sequence accessions are available at GenBank AF345988-AF345991.

Nucleic acid sequences were analyzed using a variety of computer software programs. Clustal W 1.7 was used for comparative alignment of *C. sericea* cDNA sequences. BLAST, a program available from NCBI, was used to compare *C. sericea* cDNAs to other sequences available from the database. PSORT I computer analysis was used to predict the subcellular localization of the deduced amino acid sequences. ProtScale computer analysis was used for hydropathy analysis of deduced amino acid sequences according to Kyte and Doolittle (1982).

IDENTIFICATION OF DEHYDRIN-LIKE GENE FAMILY

Complete bi-directional sequencing of four cDNAs (Rod 25, 44, 48 and 60) identified open reading frames sufficient to code for 236, 400, 440 and 567 amino acid peptides with predicted molecular weights of 25, 44, 48 and 60 kDa, respectively (Figures 1-4). These cDNAs represent a small family of closely related genes. Comparative analyses of the deduced amino acid sequences from Rod 25 and Rod 60 had 89% identity within overlapping regions (Figure 5). The deduced amino acid

FIGURE 1. Nucleotide and deduced amino acid sequence of Rod 25. The first termination codon following the open reading frame is marked by an asterisk (*). Multiple putative polyadenylation sites are underlined.

1	GCACGAGTATATCTGCTCAATATATTTCAGTGTAATTTTAAGCTTGAAAGTATGGCACAA	
	M A Q	3
61	TACGGGAACCCAACTGAAGGACTATACAAGCAACCGACTGATGTATACGGGAATCCGATT	
	Y G N P I E G L Y K Q P I D V Y G N P I	23
121	TCCCGTACCGCACAATCTGGCAACCCGGTTCAGAGAACTGCTGATGCACAATATGGGAAC	
	S R I A Q S G N P V Q R T A D A Q Y G N	43
181	CCAACTGAAGGAGTATACAAGCAACCGACTGATGTATACGGGAATCCGATTTCCCGTACC	
	P T E G V Y K Q P T D V Y G N P I S R T	63
241	CACGAATCTGGCAACCCGGTTCAGAAAACTGCTGATGCAGAATACGGGAACCCAACTGAA	
	H E S G N P V Q K T A D A E Y G N P T E	83
301	GGAGTACATAAGCACCCGACTGATGCACACGAGAATCCGCTTTCCCGTACCCACGAATCT	
	G V H K H P T D A H E N P L S R T H E S	103
361	GACAACCCAATTCAGAAAACTGCTGGTGCACAAAACAGGAACCCAACTGAAGGAGTACAT	
	D N P I Q K T A G A Q N R N P T E G V H	123
421	AAGCACCCGACTGATGCACACCGGAATCCGGTTTCCCGTACCCACGAATCTGGCAACCCG	
	K H P T D A H R N P V S R T H E S G N P	143
481	ATTCAGAAAACTGCTGATGCACAATATAGGAACCCAACTGGAGGAGTACATAAGCACCCG	
	I Q K T A D A Q Y R N P T G G V H K H P	163
541	ACTGATGCACACGGGAATCCGGTTTCCCGTACCCACGAATATGGCAACCCTATTCGAAAA	
	T D A H G N P V S R T H E Y G N P I R K	183
601	ATTGCTGATGGTCAGAGGGTTCACCCCACTGGAGGTACCACGGAAGGCTACATAACCACT	
	I A D G Q R V H P T G G T T E G Y I T T	203
661	AGTGGCACCAGCACCGGCATCGGCAACGGAGGTCGCATAGGCGGGCAGCAGCACGGCGGC	
	S G T S T G I G N G G R I G G Q Q H G G	223
721	GTTCTTCAACGCTCCGGCTCCGGTTCGAGCTCGGTAATTTAATTTTCCCCTCGTCTTTTA	
	V L Q R S G S G S S S V I *	236
781	GTTATTATACATACATATATATATATATA<u>AATATT</u>TTAAAACTTAGAAACTTCCCTATATT	
841	TTTGAATTTTGCAACTAACCGTATAG<u>AATTAA</u>ATTTCATGCCA<u>AATAAG</u>CCTAGTTTCA	
901	<u>AATATTT</u>CCTTGTCCGT<u>AAATCAC</u>TTCACCGTACCTCATGTGAAATTGCTAGTTTATTT	
961	TGTTTCAAAATAAAAAAAAAAAAAAAAAAAAC	

sequences of Rod 44 and Rod 48 were 69% identical (Figure 6). BLAST searches determined that all the open reading frames had homology to dehydrin-like proteins. As with other dehydrins, the deduced amino acid sequences were extremely hydrophilic (data not shown) and rich in glycine, proline and threonine. These three residues accounted for over 30% of the total amino acid composition of ROD 25, 44, 48 and 60 (Table 1). The predicted peptides were also deficient in tryptophan and contained reduced quantities of cysteine (Table 1). The predicted amino acid sequences contained motifs characteristic of the dehydrin family of proteins. Each contained multiple copies of the Y segment (DEYGNP) and a tract of serine residues (Figures 5 and 6). ROD 44, 48 and 60 also

FIGURE 2. Nucleotide and deduced amino acid sequence of Rod 44. The first termination codon following the open reading frame is marked by an asterisk (*). A putative polyadenylation site is underlined.

```
   1  AATTCGGCACGAGCACACACACATCCAGAGTTATAAACCAGTGTTCAACTCTTTTTCTGT
  61  GATCAATTTCTGTTTAATTTTAAGGTTGAAGGTATGGCACAATACGGAGACCAGTATGGC
                                       M  A  Q  Y  G  D  Q  Y  G      9
 121  AACCCGGTTCGCCCAACTGAAGAGGTATATGGAAGCCCGACTGATGCATACGGGAACTCG
      N  P  V  R  P  T  E  E  V  Y  G  S  P  T  D  A  Y  G  N  S     29
 181  CTTCCCCCGACCGATAAATACGGCAACCGGATTCACCAAACTGATGCACATGGCGCACAT
      L  P  P  T  D  K  Y  G  N  R  I  H  Q  T  D  A  H  G  A  H     49
 241  GGAAAGCCGATTCACAAAACTGACAAATATGGCAATCCGATTCACCCAACTGAAGAAGTA
      G  K  P  I  H  K  T  D  K  Y  G  N  P  I  H  P  T  E  E  V     69
 301  TATCGAAGCCCGACTGATGCAAACGGGAACCCGCTTACCCCATCCGATAAATACGGCAAC
      Y  R  S  P  T  D  A  N  G  N  P  L  T  P  S  D  K  Y  G  N     89
 361  CGATTCACCAAACTGATGCACATAGAAAGCCGATTCGTAAAACTGACAAATATGGCAAC
      P  I  H  Q  T  D  A  H  R  K  P  I  R  K  T  D  K  Y  G  N    109
 421  CCGGTTCACCCAACTGAAGAAGTATATCAAAGCCCGACTGATGCAAACGGGAATCCGCTT
      P  V  H  P  T  E  E  V  Y  Q  S  P  T  D  A  N  G  N  P  L    129
 481  CCTCCATCCGATAAATACGGCAACCCGATTCACCAAACTGATGCACATGGAAAGCTGATT
      P  P  S  D  K  Y  G  N  P  I  H  Q  T  D  A  H  G  K  L  I    149
 541  CGCAAAACTGACAAATATGGCAACCCGGTTCACCCAACTGAAGAAGTATATGGAAGCCCA
      R  K  T  D  K  Y  G  N  P  V  H  P  T  E  E  V  Y  G  S  P    169
 601  ACTGATGCATACGGGAACCCGCTTCCCCCAACAGATAAATATGGCAACCCAAATCGCAAA
      T  D  A  Y  G  N  P  L  P  P  T  D  K  Y  G  N  P  N  R  K    189
 661  ACAGACGAATATGGCAATCTGGTCCACCCAATTGAAAAAGTATATGGAAGCCAGACTGAT
      T  D  E  Y  G  N  L  V  H  P  I  E  K  V  Y  G  S  Q  T  D    209
 721  GCATACGGGAACCTGCTTCTCCCAACCGATAAATACGGCAACCCGATTCACCAAACTAAT
      A  Y  G  N  L  L  L  P  T  D  K  Y  G  N  P  I  H  Q  T  N    229
 781  GCACATGGAAACCAGAATCGCAAAATTGACGAATATGGAAACCCGGTTCACCCAATTGAA
      A  H  G  N  Q  N  R  K  I  D  E  Y  G  N  P  V  H  P  I  E    249
 841  AAAGTATATGGAAACCCGACTGATACATACGGGAACCCGCTTCCCCCAACCGATAAATAC
      K  V  Y  G  N  P  T  D  T  Y  G  N  P  L  P  P  T  D  K  Y    269
 901  GGCAACCCGATTCACCACACTGATGCACATGGAAACCCGATTGGCAAAACTGACAAATAT
      G  N  P  I  H  H  T  D  A  H  G  N  P  I  G  K  T  D  K  Y    289
 961  GGCAACTCGGTTCGCCCAACTGAAGAAGTATATAGGAGCCCGACTGATGCAAACGGGAAC
      G  N  S  V  R  P  T  E  E  V  Y  R  S  P  T  D  A  N  G  N    309
1021  CCGCTTCCCCCAATTGATGAATACGGCAACCCGATTCACCAAACTGATGTATATGGAAAG
      P  L  P  P  I  D  E  Y  G  N  P  I  H  Q  T  D  V  Y  G  K    329
1081  CCGATTCGCAAAACTGATGAATATGGCAACCCGGTTCACCACAGCGGCATTCTTCATCGC
      P  I  R  K  T  D  E  Y  G  N  P  V  H  H  S  G  I  L  H  R    349
1141  TCCGGCTGCTTCAGTTCTAGCTCGTCGGAGGATGATGGATATGGTGGTAGAAGGAAGAAA
      S  G  C  F  S  S  S  S  S  E  D  D  G  Y  G  G  R  R  K  K    369
```

FIGURE 2 (continued)

```
1201  CGGCTGAGGGATAAGATAAAGGAGAAGCTACGAGGTGGGAAAGAGACAGAGCAGTTGCAC
       R  L  R  D  K  I  K  E  K  L  R  G  G  K  E  T  E  Q  L  H     389
1261  ACTACAGCAACGTCCACTCCAACTGGTTACGGATAAGAAGGCTTGGAGCATCGATCATAA
       T  T  A  T  S  T  P  T  G  Y  G  *                             400
1321  AAAGAAAGGAGTCGTGGAGAAAAAGGATAAGCTTCCTGGACAACATTACTAGGTGTCTTT
1381  GCATGATGTAGCAGCTTATATATATATATGTATATGTATATGTATATGTATGTAACAAGT
1441  GTGAAAAGTACAATAATACTCCAAAAACTCGGACAAGCCTTAAAAGTACAAAAAAAAAAA
1501  AAAAAAAA
```

contained the lysine-rich motif (K segment) characteristic of dehydrins (Figures 5 and 6).

Based upon the predicted sequences, the *C. sericea* cDNA clones appear to encode a unique sub-group of Y_nSK_n dehydrins. With the exception of ROD 25, these proteins contain a single serine tract and at least one lysine rich region (Figures 5 and 6). However, the *C. sericea* ROD proteins exhibit a unique enrichment of repeated motifs similar to the Y motif (Figures 5 and 6). Previous reports have indicated that the DEYGNP motif typically occurs only one to three times near the amino terminus (Close, 1996). In addition, amino acid sequences interspersed between the Y, S, and K motifs are generally less conserved, and enriched with either glycine, alanine or proline, and polar residues (Campbell and Close, 1997). In the *C. sericea* ROD proteins, however, there are numerous repeating Y motifs distributed throughout the protein beginning near the amino terminus and extending to the poly-serine stretch close to the carboxy terminus (Figures 5 and 6). In ROD 25, 44, 48, and 60 a total of 14, 33, 33, and 35 Y segments occur, respectively (Table 2). The significance of these repeating sequences is not known.

PSORT analysis predicted nuclear localization of ROD 25 and ROD 48 and cytoplasmic localization of ROD 44 and ROD 60 (data not shown). Such predicted targeting is consistent with reports on the subcellular immunolocalization of dehydrins (Asghar et al., 1994; Godoy et al., 1994; Goday et al., 1995; Houde et al., 1995; Egerton-Warburton et al., 1997; Wisniewski et al., 1999).

We had previously shown that an isoform of the 24 kDa protein, which is abundant in cold acclimated *C. sericea*, was similar to dehydrins (Sarnighausen et al., 2002). The amino acid composition of this isoform and a partial amino acid sequence of an internal peptide fragment both exhibited similarity to dehydrins. In addition, Western blot analysis showed that the 24 kDa protein was recognized by a

FIGURE 3. Nucleotide and deduced amino acid sequence of Rod 48. The first termination codon following the open reading frame is marked by an asterisk (*). Multiple restriction sites were used for subcloning the cDNA into smaller fragments for complete sequencing. Boxes indicate restriction sites used for subcloning and sequencing.

```
   1  GACACACACATATCTAAGCGAAGGATAGAGTTATAAAACAGTGTTCAACTCTTTTGTGC
  61  GTTCAATTTCAGTTTAATTTTAAGGTTGAAGGTATGGCACAATACGGAGACCAGTATGGC
                                     M  A  Q  Y  G  D  Q  Y  G      9
 121  AACCCGGTTCGCCCAACTGAGAAAGTATATAGAAGCCCGACGGATGCATACGGGAACTCA
      N  P  V  R  P  T  E  K  V  Y  R  S  P  T  D  A  Y  G  N  S   29
 181  TTTCCCCCAACCGATGAATACAGCAACCCGATTCACCAAACTGATGCATATGGAAACTCG
      F  P  P  T  D  E  Y  S  N  P  I  H  Q  T  D  A  Y  G  N  S   49
 241  AATCGCAAAACTGACGAATATGGCAACCTGGTTCACCCACTTGAAAAAGTATATGGAAGC
      N  R  K  T  D  E  Y  G  N  L  V  H  P  L  E  K  V  Y  G  S   69
 301  CCGACTGATGCATACGGGAACCCGCTTCCCCCAACCGATAAATACGGCAACCCGCTTCAC
      P  T  D  A  Y  G  N  P  L  P  P  T  D  K  Y  G  N  P  L  H   89
 361  CAAACTGATGCATATGGAAAGCCGATTCGCAAAACTGGCGAATATGGCGACCCGACTGAT
      Q  T  D  A  Y  G  K  P  I  R  K  T  G  E  Y  G  D  P  T  D  109
 421  GCATACGGGAACCCGCTTCCCTCAACCGATAAATACGGCAACCCGATTCACCA[ATCGAT]T  ClaI
      A  Y  G  N  P  L  P  S  T  D  K  Y  G  N  P  I  H  Q  S  I  129
 481  CGCAGAACTGACGAATATGGCGACCCAGTTCACCCAATTGAAAAAGTATATAGAAGCCCG
      R  R  T  D  E  Y  G  D  P  V  H  P  I  E  K  V  Y  R  S  P  149
 541  ACTGATGCATACGGGAACCCGCTTCCCACAACCGATAAGTACGGAAACCCGATTCGCAAA
      T  D  A  Y  G  N  P  L  P  T  T  D  K  Y  G  N  P  I  R  K  169
 601  ACTGACGAATATGGCGACCCGGTTCACCCAATTGAAAAAGTATATAGAAGCCCGACTGAT
      T  D  E  Y  G  D  P  V  H  P  I  E  K  V  Y  R  S  P  T  D  189
 661  GCATACGGGAACCCGCTTTGCCGAACCGATAAATACGGCAACCCGATTCACCAAACTGAT
      A  Y  G  N  P  L  C  R  T  D  K  Y  G  N  P  I  H  Q  T  D  209
 721  ACACATGGAAACCCGATTCGCAAAACTGACAAATATGGCGACCCGATTCACCCAATTGAA
      T  H  G  N  P  I  R  K  T  D  K  Y  G  D  P  I  H  P  I  E  229
 781  AAAGTATATGGAAGCCCGACTGATGCATACATGAACCCGCTTCCTCAAACCGATAAATAC
      K  V  Y  G  S  P  T  D  A  Y  M  N  P  L  P  Q  T  D  K  Y  249
 841  GGCAACTCGATTCACCAAACTGATGCACATGGAAACCCAATTCGCAAAACTGACAAATAT
      G  N  S  I  H  Q  T  D  A  H  G  N  P  I  R  K  T  D  K  Y  269
 901  GGCAACCCAGTTCACCCAACTCAAGAAGCATATAGAAGCCCGACTGATGCATACGGGAAC
      G  N  P  V  H  P  T  Q  E  A  Y  R  S  P  I  D  A  Y  G  N  289
 961  CCGCTTCCCCCAACCGATGAATACGGCAACCCGATTCACCAAACTGATGCATATGGAAAG
      P  L  P  P  T  D  E  Y  G  N  P  I  H  Q  T  D  A  Y  G  K  309
1021  CCGATTCGCAAAACTAACGAATATGGCAACCCGGTTCACCACAC[GGTACC]CATGGGAGAC  KpnI
      P  I  R  K  T  N  E  Y  G  N  P  V  H  H  T  V  P  M  G  D  329
1081  TATGGAATCAGTGGTGCCACGGGAGCTTATGGCACGGGCACCGGAGTAACTGGCGTAGGT
      Y  G  I  S  G  A  T  G  A  Y  G  T  G  T  G  V  T  G  V  G  349
1141  GGGCAGCATCACGGCGGCATTCTTCATCACTCCGACTGCTTCAGTTCTAGCTCGTCGGAG
      G  Q  H  H  G  G  I  L  H  H  S  D  C  F  S  S  S  S  S  E  369
```

FIGURE 3 (continued)

```
1201  GATGATGGATATGGTGGTAGAAGGACGAAAGGGTTGACCGATAAGATAAAGGATAAGCTA
      D  D  G  Y  G  G  R  R  T  K  G  L  T  D  K  I  K  D  K  I   389
1261  CGAGGTGGGAAAAAGACAGAGCAGTTGCACACTACAGCAACGTCCACCACAACTGGTTAC
      R  G  G  K  K  T  E  Q  L  H  T  T  A  T  S  T  T  T  G  Y   409
1321  GGATACGAAGGTGAGGAGCATCGATATTATAAGAAGAAGGGAGTCATGGAGAAGATTATG
      G  Y  E  G  E  E  H  R  Y  Y  K  K  K  G  V  M  E  K  I  M   429
1381  GATAAGCTTCCTGGACAAGAGTCGCAACATTACTAGGTGTCTTTGCATGATGTAGCAGCA
      D  K  L  P  G  Q  E  S  Q  H  Y  *                           440
1441  TATATATATATATATATATATATATATATATGCTTAGCATTCTGGTTCCTCCAGACTCCG
1501  CCCTCTTTACCATGTAGTTGGCTATTGGTATGGTCTTTTGGTATGTAAATCAGAATCATA
1561  TGAATTGTTATTTTCTTATTTGAGCTTATTAGCATGAACAGCTTTGTGTTTGATGTCAAA
1621  AAAAAAAAAAAAAAAAA
```

polyclonal antibody that was raised against the synthetic peptide fragment (CTGEKKGIMDKIKEKLPGQH), which defines the consensus lysine-rich region of dehydrins (Close et al., 1993a). Despite the observation that the four *C. sericea* cDNAs encode dehydrin-like proteins and that these clones were isolated from an expression library using an antibody raised against the 24 kDa protein; none of these cDNAs appear to encode the 24 kDa protein that was purified and for which we have amino acid composition and partial peptide sequence data.

The functional significance of the Y, S and K dehydrin motifs is not known. The lysine-rich motif (EKKGIMDKIKEKLPG), or K segment, is highly conserved in higher and lower plants (Close, 1997). The position of hydrophobic, basic and acidic amino acids occurs with little deviation (Close et al., 1993b). Due to the highly conserved nature of this motif, it is believed that this is the most important structural component for dehydrin function (Close et al., 1993b). This lysine-rich consensus sequence has been hypothesized to form a putative amphipathic α-helix (Dure, 1993; Close et al., 1993b; Close, 1996) and is thought to interact non-specifically with nuclear and cytosolic macromolecules and cytoplasmic membrane surfaces (Close et al., 1993b; Close, 1996; 1997). The function of the Y motif is also not known. Close (1996) reported that the dehydrin Y motif had similarity to the nucleotide-binding portion of molecular chaperones (Martin et al., 1993). However, neither nucleotide binding nor molecular chaperone activity has been demonstrated for dehydrins (Close, 1996).

Plant molecular chaperones assist in the biogenesis of enzymes and structural proteins (Guy et al., 1998). An interesting feature is their capacity to bind to nascent proteins (Beckmann et al., 1990; Nelson et al.,

FIGURE 4. Nucleotide and deduced amino acid sequence of Rod 60. The first termination codon following the open reading frame is marked by an asterisk (*). A putative polyadenylation signal site is underlined. Boxes indicate restriction sites used for subcloning and sequencing.

```
1     AATTCGGCACGAGTATATCTGCTCAATATATTTCAGTGTAATTTTAAGCTTGAAACTATG
                                                              M    1
61    GCACAATACGGGAACCCAACTGAAGGACTATACAAGCAACCGACTGATGTATACGGGAAT
      A  Q  Y  G  N  P  T  E  G  L  Y  K  Q  P  T  D  V  Y  G  N   21
121   CCGATTTCCCGTACCCACGAATCTGGCAACCCGGTTCAGAGAACTGCTGATGCACAATAT
      P  I  S  R  T  H  E  S  G  N  P  V  Q  R  T  A  D  A  Q  Y   41
181   GGGAACCCAACTGAAGGAGTATACAAGCAACCGACTGATGTATACGG[GAATTC]GATTTCC EcoRI
      G  N  P  T  E  G  V  Y  K  Q  P  T  D  V  Y  G  N  S  I  S   61
241   CGTACCCACGAATCTGGCAACCCGGTTCAGAAAACTGCTGATGCACAATACGGGAACCCA
      R  T  H  E  S  G  N  P  V  Q  K  T  A  D  A  Q  Y  G  N  P   81
301   ACTGAAGGAGTATACAAGCAACCGACTGATGCACACGGGAATCCGATTTCCCGTACCCAC
      T  E  G  V  Y  K  Q  P  T  D  A  H  G  N  P  I  S  R  T  H   101
361   GAATCTGGCAACCCGGTTCAGAAAACTGCTGATGCACAATACGGGAACCCAACTGAAGGA
      E  S  G  N  P  V  Q  K  T  A  D  A  Q  Y  G  N  P  T  E  G   121
421   GTATACAAGCAACCGACTGATGCACACGGGAATCCGGTTTCCCGCACCCACGAATCTGGC
      V  Y  K  Q  P  T  D  A  H  G  N  P  V  S  R  T  H  E  S  G   141
481   AACCCGGTTCAGAAAACTGCTGATGCACAATACGGGAACCCAACTGAAGGAGTAGATAAG
      N  P  V  Q  K  T  A  D  A  Q  Y  G  N  P  T  E  G  V  D  K   161
541   CACCCGGCTGATGCACATGGGAATCCGGCTTTCCGTACCCGCGAATCTGGCAACCCGGTT
      H  P  A  D  A  H  G  N  P  A  F  R  T  R  E  S  G  N  P  V   181
601   CAGAAAACTGCTGATGCACAATACGGGAACCCAACTGAAGGAGTATACAAGCACCCGACT
      Q  K  T  A  D  A  Q  Y  G  N  P  T  E  G  V  Y  K  H  P  T   201
661   GATGCACACGGGAATCCGGTTTCCCGTACCCACGATTTTGGTAACCCGGTTCAGAAAACT
      D  A  H  G  N  P  V  S  R  T  H  D  F  G  N  P  V  Q  K  T   221
721   GCTGATGCACAATACGGGAACCCAACTGAAGGAGTAGATAAGCACCCGGCTGATGCACAT
      A  D  A  Q  Y  G  N  P  T  E  G  V  D  K  H  P  A  D  A  H   241
781   GGGAATACGGCTTTCCGTACCCGCGAATCTGGCAACCCGGTTCAGAAAACTGCTGATGCA
      G  N  T  A  F  R  T  R  E  S  G  N  P  V  Q  K  T  A  D  A   261
841   CAATACGGGAACCCAACTGAAGGAGTAGATAAGCACCCGACTGATGCACACAA[GAATTC]G EcoRI
      Q  Y  G  N  P  T  E  G  V  D  K  H  P  T  D  A  H  K  N  S   281
901   ATTTCCCGTACCCACGATTTTGGCAACCCGGTTCAGAAAACTGTTGATGCCCAATACGGG
      I  S  R  T  H  D  F  G  N  P  V  Q  K  T  V  D  A  Q  Y  G   301
961   AACCCAACTGAAGGAGTATATAAGCACCCGACTGATGCACACAGAAATTCGATTTCCCGT
      N  P  T  E  G  V  Y  K  H  P  I  D  A  H  R  N  S  I  S  R   321
1021  ACCCACAATTCTGGCAACCCGATTCAGAAAACTGCTGATGCAGAATACGGGAACCCAACT
      T  H  N  S  G  N  P  I  Q  K  T  A  D  A  E  Y  G  N  P  T   341
1081  GAAGGAGTACATAAGCACCCGACTGATGCACACGAGAATCCGCTTTCCCGTACCCACGAA
      E  G  V  H  K  H  P  I  D  A  H  E  N  P  L  S  R  T  H  E   361
```

FIGURE 4 (continued)

```
1141 TCTGACAACCCAATTCAGAAAACTGCTGGTGCACAAAACAGGAACCCAACTGAAGGAGTA
     S  D  N  P  I  Q  K  T  A  G  A  Q  N  R  N  P  T  E  G  V   381
1201 CATAAGCACCCGACTGATGCACACCGGAATCCGGTTTCCCGTACCCACGAATCTGGCAAC
     H  K  H  P  T  D  A  H  R  N  P  V  S  R  T  H  E  S  G  N   401
1261 CCGATTCAGAAAACTGCTGATGCACAATATAGGAACCCAACTGAAGGAGTACATAAGCAC
     P  I  Q  K  T  A  D  A  Q  Y  R  N  P  T  E  G  V  H  K  H   421
1321 CCGACTGATGCACACGGGAATCCGGTTTCCCGTACCCACGAATATGGCAACCCTATTCGA
     P  I  D  A  H  G  N  P  V  S  R  T  H  E  Y  G  N  P  I  R   441
1381 AAAATTGCTGATGGTCAAGGGGTTCACCCCACTGGA[GGTACC]ACGGAAGGCTACATAACC   KpnI
     K  I  A  D  G  Q  G  V  H  P  T  G  G  T  T  E  G  Y  I  T   461
1441 ACTAGTGGCACCAGCACCGGCATCGGCAACGGAGGTGCCATAGGCGGGCAGCAGCACGGC
     T  S  G  I  S  T  G  I  G  N  G  G  A  I  G  G  Q  Q  H  G   481
1501 GGCGTTCTTCAACGCTCCGGCTCCGGTTCGAGCTCGTCGGAGGATGATGGACAAGGTGGT
     G  V  I  Q  R  S  G  S  G  S  S  S  S  E  D  D  G  Q  G  G   501
1561 AGAAGGAAGAAGAAGGGGTTGACGGACAAGAAAAAGGAGAAGTTGTCAGGTGGGCGAAAG
     R  R  K  K  K  G  L  T  D  K  K  K  E  K  L  S  G  G  R  K   521
1621 CCAGCGCAGGCGTCGCACCCTACAGCGACGACCACAACTACAACTGGTCACGATATATAT
     P  A  Q  A  S  H  P  T  A  T  T  T  T  T  T  G  H  D  I  Y   541
1681 AAAGGTGGACAGCAGAATCAGGAGAAGAAAGGAGTCATGGAGAAGATCAAGGAGAAGCTT
     K  G  G  Q  Q  N  Q  E  K  K  G  V  M  E  K  I  K  E  K  L   561
1741 CCTGGACAGCATAACTAGGTCGTGTCCTTGCATGGGCAGCTTTTATATATATATATATTA
     P  G  Q  H  N  *                                             566
1801 CGTATGTAACAAGTGTGAAAAGTACAATAAATACTCCAAAAACTCGGACAAGCCCTCAAAT
1861 TCACTTTTTCTGACGGGTTATTTGTCCGGTTTAGCATTCTTGGTTAAAAAAAAAAAAAAA
```

1992; Howarth and Ougham, 1993) and prevent irreversible aggregation prior to acquisition of their native state conformation (Guy et al., 1998). Diverse heat shock proteins (HSPs) are unified by their characteristic capacity to function as molecular chaperones and to form stabilizing associations with proteins (Howarth and Ougham, 1993; Ukaji et al., 1999). HSPs have been found to accumulate during periods of cold acclimation (Anderson et al., 1994; Krishna et al., 1995; Guy et al., 1998; Ukaji et al., 1999) and Thomashow (1998) suggested that chaperone activity during freezing stress may help to prevent protein denaturation, resultant from freeze-induced dehydration. Whether ROD 25, 44, 48 and 60 play a role in protein binding and stabilization is not known. However, the presence of numerous repeated sequences within this protein family is intriguing. This large number of repeating segments has not been reported in other dehydrins. It is possible that this feature is connected to the extreme freeze-tolerance of *C. sericea*, and it would be interesting to investigate other species that can adapt to such extreme cold temperatures.

FIGURE 5. Comparison of deduced amino acid sequences from the open reading frames of Rod 25 and Rod 60. Sequences were aligned with Clustalw 1.7 computer software. Asterisks (*) indicate identical amino acids, whereas, (:) and (.) indicate highly conserved and conserved substitutions, respectively. Hyphens (-) indicate lack of homologous amino acids. DEYGNP-like motifs with at least 2 identical and 2 conserved substitutions are indicated by bold italics. Identification of conserved substitutions within DEYGNP-like motifs was performed with Clustalw 1.7 computer analysis (not shown). The serine stretch is designated by lower case bold, and the lysine rich regions are underlined. Within overlapping aligned regions, ROD 25 and ROD 60 are 89% identical.

```
ROD 60 MAQYGNPTEGLYKQPTDVYGNPISRTHESGNPVQRTADAQYGNPTEGVYKQPTDVYGNSI
ROD 25 ------------------------------------------------------------

ROD 60 SRTHESGNPVQKTADAQYGNPTEGVYKQPTDAHGNPISRTHESGNPVQKTADAQYGNPTE
ROD 25 ------------------------------------------------------------

ROD 60 GVYKQPTDAHGNPVSRTHESGNPVQKTADAQYGNPTEGVDKHPADAHGNPAFRTRESGNP
ROD 25 ------------------------------------------------------------

ROD 60 VQKTADAQYGNPTEGVYKHPTDAHGNPVSRTHDFGNPVQKTADAQYGNPTEGVDKHPADA
ROD 25 ------------------------------------------------------------

ROD 60 HGNTAFRTRESGNPVQKTADAQYGNPTEGVDKHPTDAHKNSISRTHDFGNPVQKTVDAQY
ROD 25 -------------------MAQYGNPTEGLYKQPTDVYGNPISRTAQSGNPVQRTADAQY
                          *********: *:***.: *.**** : *****:*.****

ROD 60 GNPTEGVYKHPTDAHRNSISRTHNSGNPIQKTADAEYGNPTEGVHKHPTDAHENPLSRTH
ROD 25 GNPTEGVYKQPTDVYGNPISRTHESGNPVQKTADAEYGNPTEGVHKHPTDAHENPLSRTH
       *********:***.: *.*****:****:*******************************

ROD 60 ESDNPIQKTAGAQNRNPTEGVHKHPTDAHRNPVSRTHESGNPIQKTADAQYRNPTEGVHK
ROD 25 ESDNPIQKTAGAQNRNPTEGVHKHPTDAHRNPVSRTHESGNPIQKTADAQYRNPTGGVHK
       ******************************************************* ****

ROD 60 HPTDAHGNPVSRTHEYGNPIRKIADGQGVHPTGGTTEGYITTSGTSTGIGNGGAIGGQQH
ROD 25 HPTDAHGNPVSRTHEYGNPIRKIADGQRVHPTGGTTEGYITTSGTSTGIGNGGRIGGQQH
       *************************** ************************* ******

ROD 60 GGVLQRSGSGssssEDDGQGGRRKKKGLTDKKKEKLSGGRKPAQASHPTATTTTTTGHDI
ROD 25 GGVLQRSGSGsssVI---------------------------------------------
       *************

ROD 60 YKGGQQNQEKKGVMEKIKEKLPGQHN
ROD 25 --------------------------
```

The accumulation of dehydrins during cold acclimation has been reported in several woody plant species (Muthalif and Rowland, 1994; Arora and Wisniewski, 1994; 1996; Arora et al., 1996; Salzman et al., 1996; Wisniewski et al., 1996; Welling et al., 1997; Rinne et al., 1998; 1999; Lim et al., 1999) and photoperiodic regulation of dehydrin accumulation has been described in birch (Welling et al., 1997; Rinne et al., 1998; 1999). We have observed seasonal changes in the RNA expression of Rod 44, 48 and 60. Transcripts were detected in October as tis-

FIGURE 6. Comparison of deduced amino acid sequences from the open reading frames of Rod 44 and Rod 48. Sequences were aligned with Clustalw 1.7 computer software. Asterisks (*) indicate identical amino acids, whereas, (:) and (.) indicate highly conserved and conserved substitutions, respectively. Hyphens (-) indicate lack of homologous amino acids. DEYGNP-like motifs with at least 2 identical and 2 conserved substitutions are indicated by bold italics. Identification of conserved substitutions within DEYGNP-like motifs was performed with Clustalw 1.7 computer analysis (not shown). The serine stretch is designated by lower case bold, and the lysine rich regions are underlined. Alignment of ROD 44 and ROD 48 revealed 69% identity.

```
ROD 48 MAQYGDQYGNPVRPTEKVYRSPTDAYGNSFPTTDEYSNPIHQTDAYGNSNK     KTDEYG
ROD 44 MAQYGDQYGNPVRPTEEVYGSPTDAYGNSLPPTDKYGNRIHQTDAHGAHGKPTHKTDKYG
       ****************:** **********:****:*.* ******:*    .:   ***:**

ROD 48 NLVHPLEKVYGSPTDAYGNPLPPTDKYGNPLHQTDAYGKPIRKTGEYG-----------D
ROD 44 NPIHPTEEVYRSPTDANGNPLTPSDKYGNPIHQTDAHRKPIRKTDKYGNPVHPTEEVYQS
       *  :** *:** ***** ****.*:******:*****: ******.:**           .

ROD 48 PTDAYGNPLPSTDKYGNPIHQS------IRRTDEYGDPVHPIEKVYRSPTDAYGNPLPTT
ROD 44 PTDANGNPLPPSDKYGNPIHQTDAHGKLIRKTDKYGNPVHPTEEVYGSPIDAYGNPLPPT
       **** *****.:**********:       **:**:**:**** *:** ***********.*

ROD 48 DKYGNPIRKTDEYGDPVHPIEKVYGSPTDAYGNPLCRTDKYGNPIHQTDTHGNPIRKTDK
ROD 44 DKYGNPNRKTDEYGNLVHPIEKVYGSQTDAYGNLLLPTDKYGNPIHQTNAHGNQNRKIDE
       ****** *******: *********** ****** *  ************::***  ** *:

ROD 48 YGDPIHPIEKVYGSPTDAYMNPLPQTDKYGNSIHQTDAHGNPIRKTDKYGNPVHPTQEAY
ROD 44 YGNPVHPIEKVYGNPTDTYGNPLPPTDKYGNPIHHTDAHGNPIGKTDKYGNSVRPTEEVY
       **:*:*********.***:* **** ******.**:********* *******.*:**:*.*

ROD 48 RSPTDAYGNPLPPTDEYGNPIHQTDAYGKPIRKTNEYGNPVHHTVPMGDYGISGATGAYG
ROD 44 RSPTDANGNPLPPIDEYGNPIHQTDVYGKPIRKTDEYGNPVHHS----------------
       ****** ****** ************.*********:*********:

ROD 48 TGTGVTGVGGQHHGGILHHSDCFsssssEDDGYGGRRTKGLTDKIKDKLRGGKKTEQLHT
ROD 44 ---------------GILHRSGCFsssssEDDGYGGRRKKRLRDKIKEKLRGGKETEQLHT
                      ****:*.*****************.* * ****:******:******

ROD 48 TATSTTTGYGYEGEEHRYYKKKGVMEKIMDKLPGQESQHY
ROD 44 TATSTPTGYG------------------------------
       *****.****
```

sues were cold acclimating and during January when freeze-tolerance was maximal (Figure 7). Expression was not detected during summer in non-acclimated tissues. Preliminary experiments also showed that Rod 25, 44, 48 and 60 are up-regulated in response to decreasing day length and controlled short day photoperiods (data not shown).

CONCLUSIONS

Dehydrins are a family of hydrophilic proteins that accumulate in response to desiccation-stress. These proteins contain characteristic motifs and are found in a wide variety of higher and lower plant species

TABLE 1. Predicted amino acid composition of the deduced peptides of Rod 25, 44, 48, and 60.

Amino Acid	ROD 25% Comp	ROD 44% Comp	ROD 48% Comp	ROD 60% Comp
Ala	5.9	3.5	3.6	7.2
Arg	5.5	4.3	3.9	4.2
Asn	7.2	7.8	5.2	7.2
Asp	4.2	7.5	8.2	5.8
Cys	0.0	0.3	0.5	0.0
Gln	6.4	2.8	3.2	6.3
Glu	4.7	5.0	4.5	5.1
Gly	13.6	10.8	12.0	12.2
His	6.4	5.8	5.0	6.3
Ile	4.2	4.5	4.3	2.6
Leu	1.3	3.8	3.2	1.1
Lys	3.8	7.3	7.0	6.7
Met	0.4	0.3	1.1	0.4
Phe	0.0	0.3	0.5	0.7
Pro	8.9	11.8	10.5	8.5
Ser	6.8	4.5	4.8	5.6
Thr	11.0	9.0	10.0	10.8
Trp	0.0	0.0	0.0	0.0
Tyr	4.2	7.8	9.1	3.9
Val	5.5	3.8	3.4	5.1

TABLE 2. Analysis of repeated motifs found in the deduced open reading frame of Rod 25, 44, 48, and 60. Two types of motifs, similar to DEYGNP and DAQYGNP were repeated throughout the open reading frames. The total number of amino acids was tallied at each position within all repeats and listed in order from greatest to least common at each position of the repeat.

ROD 25

D_3	A_4	Q_3	Y_4	G_3	N_4	P_4
M_1		E_1		R_1		

D_5	E_4	S_4	G_7	N_{10}	P_{10}
H_4	A_3	Y_3	R_1		
A_1	V_2	H_3	D_1		
	Q_1		E_1		

ROD 60

D_{10}	A_{11}	Q_{10}	Y_{11}	G_{10}	N_{11}	P_{11}
M_1		E_1		R_1		

D_{12}	A_{10}	H_{10}	G_{19}	N_{24}	P_{20}
H_{10}	E_9	S_9	R_2		S_3
R_2	D_2	Y_3	D_1		T_1
	V_2	F_2	E_1		
	N_1		K_1		

ROD 44

D_{26}	A_{10}	Y_{26}	G_{30}	N_{25}	P_{25}
E_4	K_{10}	H_4	R_3	S_5	L_3
K_2	V_7	N_3		K_3	S_2
N_1	E_4				R_1
	Q_1				Q_2
	T_1				

ROD 48

D_{25}	A_{12}	Y_{31}	G_{28}	N_{21}	P_{29}
K_5	K_7	H_2	R_3	S_6	S_3
N_1	E_7		M_1	D_4	L_1
G_1	V_5		S_1	K_2	
E_1	Q_1				
	T_1				

FIGURE 7. Seasonal changes in expression of Rod 44, 48, 60 in *C. sericea* xylem. Total RNA was isolated from lyophilized *C. sericea* xylem tissue harvested in January (acclimated), June (non-acclimated), and October (acclimated). RNA blots were hybridized with radiolabeled probes prepared from Rod 44, 48, 60 cDNAs.

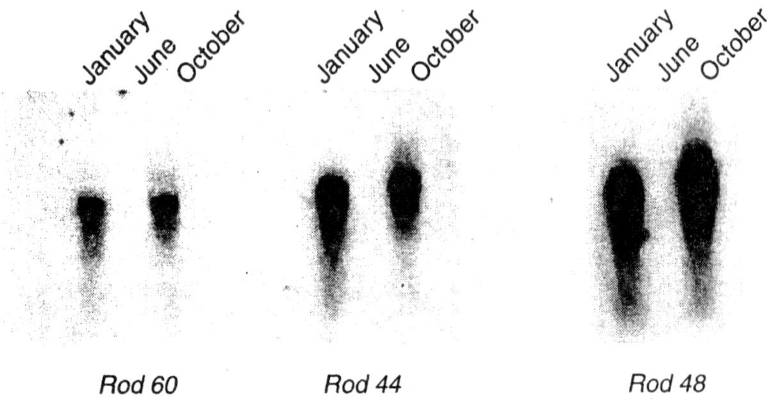

(Close, 1997). In this paper, we describe the identification and characterization of 4 cDNAs that encode a family of dehydrin-like proteins in the extremely cold tolerant shrub, *Cornus sericea.* This family of proteins appears unique among dehydrins due to the high number of repetitive Y motifs. The biological significance of this structural feature is not known but is intriguing and warrants further investigation.

REFERENCES

Alwine, J.C., D.J. Kemp and G.R. Stark. (1977). Method for detection of specific RNAs in agarose gels by transfer to diazobenzyloxymethyl-paper and hybridization of DNA probes. *Proceedings of the National Academy of Science USA* 74:5350-5354.

Anderson, J.V., Q-B. Li, D.W. Haskell and C.L. Guy. (1994). Structural organization of the spinach endoplasmic reticulum-luminal 70-kilodalton heat-shock cognate gene and expression of 70-kilodalton heat-shock genes during cold acclimation. *Plant Physiol*ogy 104: 1359-1370.

Arora, R. and M.E. Wisniewski. (1994). Cold acclimation in genetically related (sibling) deciduous and evergreen peach (*Prunus persica* [L.] Batsch). II. A 60-kilodalton bark protein in cold-acclimated tissues of peach is heat stable and related to the dehydrin family of proteins. *Plant Physiology* 105: 95-101.

Arora, R. and M. Wisniewski. (1996). Accumulation of a 60-kD dehydrin protein in peach xylem tissues and its relationship to cold acclimation. *HortScience* 31: 923-925.

Arora, R., M. Wisniewski and L.J. Rowland. (1996). Cold acclimation and alterations in dehydrin-like and bark storage proteins in the leaves of sibling deciduous and evergreen peach. *Journal of the American Society for Horticultural Science* 121: 915-919.

Asghar, R., R.D. Fenton, D.A. DeMason and T.J. Close. (1994). Nuclear and cytoplasmic localization of maize embryo and aleurone dehydrin. *Protoplasma* 177: 87-94.

Aviv, H. and P. Leder. (1972). Purification of biologically active globin messenger RNA by chromatography on oligothymidilic acid-cellulose. *Proceedings of the National Academy of Science U.S.A.* 69: 1408-1412.

Baker, J.C., C. Steele and L. Dure III. (1988). Sequence and characterization of 6 Lea proteins and their genes from cotton. *Plant Molecular Biology* 11: 277-291.

Beckmann, R.P., L.A. Mizzen and W.J. Welch. (1990). Interaction of Hsp 70 with newly synthesized proteins: Implications for protein folding and assembly. *Science* 248: 850-854.

Burke, M.J., L.V. Gusta, H.A. Quamme, C.J. Weiser and P.H. Li. (1976). Freezing and injury in plants. *Annual Review of Plant Physiol*ogy 27: 507-528.

Campbell, S.A. and T.J. Close. (1997). Dehydrins: Genes, proteins, and associations with phenotypic traits. *New Phytol*ogy 137: 61-74.

Chang, S., J. Puryear and J. Cairney. (1993). A simple and efficient method for isolating RNA from pine trees. *Plant Molecular Biology Reporter* 11: 113-116.

Close, T.J., R.D. Fenton and F. Moonan. (1993a). A view of plant dehydrins using antibodies specific to the carboxy terminal peptide. *Plant Molecular Biol*ogy 23: 279-286.

Close, T.J., R.D. Fenton, A. Yang, R. Asghar, D.A. DeMason, D.E. Crone, N.C. Meyer and F. Moonan. (1993b). Dehydrin: The protein. In *Plant responses to cellular dehydration during environmental stress*, eds. T.J. Close and E.A. Bray. American Society of Plant Physiologists. Rockville, pp. 104-118.

Close, T.J. (1996). Dehydrins: Emergence of a biochemical role of a family of plant dehydration proteins. *Physiologia Plantarum* 97: 795-803.

Close, T.J. (1997). Dehydrins: A commonality in the response of plants to dehydration and low temperature. *Physiologia Plantarum* 100: 291-296.

Dure, L. III, S.C. Greenway and G.A. Galau. (1981). Developmental biochemistry of cottonseed embryogenesis and germination: XIV. Changing mRNA populations as shown by *in vivo* and *in vitro* protein synthesis. *Biochemistry* 20: 4162-4168.

Dure, L. III. (1993). A repeating 11-mer amino acid motif and plant desiccation. *Plant Journal* 3: 363-369.

Egerton-Warburton, L.M., R.A. Balsamo and T.J. Close. (1997). Temporal accumulation and ultrastructural localization of dehydrins in *Zea mays*. *Physiologia Plantarum* 101: 545-555.

Goday, A., A.B. Jensen, F.A. Culianez-Macia, M.M. Alba, M. Figueras, J. Serratosa, M. Torrent and M. Pages. (1994). The maize abscisic acid-responsive protein Rab17 is located in the nucleus and interacts with nuclear localization signals. *Plant Cell* 6: 351-360.

Godoy, J., R. Lunar, S. Torres-Schumann, J. Moreno, R.M. Rodrigo and J.A. Pintor-Toro (1994). Expression, tissue distribution and subcellular localization of dehydrin TAS14 in salt-stressed tomato plants. *Plant Molecular Biology* 26: 1921-1934.

Guy, C. (1990). Cold acclimation and freezing stress tolerance: Role of protein metabolism. *Annual Review of Plant Physiology and Plant Molecular Biology* 41: 187-223.

Guy, C., D. Haskell and Q-B. Li. (1998). Association of proteins with the stress 70 molecular chaperones at low temperature: Evidence for the existence of cold labile proteins in spinach. *Cryobiology* 36: 301-314.

Houde, M., C. Daniel, M. Lachapelle, F. Allard, S. Laliberte and F. Sarhan. (1995). Immunolocalization of freezing-tolerance-associated proteins in the cytoplasm and nucleoplasm of wheat crown tissues. *Plant Journal* 8: 583-593.

Howarth, C.J. and H.J. Ougham. (1993). Gene expression under temperature stress. *New Phytologist* 125: 1-26.

Krishna, P., M. Sacco, J.F. Cherutti and S. Hill. (1995). Cold-induced accumulation of hsp90 transcripts in *Brassica napus*. *Plant Physiology* 107: 915-923.

Kyte, J. and R.F. Doolittle. (1982). A simple method for displaying the hydropathic character of a protein. *Journal of Molecular Biology* 157: 105-132.

Levitt, J. (1980). *Responses of plants to environmental stresses. Chilling freezing and high temperature stresses.* 2nd edn., vol 1. Academic Press, New York.

Lim, C.C., S.L. Krebs and R. Arora. (1999). A 25-kDa dehydrin associated with genotype- and age-dependent leaf freezing tolerance in *Rhododendron*: A genetic marker for cold hardiness? *Theoretical Applied Genetics* 99: 912-920.

Martin, J., S. Geromanos, P. Tempst and F.U. Hartl. (1993). Identification of nucleotide binding regions in the chaperonin proteins GroEL and GroES. *Nature* 366: 279-282.

Muthalif, M.M. and L.J. Rowland. (1994). Identification of dehydrin-like proteins responsive to chilling in floral buds of blueberry (*Vaccinium* section *Cyanococcus*). *Plant Physiology* 104:1439-1447.

Nelson, R.J., T. Ziegelhoffer, C. Nicolet, M. Werner-Washburne and E.A. Craig. (1992). The translation machinery and 70 kD heat shock protein cooperate in protein synthesis. *Cell* 71: 97-105.

Pearce, R.S. (1999). Molecular analysis of acclimation to cold. *Plant Growth Regulation* 29: 47-76.

Rinne, P.L.H., A. Welling and P. Kaikuranta. (1998). Onset of freezing tolerance in birch (*Betula pubescens* Ehrh.) involves LEA proteins and osmoregulation and is impaired in an ABA-deficient genotype. *Plant, Cell, Environment* 21: 601-611.

Rinne, P.L.H., P.L.M. Kaikuranta, L.H.W. van der Plas and C. van der Schoot. (1999). Dehydrins in cold-acclimated apices of birch (*Betula pubescens* Ehrh.): Production, localization and potential role in rescuing enzyme function during dehydration. *Planta* 209: 377-388.

Salzman, R.A., R.A. Bressan, P.M. Hasegawa, E.N. Ashworth and B.P. Bordelon. (1996). Programmed accumulation of LEA-like proteins during desiccation and cold acclimation of over-wintering grape buds. *Plant, Cell, Environment* 19: 713-720.

Sanger, F., S. Nicklen and A.R. Coulson. (1977). DNA sequencing with chain-terminating inhibitors. *Proceedings of the National Academy of Sciences USA* 74: 5463-5467.

Sarnighausen, E., D.T. Karlson and E.N. Ashworth. (2002). Seasonal regulation of a 24 kD protein from red-osier dogwood (*Cornus sericea* L.) xylem. *Tree Physiology* 22: 423-430.

Thomashow, M.F. (1990). Molecular genetics of cold acclimation in higher plants. *Advances in Genetics* 28: 99-131.

Thomashow, M.F. (1998). Role of cold-responsive genes in plant freezing tolerance. *Plant Physiology* 118: 1-7.

Thomashow, M.F. (1999). Plant cold acclimation: Freezing tolerance genes and regulatory mechanisms. *Annual Review of Plant Physiology and Plant Molecular Biology* 50: 571-599.

Ukaji, N., C. Kuwabara, D. Takezawa, K. Arakawa, S. Yoshida and S. Fujikawa (1999). Accumulation of small heat-shock protein homologs in the endoplasmic reticulum of cortical parenchyma cells in mulberry in association with seasonal cold acclimation. *Plant Physiology* 120: 481-489.

Weiser, C.J. (1970). Cold resistance and injury in woody plants. *Science* 169: 1269-1278.

Welling, A., P. Kaikuranta and P. Rinne (1997). Photoperiodic induction of dormancy and freezing tolerance in *Betula pubescens*. Involvement of ABA and dehydrins. *Physiologia Plantarum* 100: 119-125.

Wisniewski, M., T.J. Close, T. Artlip and R. Arora. (1996). Seasonal patterns of dehydrins and 70-kDa heat-shock proteins in bark tissues of eight species of woody plants. *Physiologia Plantarum* 96: 496-505.

Wisniewski, M., R. Webb, R. Balsamo, T.J. Close, X. Yu and M. Griffith. (1999). Purification, immunolocalization, cryoprotective, and antifreeze activity of PCA60: A dehydrin from peach (*Prunus persica*). *Physiologia Plantarum* 105: 600-608.

Wood, A.J. (1994). Molecular, biochemical, and physiological responses of sorghum to pre-flowering drought stress: The role of dehydrins, glycinebetaine and osmotic adjustment as protective mechanisms, PhD dissertation, Purdue University.

Inheritance of Cold Hardiness and Dehydrin Genes in Diploid Mapping Populations of Blueberry

Ganesh R. Panta
Lisa J. Rowland
Rajeev Arora
Elizabeth L. Ogden
Chon-Chong Lim

SUMMARY. Studies of herbaceous plants suggest that cold hardiness is a complex, quantitatively inherited trait. Although development of

Ganesh R. Panta is affiliated with the Department of Pharmacology, University of Tennessee Health Science Center, Memphis, TN 38163.

Lisa J. Rowland and Elizabeth L. Ogden are affiliated with the Fruit Laboratory, Henry A. Wallace Beltsville Agricultural Research Center, U.S. Department of Agriculture, Agricultural Research Service, Beltsville, MD 20705.

Rajeev Arora is affiliated with the Department of Horticulture, Iowa State University, Ames, IA 50011.

Chon-Chong Lim is affiliated with Soft Landscaping, CyPark Sdn Bnd No. 15 Jalan 16/11, 46350 Petaling Jaya, Malaysia.

Address correspondence to: Lisa J. Rowland, USDA-ARS, Fruit Laboratory, 10300 Baltimore Avenue, Building 010A, BARC-West, Beltsville, MD 20705 (E-mail: rowlandj@ba.ars.usda.gov).

Journal paper No. J-19832 of the Iowa Agriculture and Home Economics Experiment Station, Ames, Iowa, Project No. 3601, and, in part, supported by Hatch Act and State of Iowa funds.

[Haworth co-indexing entry note]: "Inheritance of Cold Hardiness and Dehydrin Genes in Diploid Mapping Populations of Blueberry." Panta, Ganesh R. et al. Co-published simultaneously in *Journal of Crop Improvement* (Food Products Press, an imprint of The Haworth Press, Inc.) Vol. 10, No. 1/2 (#19/20), 2004, pp. 37-52; and: *Adaptations and Responses of Woody Plants to Environmental Stresses* (ed: Rajeev Arora) Food Products Press, an imprint of The Haworth Press, Inc., 2004, pp. 37-52. Single or multiple copies of this article are available for a fee from The Haworth Document Delivery Service [1-800-HAWORTH, 9:00 a.m. - 5:00 p.m. (EST). E-mail address: docdelivery@haworthpress.com].

http://www.haworthpress.com/web/JCRIP

Digital Object Identifer: 10.1300/J411v10n01_04

cold hardiness is an integral part of the life cycle of woody perennial plants, studies on the genetic control of cold hardiness in woody perennials are scarce. A better understanding of the genetic control of cold hardiness would be valuable for developing more effective strategies to increase cold hardiness and, hence, climatic adaptation of woody perennial crops. In blueberry, three major dehydrins of 65, 60, and 14 kDa have been found to increase with cold acclimation and decrease with deacclimation. A comparison of these dehydrin levels among various blueberry cultivars and selections has revealed their level of accumulation to be closely associated with cold hardiness level. Efforts are underway to isolate and map the dehydrin genes of blueberry utilizing blueberry populations that segregate for cold hardiness in order to determine if the dehydrin genes map to or co-segregate with QTLs controlling cold hardiness. Progress has been made toward this goal. Cold hardiness levels were determined for a portion of the blueberry mapping populations (derived from testcrosses of *Vaccinium darrowi* Camp × *V. caesariense* Mackenz. F_1s to another *V. darrowi* and another *V. caesariense*) using a laboratory controlled freeze-thaw regime, followed by visual assessment of injury to floral buds. As expected, the *V. darrowi* and *V. caesariense* parents were found to differ significantly in terms of cold hardiness levels (LT_{50}s of −13°C and −20°C, respectively). Mean cold hardiness level of F_1s (LT_{50} of −14.7°C) was skewed toward the *V. darrowi* parents suggesting that cold hardiness is a partially recessive trait. The sequence of a 2.0 kb cDNA clone, which encodes the 60 kDa blueberry dehydrin, was used to map a dehydrin-related gene to current linkage group 12 of the *V. caesariense* testcross population. A preliminary comparison of the segregation pattern of the dehydrin-related gene to that of the cold hardiness trait suggests that the marker does not segregate with cold hardiness.

KEYWORDS. Blueberry, cold acclimation, cold hardiness, dehydrin, freezing tolerance, genetics, mapping, *Vaccinium*

INTRODUCTION

The capacity of woody perennial plants to survive winter freezes is dependent on two phenological events which occur during late summer or early fall: (1) the onset of endodormancy and (2) an increase in cold hardiness upon exposure to low non-freezing temperatures called cold acclimation. Exposure to low temperature is also required for breaking endodormancy and resumption of growth the following spring. This low temperature or chilling requirement helps prevent growth during transitory periods of warm temperatures throughout a large portion of the winter, assisting to synchronize growth with exposure to favorable temperatures. Cold acclimation, on the other hand, enables plants to withstand subfreezing temperatures during the winter. Some plants which would be killed by temperatures slightly below 0°C during the summer and early fall may survive temperatures as low as − 196°C during the winter when fully cold acclimated (Sutinen et al., 1992). Consequently, chilling requirement and cold hardiness levels are major factors determining to what degree temperate-zone fruit crops will survive the winter and early spring without damage to shoot and flower buds.

Quantitative trait loci (QTLs) are defined as two or more genetic loci that contribute cooperatively for the establishment of a specific phenotype. Genetic evidence for cold hardiness being a complex quantitative trait has been reported for numerous plants, including apple (Fejer, 1976), alfalfa (Perry et al., 1987), barley (Hayes et al., 1993), *Eucalyptus* (Tibbits et al., 1991; Byrne et al.,1997), potato (Stone et al., 1993), wheat (Sutka, 1981; Brule-Babel and Fowler, 1988; Roberts, 1990; Galiba et al., 1995), rye (Brule-Babel and Fowler, 1989), and oilseed *Brassica* (Teutonico et al., 1995). Studies on the gene action of cold hardiness in wheat and potato indicate that cold hardiness has additive and dominance components and that frost sensitivity is a partially dominant trait (Sutka, 1981; Stone et al., 1993). With respect to recent molecular genetic research on cold hardiness, Jaglo-Ottosen et al. (1998) isolated the regulatory gene *CBF1*, which is responsible for induction of a family of cold-regulated or COR proteins (some of which are members of a family of proteins known as dehydrins) in *Arabidopsis*. CBF1 binds a DNA regulatory element present in the promoter region of many COR genes and stimulates transcription in response to low temperature. By making a construct of the *CBF1* gene for its constitutive overexpression, researchers were able to develop transgenic plants with elevated COR gene expression in the absence of low temperature treatment. Transgenic plants that were not cold acclimated, but were overexpressing CBF1, were shown to be more frost-tolerant than nonacclimated control plants and about as tolerant as cold acclimated plants.

Despite the critical role cold hardiness plays in determining the winter and early spring survival of woody perennial plants, few studies have addressed the genetic or molecular genetic control of cold hardiness in woody perennials. This scarcity of information is due to many factors including long generation times, inbreeding depression, and heterozygosity of various traits in parental populations. Despite these difficulties, limited genetic studies in apple (Watkins and Spangelo, 1970; Fejer, 1976) and *Eucalyptus* (Tibbits et al., 1991) have been performed and suggest that cold hardiness is inherited in an additive manner.

The United States is the world's leading blueberry (*Vaccinium* spp.) producer. A survey of blueberry research and extension scientists in the United States has identified (1) lack of cold hardiness and (2) susceptibility to spring frosts as the most important genetic limitations of current blueberry cultivars (Moore, 1993). Studies on the genetics of cold hardiness in blueberry, however, are lacking. One study on the heritability of winter injury in the field found heritability estimates over three years to be low although individual year estimates were higher. In addition, general combining ability variance components were consistently larger than specific combining ability suggesting that additive variance is more important than nonadditive variance (Fear et al., 1985). However, because winter survival in the field is determined by multiple factors, not only cold hardiness in the cold acclimated state, the authors concluded that screening for cold hardiness under controlled conditions would be desirable before concrete conclusions could be drawn concerning additive, dominance, and epistatic interactions.

We have been using a combination of molecular, genetic, and physiological approaches to investigate genetic controls of both chilling requirement and cold hardiness in the cold acclimated state in blueberry with the ultimate goal of identifying markers for use in marker-assisted breeding approaches. Marker-assisted strategies would be particularly useful in the case of woody perennials like blueberry for following such complex phenotypes as cold hardiness and chilling requirement in breeding populations. Previously, we identified three major proteins of 65, 60, and 14 kDa in blueberry floral buds, whose levels correlate well with cold hardiness levels (Muthalif and Rowland, 1994; Arora et al., 1997). Further study revealed that these proteins are dehydrins (Muthalif and Rowland, 1994). Dehydrins are heat-stable, glycine-rich plant proteins that are induced in response to dehydration stress, such as drought, cold, and seed maturation, and also in response to abscisic acid (Close, 1996). Efforts are underway in our laboratory to isolate and map the dehydrin genes of blueberry utilizing blueberry populations that segregate for cold hardiness in order to determine if the dehydrin genes map to or co-segregate with QTLs controlling cold hardiness.

A 2.0 kb cDNA, which encodes the 60 kDa dehydrin from blueberry, has been cloned and sequenced (Levi et al., 1999). Here, we report using the DNA sequence of the cDNA clone to map a dehydrin or dehydrin-related gene in our diploid blueberry population segregating for chilling requirement and cold hardiness. In addition, portions of our mapping populations were evaluated for cold hardiness in the cold acclimated state and mean cold hardiness (LT_{50}s) of parent, F_1, and testcross populations were calculated and analyzed for significant differences and degree of dominance. Furthermore, a preliminary comparison was made of the segregation pattern of the dehydrin-related gene to that of the cold hardiness trait.

MATERIALS AND METHODS

Plant Material

The blueberry populations being used for mapping QTLs controlling chilling requirement and cold hardiness, and for mapping dehydrins, have been described previously in detail (Rowland et al., 1999). Briefly, two diploid testcross populations were generated by crossing *V. darrowi* × *V. caesariense* hybrids [Fla4B × W85-20 (#5, #6, and #10)] back to another *V. darrowi* and another *V. caesariense* selection, NJ88-13-15 and W85-23, respectively. According to the more recent classification of blueberry species, *V. caesariense* is now generally referred to as diploid *V. corymbosum* L. (Galletta and Ballington, 1996). However, to avoid confusion with tetraploid *V. corymbosum*, here we have used the older *V. caesariense* terminology. *V. darrowi* parents have a lower chilling requirement and are more cold sensitive than *V. caesariense* parents. For mapping purposes, segregation of the dehydrin-related gene was followed in a total of 62 plants of the testcross populations. For cold hardiness studies, 95 plants were evaluated: original and testcross *V. darrowi* parents, Fla4B and NJ88-13-15 (called P_1 population); original and testcross *V. caesariense* parents, W85-20 and W85-23 (P_2); 9 Fla4B × W85-20 F_1s (including #5, #6, and #10, as well as 6 other F_1s); 40 F_1 × NJ88-13-15 testcross plants (TC_1); and 42 F_1 × W85-23 testcross plants (TC_2).

Cold Hardiness Determinations

Four-to-five year-old potted plants (3-7 clones of each individual plant) were transferred from the greenhouse to a cold room in November and were cold acclimated by exposure to 4°C for 4 weeks at a photoperiod of 10/14 h (day/night) with 100 μmol $m^{-2}s^{-1}$ photosynthetic photon-flux density. Freezing treatment of floral buds was as described by Arora et al. (2000). This freeze-thaw test entailed placing three 5-6 cm long shoots with 3-8 attached buds per treatment temperature (−6°C to −24°C at 2°C increments) from each plant in test tubes (one shoot/tube) with 0.5 ml water. Shoots were arranged in a completely randomized block design. The samples were subjected to controlled freezing in a glycol bath (Model 2325, Forma Scientific, Marietta, OH). Ice nucleation was initiated at −1°C, and samples further cooled at 1°C/hr down to −4°C, 1°C/30 min down to −8°C, and 2°C/30 min down to −24°C. Samples were thawed overnight at 4°C followed by a 24-h incubation at 23°C, at which time buds were dissected and observed for visual browning (Fear and Lawson, 1987; Flinn and Ashworth, 1994). Buds were rated for 0-100% browning and the cold hardiness level (LT_{50}) defined as the temperature causing 50% injury/browning.

DNA Extraction and PCR

DNA was extracted from blueberry leaf tissue using a CTAB (hexadecyltrimethyl-ammonium bromide) procedure (Doyle and Doyle, 1987) as modified by Rowland and Ngyen (1993) but with the omission of the polyethylene glycol precipitation. Reaction conditions for PCR were according to Levi et al. (1993) except that 0.1% w/v BSA was used in place of the gelatin (Stommel et al., 1997). Reactions using *bbdhn1* primers were run for 40 cycles of 92.5°C (40 sec), 58°C (70 sec), and 72°C (2 min) in a PTC-100 thermal cycler (MJ Research, Watertown, MA). Amplified DNA fragments were separated through 1.4% agarose gels along with molecular weight standards (1 kb ladder, Life Technologies, Gaithersburg, MD). DNA fragments were stained with ethidium bromide and photographs taken using the Eagle Eye II still video system (Stratagene, CA).

Linkage Analysis of Dehydrin-Related Gene

Sequence of the 2.0 kb *bbdhn1* cDNA was analyzed to design primer pairs for PCR amplification using MacVector 5.0 software (Scientific Imaging System, New Haven, CT). Selected primer sequences were sent to Informagen, Inc. (Newington, NH) for oligonucleotide synthesis. The dehydrin-related marker was placed on the current blueberry

RAPD marker map (Rowland et al., 1999) using segregation data and the computer program MAPMAKER (Lander et al., 1987).

RESULTS AND DISCUSSION

Cold Hardiness Determinations

Mean cold hardiness (LT_{50}) values of the five blueberry populations (two parental, one F_1, and two testcross populations) analyzed are presented in Table 1. Mean cold hardiness of the *V. darrowi* and *V. caesariense* parent populations differed significantly, with LT_{50}s of −13.0°C and −20.0°C, respectively. The mean cold hardiness of the F_1 population (−14.7°C) was closer to that of the *V. darrowi* parent than to the *V. caesariense* parent population. According to Falconer (1989), degree of dominance is calculated as h/d, where h is the difference between the F_1 population and parental mean (deviation due to the dominance effect) and d is the difference between the parental population and their mean (deviation due to the additive effect). In this case, $h/d = |(-14.7°C) - (-16.5°C)|/|(-13.0°C) - (-16.5°C)| = 0.51$. Theoretically, if h/d = 0, then the variance is due to completely additive effects. If h/d = 1, then the variance is due to a complete dominance effect. A h/d value of 0.51 suggests that the variance in cold hardiness may be explained by both additive and dominance effects. This, along with the skewed cold hardiness mean of the F_1 population toward the *V. darrowi* parent, suggests that cold hardiness (as opposed to cold sensitivity) is a partially recessive trait. These results are in accordance with other reports from herba-

TABLE 1. Mean cold hardiness (LT_{50},°C) of parental, F_1, and testcross generations of a cross between *V. darrowi* and *V. caesariense*.

Generations	Sample size (n)	LT_{50} (°C) Means ± SE
P_1: *V. darrowi* (Fla4B, NJ88-13-15)	2	−13.0 ± 1.00
P_2: *V. caesariense* (W85-20, W85-23)	2	−20.0 ± 1.00
F_1: (Fla4B × W85-20)	9	−14.7 ± 0.33
TC_1: F1 × *V. darrowi* (NJ88-13-15)	40	−14.1 ± 0.38
TC_2: F1 × *V. caesariense* (W85-23)	42	−17.9 ± 0.36

ceous and woody plant genera, namely *Solanum* (Stone et al., 1993), wheat (Sutka, 1981), and *Eucalyptus* (Tibbits et al., 1991). In addition, results from our own previously reported generation means analysis of the cold hardiness data revealed that cold hardiness in blueberry is controlled largely by additive gene effects and, to a lesser degree, by dominance gene effects. Testing of various genetic models also indicated that cold hardiness inheritance could be adequately explained by a simple additive-dominance model of gene action (Arora et al., 2000). Previous analyses of results from reciprocal crosses also indicated that there is no significant maternal effect (cytoplasmic) on cold hardiness (Arora et al., 2000).

The $F_1 \times$ *V. darrowi* testcross population had a mean cold hardiness of −14.1°C whereas the $F_1 \times$ *V. caesariense* testcross population had a mean cold hardiness of −17.9°C. The cold hardiness levels of the testcross populations had continuous and normal (bell-shaped) distributions (data not shown), which is as expected for a multigenic trait. Most of the individuals, with a few exceptions, were distributed between the parental values. In both testcross populations, there were a few individuals recovered which had the same cold hardiness levels as the recurrent parents. Because the parental phenotypes were fully recovered in the relatively small populations of 40 and 42 plants, results suggest that cold hardiness may be controlled by relatively few genes. The probability that backcross or, in this case, testcross segregants will have cold hardiness levels the same as the recurrent parent is $(1/2)^n$, where n = number of gene pairs, assuming additive gene effects. In each of the testcross populations, approximately 5 out of 40 progeny or 1/8 [$= (1/2)^3$] had the same cold hardiness level as the recurrent parent, indicating that as few as 3 genes may be responsible for the cold hardiness variation observed in these populations.

Mapping Dehydrin-Related Gene

The two testcross populations described above, which segregated for both chilling requirement and cold hardiness, are being utilized to construct genetic linkage maps of blueberry, with the ultimate goal of mapping QTLs controlling chilling requirement and cold hardiness. Testcross, rather than true backcross populations, are being used because diploid blueberry tolerates very little inbreeding. The *V. darrowi* clones, Fla4B (original parent) and NJ88-13-15 (testcross parent), are evergreen, low-chilling, cold-sensitive, lowbush plants whereas the *V. caesariense* clones, W85-20 (original parent) and W85-23 (testcross parent), are deciduous, high-chilling, cold-hardy, highbush plants. Fla4B and NJ88-13-15 are wild selections collected from Florida; W85-20 and W85-23 are wild selections from New Jersey. Also, Fla4B has been used extensively in breeding programs to introduce low chilling requirement into the cultivated highbush (*V. corymbosum* L.) background. F_1s were obtained by crossing Fla4B with W85-20, using Fla4B as the female parent. To generate the mapping populations, three of the F_1s were crossed to testcross parents NJ88-13-15 and W85-23. Although F_1s were mainly used as female parents, some reciprocal crosses were made as well.

To map blueberry dehydrin gene *bbdhn1* [represented by a 2.0 kb cDNA clone, which encodes the 60 kDa dehydrin (Levi et al., 1999)], or possibly a related dehydrin gene, primers were designed which should result in amplification in PCRs of the entire coding region of the *bbdhn1* gene (Figure 1). The forward primer was based on the 49-73 base sequence and the reverse primer on the 1305-1281 base sequence.

FIGURE 1. PCR primers designed from the *bbdhn1* cDNA sequence. Twenty-five nucleotide long primers were selected such that they had 45-55% GC content, T_m of 55 to 60°C, and 3′ end of G or C. The forward (49-73 bases) and reverse (1305-1281 bases) primer sequences are shown in bold face. Start (ATG) and stop (TAG) codons are also in bold face.

1 TTGATGTTAACTTATTTTGACGGTAGATGAGGAAATTAATTTATTCAA**GGAAAGAGAAGG**

61 **CTCATAGGCGTTG**GGATTGTGATATTGACGTTTCAATTGACACTTGCTAACTTCACTGCG

121 CCTCGTACTCGTATAGTCGCATCCAAGGCTATTAACGAACTTTCACGTCCCCCTCAAACG

181 AATAATCCTCTATTTTGGCGCCTATAAAACGATACCCTTGGACGCTAGTTCTCTTCACAG

241 CCAATCCCCAATCCACGATTCCCTTTTATATTACAGAAAACAGAGAGTGAGAGAAACAAA

301 CAGAGAGGAGAGACAG**AAAATG**GCAGGAATCATGAACAAGATTGGCCAGACCCTCCCTTG

361 TGGAGGCAACAAGGAAGAGGACAAGTACAAGGGAGGAGAGCAGCATCAGCAGCAGCAGTA

421 CAACAAACCAGGTCAGCACCAGGGCGAGTCCAGGCAGGACCAGCGCAAGGAAGGGGGAGG

481 GCTTATGGGCGAGTTGAAGGATAAGATCCATGGCGGGGGTGTCGGTACTGATGGCGGCGG

541 TGTTGGTAGTGTTGATCCGCATGGTCACCAGGGCGGCAAATACAGGCATGATCAACAACA

601 CAGCGAGTACGTGCAGGATCAGCAGCCCGGCGGATATAGGCAGGATCAGCGCAAGGAAGG

661 GGGAGGGCTTATGGACAAGGTGAAGGATAAAATCCATGGCGGCGGTGTTGGTAGTGCTGA

721 CCAGCACCAGGACGGATACAAGCAGGATCAGCTGCATGGCGGGTACAAGCAGGATCAGCA

781 GCTTGGCGGGTATAGGCAGGATCAGCACAAGGAAGGCGGAGGGCTTATCTACAAGGTGAA

841 GGACAAGATCCATGGCGGCGATGGTGGTAGTGCTGATCAGCACCAGGGCATATACGGGCA

901 GGATCAACAGCTCGGCGGGTACAGGCAGGATCAACAGCTCGGCGGGTACAGGCAGGATCA

961 ACAGCATGGCGAGTACAAGCAGGATCAACGCAAGGAAGGGGGAGGGCTTATGGACAAGGT

1021 GAAAGATAAGATCCCTGGCGGCAATGGTGGTAGTGCTGCTGATCAGCACCAGGGCGTATA

1081 CGGGCAGGATCAACAGCTCGGCGGGTACAGGCAGGATCAACAGCGTGGCGAGTACAAGCA

1141 GGATCAACGCAAGGAAGGGGGAGGGCTTATGGACAAGGTGAAAGATACGATCCATGGCGG

1201 CGCTGGTGGTGGTGCTGATAAGCACCGCGGCGAGTACAAGCAGGATCAGTACCGCGGTGA

1261 T**TAG**ATTGATTCATCATCAC**GTCTTCTTTTTCGCCTCTACGACTC**TTTAGGTAACTACAT

1321 ACACATATTTGTATATAGATATGCAAACAACCGATGTATGTATGTATGTTATTTTTTTTT

1381 TTTCCTTATTCTGAATTTTTTTTTTTTCCAGGTGTGGATAGAGTCCCATGGGAGATGCAAA

1441 TAAGTGGGGGAAAAGGAATGAGAGTGTGTTTTTCTCTTCAGTATCTATTTTGTAGTGTTA

1501 ATATACGAGTTGTAATCTGAGTGGGAGTTCAATATAATGTCTCTTTTATATGTTTGAGTA

1561 TAAGAAATGTTGCAGAAATTCTTGTCTATGCCTTCTTTTTTCTTTTCTTTTTTTTTTTAT

1621 CTAATTCTTCTCCATCTTTTTTTTTTTGGATCCGAAGCCCTGTTTCATTAAGCATGTTCTG

1681 CAAAAGCAGCAAACCGACTCATTCTCAAAATGGCTTAATGTGATTTACAAGACAGACCAT

1741 TATCTCACAATACATCAACCAAATTTTATTTTGCCAATAAGTAAATTCGTGAAACATCCC

1801 TAATTTGAACTTGTGAACTTAGGTGTTGTTTGCGCACATAGATTATCTATAACCGTGTAA

1861 GTGATATCAATCACGTAATTTGAACTAGCAAATTGACATGCCTTGACAAATAAGTGATTT

1921 GAATCATGCAATCTAGCGAATTGCCTGTGTTTGTAGGCACGTCACAATACAATTTGGC

First, PCRs were performed using DNA from the original parents, testcross parents, and F_1s to determine if a polymorphic marker would be generated by the above primers that could be used for mapping. Two intense bands of 2.0 and 1.8 kb were amplified in PCR (Figure 2). The 2.0 kb band was present in all plant DNA samples whereas the 1.8 kb band was present in one original parent (Fla4B), absent in the other original parent (W85-20), absent in both testcross parents (NJ88-13-15 and W85-23), and present in two of the three F_1s (#6 and #10). Hence, the 1.8 kb marker was a good candidate for mapping in subpopulations derived from F_1s #6 and #10. A very faint band of 1.3 kb, not readily visible in the gel photograph shown in Figure 2, was also present in all plant DNA samples, and, therefore, could not be mapped.

The segregation pattern of the 1.8 kb marker is shown in Figure 3. Overall, the 1.8 kb marker segregated in the subpopulations at a ratio of 23:39 (+:−), which was significantly different from the expected 1:1 ratio ($\chi^2_{0.05,1df} = 3.84 < 4.12$). Skewed segregation is common in woody perennials (reviewed by Torres et al., 1985), especially in wide crosses (Weeden, 1989). Our mapping population was obtained from inter-

FIGURE 2. PCR amplification products from original parents Fla4B and W85-20, testcross parents NJ88-13-15 and W85-23, and F1s #5, 6, and 10 of blueberry mapping populations using the *bbdhn1*-derived primers shown in Figure 1. 1, 2: Each reaction was performed in duplicate. Amplification products of 2.0 and 1.8 kb are indicated.

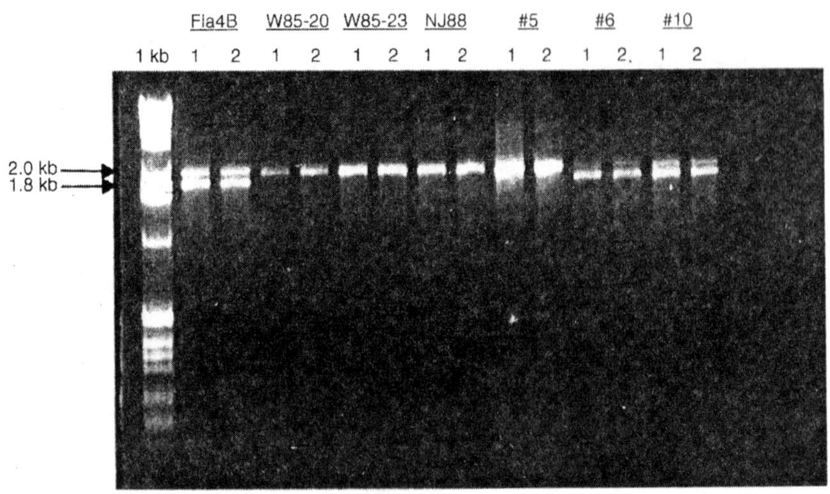

FIGURE 3. Segregation pattern of the polymorphic 1.8 kb PCR marker in a mapping subpopulation. Here, the 1.8 kb marker is present in 13 of 28 plants. One reaction (#27) did not work on the gel shown. Forward and reverse primers designed from the *bbdhn1* cDNA sequence were used to amplify the genomic DNA.

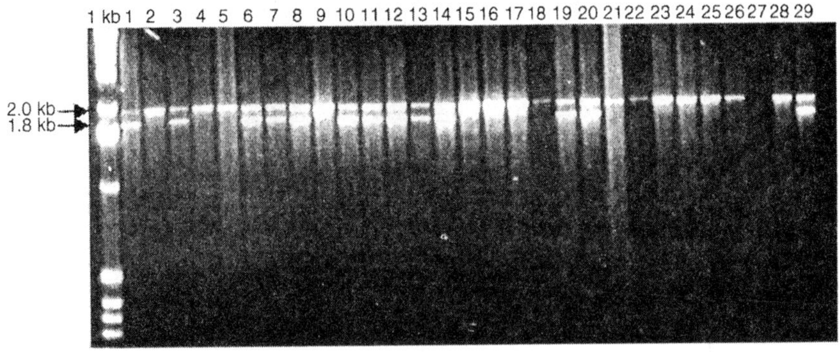

specific crosses; hence, the distorted segregation was not surprising. Skewedness in segregation ratios of molecular markers is not thought to result from pleiotropy or epistasis, but to reflect an abnormal segregation (Weeden, 1989). The skewedness does not prevent the use of such loci in mapping studies (Durham et al., 1992).

To date ~70 RAPD markers have been mapped in the *V. caesariense* testcross population spanning 795.0 cM, in 20 linkage groups (Rowland et al., 1999 and unpublished). Using segregation data for the 1.8 kb marker, designated D100, it was placed on the *V. caesariense* genetic linkage map. As shown in Figure 4, the marker mapped to linkage group 12 (if numbered from longest to shortest), 7.2 cM from the RAPD marker OPAA1b. To date, the marker does not show linkage to any of the RAPD markers mapped in the *V. darrowi* testcross population. Efforts are currently underway in our laboratory to determine if the 1.8 kb marker is actually an allele of *bbdhn1* or a related dehydrin. The 2.0, 1.8, and 1.3 kb fragments from Fla4B have been cloned, as have the 2.0 and 1.3 kb fragments from W85-20; and all are being sequenced. The expected size of the fragment amplified from *bbdhn1*, if the gene contains no introns, is 1256 bp. However, *bbdhn1* may contain one or more introns or the *bbdhn1* alleles from Fla4B and W85-20 may be larger than those from 'Bluecrop', the cultivar from which *bbdhn1* was cloned. Sequencing the fragments should clarify this.

FIGURE 4. Current RAPD-based genetic linkage map of blueberry population derived from a cross between diploid *V. darrowi* × *V. caesariense* and *V. caesariense*. Marker names and distances between adjacent markers are shown to the right of each linkage group. If no distance is indicated between markers, then no recombination was detected between those markers. Total length in cM is indicated above each linkage group. The 1.8 kb dehydrin-related marker is designated D100. Reprinted with permission from the publisher of *HortScience* for the Figure 3 of the article by Rowland et al. 1999, 34(7):1185-1191.

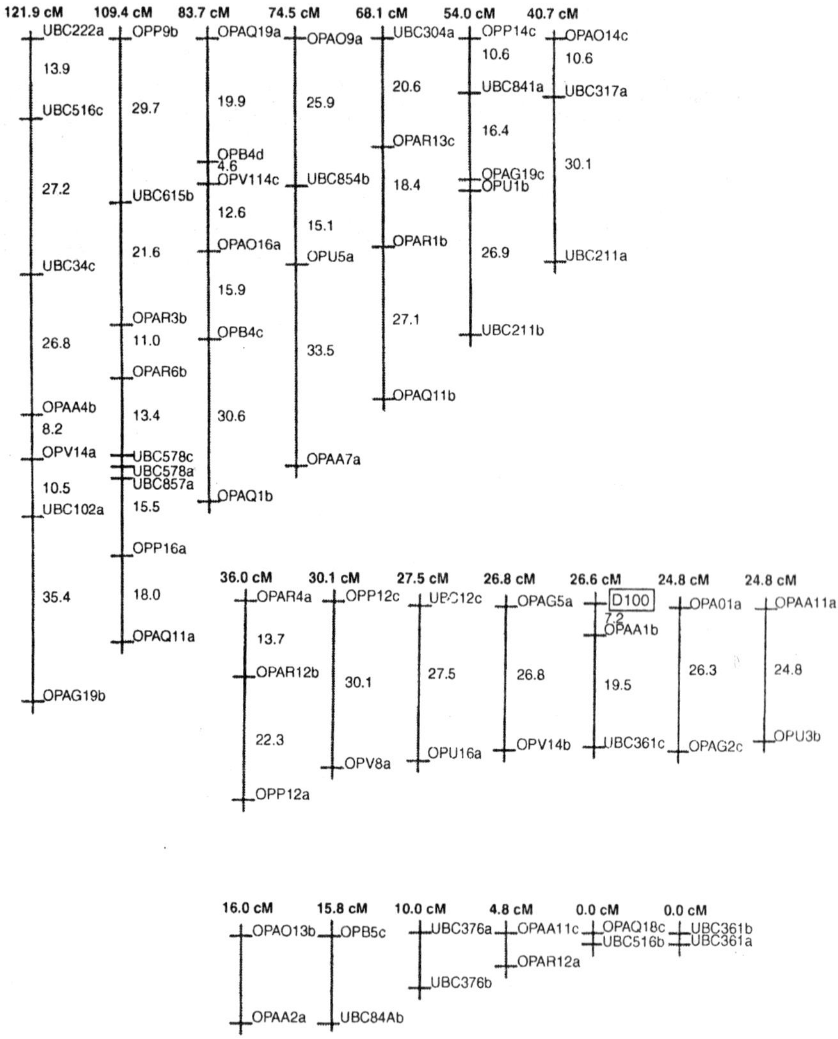

Comparison of the Segregation Pattern of the Dehydrin-Related Marker and Cold Hardiness

Although cold hardiness in plants is a multigenic trait, a preliminary comparison was made of the segregation pattern of the 1.8 kb marker to that of the cold hardiness trait. The subpopulations of the mapping populations derived from F_1s #6 and #10 crossed to testcross parents NJ88-13-15 and W85-23 were divided into two groups, those that carry the 1.8 kb marker and those that do not. The distribution of cold hardiness was examined and the mean cold hardiness determined for each group. Distribution of cold hardiness appeared continuous and normal for each group. Based on cold hardiness and marker segregation data that we have available at this time, mean cold hardiness for the NJ88-13-15 testcross subpopulation that carries the 1.8 kb marker was −15.0°C and mean cold hardiness for the NJ88-13-15 testcross subpopulation that does not carry the 1.8 kb marker was −15.2°C. Mean cold hardiness for the W85-23 testcross subpopulation that carries the 1.8 kb marker was −18.3°C while mean cold hardiness for the W85-23 testcross subpopulation that does not carry the 1.8 kb marker was −18.5°C. The mean cold hardiness of each group (based on presence or absence of the marker) was not significantly different, suggesting that inheritance of the 1.8 kb marker may not be a factor in determination of cold hardiness. These results are preliminary, based on small population sizes, and may change as cold hardiness and marker segregation data become available for more plants. Also, as stated earlier, cold hardiness is determined by multiple genes; thus, we do not expect to see an absolute correlation between segregation of cold hardiness and any one gene. In addition, if expression levels of some of the genes determining cold hardiness are controlled by one regulatory gene, like the *CBF1* gene in *Arabidopsis* (Jaglo-Ottosen et al., 1998), inheritance of cold hardiness may correlate with inheritance of the regulatory gene rather than the genes with the more direct effect on cold hardiness. Finally, the 1.8 kb fragment may not be an allele of *bbdhn1* but an allele of a related gene that may or may not even be expressed and may not play a role in cold hardiness determination. Sequencing the various fragments amplified from the parent plants should enable us to determine if the 1.8 kb fragment is a *bbdhn1* allele.

CONCLUSIONS

Cold hardiness in plants is a complex quantitative trait. Mapping QTLs controlling cold hardiness and identification of molecular markers associated with major cold hardiness QTLs would aid in breeding plants for greater climatic adaptation. Plants could be evaluated at the seedling stage for the likelihood of having an appropriate level of cold hardiness by the presence or absence of specific DNA markers. In blueberry, levels of three major dehydrins of 65, 60, and 14 kDa increase with chill unit accumulation and cold acclimation. Their level of accumulation is closely associated with cold hardiness level. Progress has been made toward mapping QTLs controlling cold hardiness and mapping dehydrins in blueberry populations that segregate for cold hardiness and chilling requirement. Evaluation of portions of the mapping populations for cold hardiness reveal that the two parent populations differ significantly in terms of their mean cold hardiness levels, as expected. Mean cold hardiness of the F_1 population is skewed toward that of the *V. darrowi* parent population suggesting that cold hardiness is a partially recessive trait. Distribution of cold hardiness in the testcross populations is continuous and normal, as expected for a multigenic trait. The number of segregants in the testcross populations with parental-like phenotypes suggests that cold hardiness may be controlled by relatively few genes. Sequence of the *bbdhn1* gene, encoding the 60 kDa dehydrin, was used to map a 1.8 kb fragment in the *V. caesariense* testcross population. The fragment may be an allele of *bbdhn1* or may be derived from a related gene. Preliminary analyses suggest that the marker does not segregate with cold hardiness.

REFERENCES

Arora, R., L.J. Rowland and G.R. Panta. (1997). Cold hardiness and dormancy transitions in blueberry and their association with accumulation of dehydrin-like proteins. *Physiol. Plant.* 101:8-16.

Arora, R., L.J. Rowland, J.S. Lehman, C.C. Lim, G.R. Panta and N. Vorsa. (2000). Genetic analysis of freezing tolerance in blueberry (*Vaccinium* section *Cyanococcus*). *Theor. Appl. Genet.* 100:690-696.

Brule-Babel, A.L. and D.B. Fowler. (1988). Genetic control of cold hardiness and vernalization requirement in winter wheat. *Crop Sci.* 28:879-884.

Brule-Babel, A.L. and D.B. Fowler. (1989). Genetic control of cold hardiness and vernalization requirement in rye. *Genome* 32:19-23.

Byrne, M., J.C. Murrell, J.V. Owen, E.R. Williams and G.F. Moran. (1997). Mapping of quantitative trait loci influencing frost tolerance in *Eucalyptus nitens*. *Theor. Appl. Genet.* 95:975-979.

Close, T.J. (1996). Dehydrins: emergence of a biochemical role of a family of plant desiccation proteins. *Physiol. Plant.* 97:795-803.

Doyle, J.J. and J.L. Doyle. (1987). A rapid DNA isolation procedure for small quantities of fresh leaf tissue. *Phytochem. Bull.* 19:11-15.

Durham, R.E., P.C. Liou, F.G. Gmitter, Jr. and G.A. Moore. (1992). Linkage of restriction fragment length polymorphisms and isozymes in *Citrus*. *Theor. Appl. Genet.* 84:39-48.

Falconer, D.S. (1989). Introduction to quantitative genetics, Ed 3. Longman Scientific and Technical, UK. p. 438.

Fear, C.D. and V.F. Lawson. (1987). Cold injury to flower buds and shoots of blueberry cultivars following extreme low fall temperatures. *Fruit Varieties J.* 41:148-150.

Fear, C.D., F.I. Lauer, J.J. Luby, R.L. Stucker and C. Stushnoff. (1985). Genetic components of variance for winter injury, fall growth cessation, and off-season flowering in blueberry progenies. *J. Amer. Soc. Hort. Sci.* 110:262-266.

Fejer, S.O. (1976). Combining ability and correlations of winter survival, electrical impedance and morphology in juvenile apple trees. *Can. J. Plant Sci.* 56:303-309.

Flinn, C.L. and E.N. Ashworth. (1994). Blueberry flower hardiness is not estimated by differential thermal analysis. *J. Amer. Soc. Hort. Sci.* 119:295-298.

Galiba, G., S.A. Quarrie, J. Sutka, A. Morgounov and J.W. Snape. (1995). RFLP mapping of the vernalization (*Vrn1*) and frost resistance (*Fr1*) genes on chromosome 5A of wheat. *Theor. Appl. Genet.* 90:1174-1179.

Galletta, G.J. and J.R. Ballington. (1996). Blueberries, cranberries and lingonberries. In: J. Janick and J.N. Moore (eds.) Fruit Breeding. Vol. II, Vine and Small Fruit Crops. Prentice Hall, New York, pp. 1-107.

Hayes, P.M., T.K. Blake, T.H.H. Chen, S. Tragoonrung, F. Chen, A. Pan and B. Liu. (1993). Quantitative trait loci on barley (*Hordeum vulgare*) chromosome 7 associated with components of winter hardiness. *Genome* 36:66-71.

Jaglo-Ottosen, K.R., S.J. Gilmour, D.G. Zarka, O. Schabenberger and M.F. Thomashow. (1998). Arabidopsis *CBF1* overexpression induces *COR* genes and enhances freezing tolerance. *Science* 280:104-106.

Lander, E.S., P. Green, J. Abrahamson, A. Barlow, M.J. Daly, S.E. Lincoln and L. Newburg. (1987). MAPMAKER, an interactive computer package for constructing primary genetic linkage maps of experimental and natural populations. *Genomics* 1:185-199.

Levi, A., G.R. Panta, C.M. Parmentier, M.M. Muthalif, R. Arora, S. Shanker and L.J. Rowland. (1999). Complementary DNA cloning, sequencing, and expression of an unusual dehydrin from blueberry floral buds. *Physiol. Plant.* 107:98-109.

Levi, A., L.J. Rowland and J.S. Hartung. (1993). Production of reliable randomly amplified polymorphic DNA (RAPD) markers from DNA of woody plants. *HortScience* 28:1188-1190.

Moore, J.N. (1993). The blueberry industry of North America. *Acta Hort.* 346:15-26.

Muthalif, M.M. and L.J. Rowland. (1994). Identification of dehydrin-like proteins responsive to chilling in floral buds of blueberry (*Vaccinium*, section *Cyanococcus*). *Plant Physiol.* 104:1439-1447.

Perry, M.C., M.S. Mcintosh, W.J. Wiebold and M. Welterlen. (1987). Genetic analysis of cold hardiness and dormancy in alfalfa. *Genome* 29:144-149.

Roberts, D.W.A. (1990). Identification of loci on chromosome 5A of wheat involved in control of cold hardiness, vernalization, leaf length, rosette growth habit, and height of hardened plants. *Genome* 33:247-259.

Rowland, L.J. and B. Ngyen. (1993). Use of PEG for purification of DNA from leaf tissue of woody plants. *BioTechniques* 14:735-736.

Rowland, L.J., E.L. Ogden, R. Arora, C.C. Lim, J.S. Lehman, A. Levi and G.R. Panta. (1999). Use of blueberry to study genetic control of chilling requirement and cold hardiness in woody perennials. *HortScience* 34:1185-1191.

Stommel, J.R., G.R. Panta, A. Levi and L.J. Rowland. (1997). Effects of gelatin and BSA on the amplification reaction for generating RAPD. *Biotechniques* 22:1064-1066.

Stone, J.M., J.P. Palta, J.B. Bamberg, L.S. Weiss and J.F. Harbage. (1993). Inheritance of freezing resistance in tuber bearing *Solanum* species: Evidence for independent genetic control of nonacclimated freezing tolerance and cold acclimation capacity. *Proc. Natl. Acad. Sci. USA* 90:7869-7873.

Sutka, J. (1981). Genetic studies of frost resistance in wheat. *Theor. Appl. Genet.* 59:145-152.

Sutinen, M.L., J.P. Palta and P.B. Reich. (1992). Seasonal differences in freezing stress resistance of needles of *Pinus nigra* and *Pinus resinosa*: Evaluation of the electrolyte leakage method. *Tree Physiol.* 11:241-254.

Teutonico, R.A., B. Yandell, J.M. Satagopan, M.E. Ferreira, J.P. Palta and T.C. Osborn. (1995). Genetic analysis and mapping of genes controlling freezing tolerance in oilseed *Brassica. Mol. Breed.* 1:329-339.

Tibbits, W.N., B.M. Potts and M.H. Savva. (1991). Inheritance of freezing resistance in interspecific F_1 hybrid of *Eucalyptus. Theor. Appl. Genet.* 83:126-135.

Torres, A.M., T. Mau-Lastovicka, T.E. Williams and R.K. Soost. (1985). Segregation distortion and linkage of *Citrus* and *Poncirus* isozyme genes. *J. Hered.* 76: 289-294.

Watkins, R. and L.P.S. Spangelo. (1970). Components of genetic variance for plant survival and vigor of apple trees. *Theor. Appl. Genet.* 40:195-203.

Weeden, N.F. (1989). Application of isozymes in plant breeding. In: J. Janick (ed.) Plant Breeding Reviews, vol. 6, Timber Press, Portland, OR, pp. 11-54.

Molecular Genetic and Physiological Analysis of the Cold-Responsive Dehydrins of Blueberry

L. J. Rowland
G. R. Panta
S. Mehra
C. Parmentier-Line

SUMMARY. Blueberry plants, like many perennial plants, must undergo physiological changes in order to survive winter. Lack of winter hardiness and susceptibility to spring frosts have been identified as two

---

L. J. Rowland is Research Geneticist (Plants), Fruit Laboratory, U.S. Department of Agriculture, Agricultural Research Service, Beltsville, MD.

G. R. Panta is affiliated with the Department of Pharmacology, University of Tennessee Health Science Center, Memphis, TN 38163.

S. Mehra is affiliated with the Fruit Laboratory, Henry A. Wallace Beltsville Agricultural Research Center, U.S. Department of Agriculture, Agricultural Research Service, Beltsville, MD 20705.

C. Parmentier-Line is affiliated with the Department of Natural Resource Sciences and LARC, University of Maryland, College Park, MD 20742.

Address correspondence to: L. J. Rowland, USDA-ARS, Fruit Laboratory, 10300 Baltimore Avenue, Building 010A, BARC-West, Beltsville, MD 20705 (E-mail: rowlandj@ba.ars.usda.gov).

The contributions of previous work by numerous colleagues who are cited in this paper are gratefully acknowledged.

[Haworth co-indexing entry note]: "Molecular Genetic and Physiological Analysis of the Cold-Responsive Dehydrins of Blueberry." Rowland, L. J. et al. Co-published simultaneously in *Journal of Crop Improvement* (Food Products Press, an imprint of The Haworth Press, Inc.) Vol. 10, No. 1/2 (#19/20), 2004, pp. 53-76; and: *Adaptations and Responses of Woody Plants to Environmental Stresses* (ed: Rajeev Arora) Food Products Press, an imprint of The Haworth Press, Inc., 2004, pp. 53-76. Single or multiple copies of this article are available for a fee from The Haworth Document Delivery Service [1-800-HAWORTH, 9:00 a.m. - 5:00 p.m. (EST). E-mail address: docdelivery@haworthpress.com].

http://www.haworthpress.com/web/JCRIP

Digital Object Identifer: 10.1300/J411v10n01_05

of the most important genetic limitations of current blueberry cultivars. These traits are influenced by several factors such as the level of cold hardiness reached while plants are in the cold acclimated state and the chilling requirement or amount of low-temperature exposure required for breaking dormancy. How these factors are controlled genetically and how they interact with each other is not well understood. For these reasons, we have been using a combination of molecular, genetic, and physiological approaches to investigate genetic controls of chilling requirement, cold hardiness, and related factors in blueberry. Previously, we identified three dehydrins of 65, 60, and 14 kDa as the predominant proteins present in cold acclimated blueberry floral buds. Dehydrins are a group of heat-stable, glycine-rich plant proteins that are induced by environmental stimuli that have a dehydrative component, such as drought and low temperature. Levels of the blueberry dehydrins increase with cold acclimation and decrease with deacclimation and resumption of growth.

Expression studies with whole plants indicate that blueberry dehydrins are induced by cold stress in all organs examined including floral buds, leaves, stems, and roots, and by drought stress in primarily stems. Although dehydrin accumulation correlates positively with cold hardiness levels, it does not correspond precisely to the degree of drought tolerance or drought avoidance. Our studies of dehydrin expression in cell suspension cultures indicate that cell cultures are not a good system for studying blueberry dehydrin expression.

Peptide sequence information from the blueberry dehydrins has been used to prepare degenerate primers and amplify a portion of a gene encoding a dehydrin. This amplification product has been used to screen our cDNA library, prepared from RNA from cold acclimated blueberry floral buds, and has resulted in the isolation of a full-length cDNA clone thought to encode the 60 kDa dehydrin. The gene represented by this clone has been designated *bbdhn1*. This clone has now been used as a probe to further screen the cDNA library and has resulted, to date, in the isolation of four partial-length dehydrin cDNAs. All have been completely sequenced and the sequences compared to each other and to that of the *bbdhn1* cDNA. The sequences are identical at the 3′ end and diverge more and more as they approach the 5′ end. Whether the cDNAs represent different genes or a combination of different alleles and different genes remains to be determined. Efforts are currently underway to complete cloning and sequencing the remaining unique dehydrin cDNAs, as well as isolate and characterize cDNA clones representing other cold-responsive messages from blueberry. *[Article copies available for a fee from The Haworth Document Delivery Service: 1-800-HAWORTH. E-mail address: <docdelivery@haworthpress.com> Website: <http://www.HaworthPress.com>*

KEYWORDS. Blueberry, chilling requirement, cloning, cold acclimation, cold hardiness, cold-responsive genes, cold-responsive proteins, cold tolerance, dehydrins, dormancy, molecular biology, molecular genetics, physiology, *Vaccinium*

INTRODUCTION

Annually, woody perennial plants of the temperate zone must undergo physiological changes in order to survive winter. As part of this annual growth cycle, by late summer to mid autumn plants enter a state of dormancy known as endodormancy and begin to develop cold hardiness, referred to as the first stage of cold acclimation (Nissila and Fuchigami, 1978; Sakai and Larcher, 1987). Here, the prefix "endo" refers to the inhibition of bud growth which is maintained within the bud itself, as distinguished from paradormancy which is controlled by a factor outside the bud (for example, apical dominance), and ecodormancy, which is imposed by environmental factors (Lang, 1987; Lang et al., 1987). Endodormancy and this first stage of cold acclimation are thought to be triggered or enhanced by short photoperiods (Downs and Borthwick, 1957; Nitsch, 1957; Perry, 1971; Vince-Prue, 1975; Nooden and Weber, 1978). Endodormancy is characterized by a chilling requirement, i.e., exposure to an accumulated number of hours of low, nonfreezing temperatures in order for budbreak to occur the following spring (Lang et al., 1987). Thus, the chilling requirement serves to synchronize a plant's growth upon exposure to favorable environmental conditions. During winter months, while buds are fully endodormant and then ecodormant, there is a further increase in cold hardiness to reach maximum hardiness. This is the second stage of cold acclimation (Fuchigami et al., 1982). Some plants that would be killed by temperatures slightly below 0°C during the summer and early fall may survive temperatures as low as − 196°C during the winter when fully cold acclimated (Sutinen et al., 1992). These physiological changes culminate, upon the return of warmer temperatures, in resumption of growth and fully deacclimated plants.

The United States is the world's leading producer of blueberries. In a survey of blueberry research and extension scientists in the United States, lack of winter hardiness and susceptibility to spring frosts were identified as the most important genetic limitations of current cultivars (Moore, 1993). The expression of these traits are due to the interaction of several components including how quickly plants cold acclimate in the fall and deacclimate in the spring, the level of cold hardiness reached while plants are in the cold acclimated state, timing of plants entering and breaking dormancy, depth of dormancy, the chilling requirement, and the heat units required for resumption of growth in the spring. How these factors are controlled genetically and how they interact with each other is not well understood. One of the reasons for this is the difficulty in conducting genetic research in woody perennials. Factors such as long generation times, high levels of heterozygosity, and problems associated with inbreeding depression often render use of recombinant inbred lines and, sometimes, even true backcross and F_2 populations impossible for genetic and mapping studies. In terms of molecular genetic research, procedures that are routine in many herbaceous species, such as DNA, RNA, and protein extractions, often times are more difficult in woody plant species. Despite these difficulties and because of the importance of cold hardiness and dormancy-related traits to fruit crop industries, we are using a combination of molecular, genetic, and physiological approaches to investigate genetic controls of chilling requirement, cold hardiness, and related factors in blueberry.

This article will focus on describing molecular and physiological work related to the investigation of cold hardiness in acclimated blueberry. We will summarize our past work as well as present new findings. Our classical genetic and mapping work on control of chilling requirement and cold hardiness in blueberry has been recently reviewed elsewhere (Rowland et al., 1999).

IDENTIFICATION AND CHARACTERIZATION OF PROTEINS RESPONSIVE TO LOW TEMPERATURE IN BLUEBERRY

Identification of Chilling-Responsive Proteins in Floral Buds of Blueberry

Initially, to identify proteins responsive to low temperature exposure or "chilling," we examined changes in protein levels associated with chill unit accumulation in floral buds of two blueberry cultivars with different chilling requirements and levels of cold hardiness (Muthalif and Rowland, 1994a). The cultivars used were a high chilling, cold tolerant *Vaccinium corymbosum* L. cultivar 'Bluecrop' (chilling requirement of about 1200 chill units) and a lower chilling, more cold sensitive *V. ashei* Reade cultivar 'Tifblue' (chilling requirement of about 600

FIGURE 1. Profiles of total proteins from high chilling, cold tolerant *V. corymbosum* cultivar 'Bluecrop'. One hundred micrograms of total proteins, extracted from floral buds of field plants of 'Bluecrop' collected at various times during chill unit accumulation, were fractionated on a 12.5% gel by SDS-PAGE. Chill units are indicated by the number above each lane. In the far left lane are molecular mass markers. Arrows to the right mark the 65, 60, and 14 kDa polypeptides that accumulate with chilling.

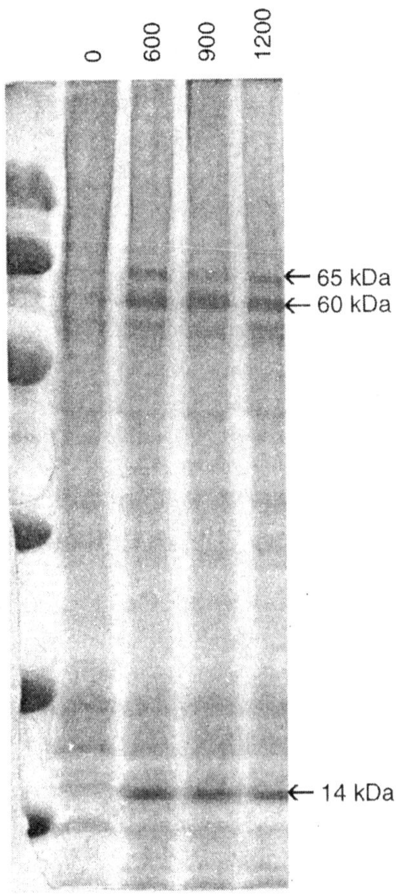

chill units). From profiles of total and soluble proteins (Figure 1), the levels of three proteins of 65, 60, and 14 kDa were observed to increase with chill unit accumulation such that they become the predominant proteins visible on one-dimensional sodium dodecyl sulfate (SDS)-polyacrylamide gels in both cultivars (Muthalif and Rowland, 1994a,b).

The level of an additional 10 kDa protein was observed to increase in 'Tifblue' as well with chilling.

Densitometric scans of the protein gels indicated that levels of the proteins increased dramatically during the initial 300 chill units, reaching a maximum level by about 600-900 chill units. Protein levels declined sharply with exposure to higher temperatures and resumption of growth, reaching levels nearly equal to or below the 0-chill unit level by the time of budbreak. Additionally, the magnitude of the overall changes observed in 'Tifblue' were not as dramatic as those observed in 'Bluecrop', with the exception of the 10 kDa protein which was not detected in 'Bluecrop' (Muthalif and Rowland, 1994a).

Categorization of Chilling-Responsive Proteins as Dehydrins

Further characterization of the chilling-induced proteins revealed them to be members of a family of proteins known as dehydrins. Dehydrins are a group of heat-stable, glycine-rich plant proteins that are induced by environmental stimuli that have a dehydrative component including drought, low temperature, salinity, and seed maturation (Close, 1996). Also indicative of dehydrins is the presence of a highly conserved lysine-rich amino acid sequence (consensus sequence EKKGIMDKIKEKLPG) referred to as the K segment, which is often repeated several times (Close, 1996). The K segments are predicted to form amphipathic α-helices (Close, 1997) and amphipathic α-helices may have a role in stabilizing cell membranes against freezing damage (Thomashow, 1999). The categorization of chilling-responsive proteins in blueberry as dehydrins was based on several factors, including their reaction to antiserum raised against the dehydrin-specific consensus peptide or K segment, their boiling-stability, and the similarity in amino acid composition of selected sequenced peptides from the chilling-responsive proteins to dehydrins (Muthalif and Rowland, 1994a).

Association of Blueberry Dehydrins with Cold Acclimation

To determine if levels of the chilling-induced dehydrins were associated with levels of cold hardiness, freezing tolerance of floral buds of 'Bluecrop' and 'Tifblue' was determined about every 300 chill units and compared to protein profiles (Muthalif and Rowland, 1994a). Bud cold hardiness was evaluated using a laboratory controlled freeze-thaw regime, followed by a visual assessment of injury. Cold hardiness or freezing tolerance (LT_{50}) was defined as the temperature causing 50% of the flower buds to be injured. For the two cultivars tested, levels of the dehydrins did appear to correlate with cold hardiness levels. The largest increase in the dehydrin levels (during the initial 300 chill units) coincided with the largest increase in the level of cold hardiness and the most dramatic decline in cold hardiness occurred with the resumption of growth, as did the decline in levels of the dehydrins. Also, maximum level of cold hardiness was higher in 'Bluecrop' (LT_{50} of −29°C for field plants) than 'Tifblue' (LT_{50} of −22°C) as was maximum level of the dehydrins.

The most cold hardy cultivar in this initial study, 'Bluecrop', is also the cultivar with the highest chilling requirement. Since cold acclimation and development of dormancy, as well as deacclimation and release from dormancy, occur simultaneously in woody perennials, it is impossible from the work described above to conclude unequivocally that dehydrins are more closely associated with cold acclimation than with dormancy transitions. Consequently, in a follow-up study, we used a novel strategy to separate cold acclimation/deacclimation and dormancy transitions in blueberry floral buds in order to independently study the relationship of dehydrin expression to these two processes (Arora et al., 1997). In this study, greenhouse-grown plants of three cultivars, 'Bluecrop', 'Tifblue', and 'Gulfcoast' (*V. corymbosum*) were used, having chilling requirements of approximately 1400, 900, and 600 chill units, respectively, when assessed by chilling plants constantly at 4°C under controlled conditions (where 1 chill unit = 1 hour at 4°C). These chilling requirements are somewhat higher than those reported for field-grown plants of these cultivars (where 1 chill unit was defined as 1 hour of exposure to 0-7°C [Muthalif and Rowland, 1994a]). This is consistent with other studies comparing chilling requirements under natural versus artificial conditions.

In this study, the three cultivars were first exposed to 4°C for long enough to satisfy one-half of their respective chilling requirements. This treatment resulted in cold acclimation. A fraction of the dormant and cold hardy plants were then given a temperature treatment of 15°C day/12°C night for 2 weeks which resulted in deacclimation but was dormancy-neutral, i.e., did not negate the chill units already accumulated. The remainder of the plants continued storage at 4°C until and beyond satisfaction of their chilling requirements. Before and after each treatment, cold hardiness and dormancy status of floral buds were determined. In addition, proteins were extracted and analyzed by one-dimensional SDS-PAGE followed by immunoblotting with anti-dehydrin antiserum. We found first that in plants given the deacclimation/dormancy-neutral treatment the dehydrin levels decreased to the pre-chilling level. Densitometric scans of protein gels indicated a good correlation between the abundance of dehydrins in cold acclimated plants and the degree of cold hardiness in the three cultivars. In addition, levels of the dehydrin proteins and cold hardiness levels remained about the same between 100% and >100% satisfaction of the chilling requirements. Taken together, these results indicated that changes in dehydrin levels are associated with cold hardiness transitions rather than with changes in dormancy status.

Effect of Cold and Drought Stress on Accumulation of Blueberry Dehydrins in Leaves, Stems, and Roots

To determine if the dehydrins are induced in other organs in response to low temperature treatment and in response to drought stress, accumulation of dehydrins was also examined in leaves, stems, and roots of various blueberry genotypes (*V. ashei* cultivar 'Climax' for low temperature treatments and *V. ashei* cultivar 'Premier' for drought treatments, and *V. corymbosum* cultivar 'Bluecrop' and *V. darrowi* Camp selection Fla4B for both low temperature and drought treatments) by SDS-polyacrylamide gel electrophoresis followed by immunoblotting with antidehydrin antiserum (Panta et al., 2001). Cold treatment involved placing plants in a cold room maintained at 4°C for 5 weeks; drought stress was imposed by withholding water from potted, greenhouse-grown plants for over 30 days. Relative water content of shoots was determined periodically throughout the drought treatment.

Dehydrins were found to accumulate with both cold and drought stress but their molecular masses varied depending upon blueberry genotype. Major 65 and 60 kDa dehydrins accumulated in the *V. corymbosum* cultivar 'Bluecrop' and the two *V. ashei* cultivars, 'Climax' and 'Premier'. However, 60 and 54 kDa dehydrins accumulated in the *V. darrowi* selection Fla4B. The 14 kDa dehydrin accumulated to low levels in response to cold stress in 'Bluecrop' but did not appear to accumulate in response to drought stress in any of the genotypes.

Blueberry dehydrins accumulated to higher levels in floral buds, stems, and roots than in leaves with cold stress and to higher levels in stems than in either roots or leaves with drought stress. In the cold stress experiment described above, the level of dehydrin accumulation correlated with expected level of plant cold hardiness in the three genotypes examined. With drought stress, dehydrins accumulated prior to significant changes in relative water content, and dehydrin levels did not appear to be closely correlated with relative water content either among or within genotypes, suggesting that dehydrins may not play a major role in determination of drought tolerance in blueberry and/or they may not respond to desiccation *per se* but, perhaps, to associated rises in ABA levels. This ABA-responsiveness, rather than desiccation responsiveness, has been shown to be the case with many drought-inducible proteins (Shinozaki and Yamaguchi-Shinozaki, 2000).

Effect of Photoperiod on Expression of Blueberry Dehydrins in Leaves and Stems

Although we do not have direct evidence for short photoperiods inducing accumulation of blueberry dehydrins, we do have some indirect evidence that supports this idea. In an experiment comparing the effect of cold treatment combined with total dark treatment to cold treatment combined with a 10 hour light/14 hour dark photoperiod on dehydrin induction in leaves and stems, dehydrins were induced to higher levels with the total dark treatment (Panta et al., 2001). This suggests that dehydrins may be responsive to changes in photoperiod. Decreasing photoperiod and temperature are the environmental signals for plant cold acclimation. Since dehydrins are associated with plant cold hardiness, a combination of cold and no light (or daylength shorter than 10 hours of light) might induce some dehydrins to a higher level than cold and 10 hours of light, as seen in stems and leaves. Also, in light, some leaf nitrogen might be diverted to photosynthetic proteins, while in dark, all available nitrogen can be used for other proteins. Alternatively, total darkness itself may constitute another type of "stress" that results in higher dehydrin levels. Whether the blueberry dehydrins respond to changes in photoperiod alone (in the absence of cold treatment) is unknown at this time.

Expression of Blueberry Dehydrins in Cell Suspension Cultures versus Whole Plants

Low temperature treatment (Hellergren, 1983; Wallner et al., 1986; Arora and Wisniewski, 1995) and ABA applied at warm temperatures (Tremblay et al., 1992) have been shown to result in cold acclimation of cell suspension cultures derived from some woody plants. Cell suspension culture systems offer some advantages over whole plants for gene expression studies. Treatment of whole plants with ABA can be problematic possibly due to inadequate uptake, rapid metabolism, and/or microbial degradation (Chen and Gusta, 1983). Besides allowing for more control over treatments with ABA and other chemicals, a cell suspension culture system for blueberry would allow experiments to be more easily repeated and to be carried out year-round without concern for differences in developmental or physiological stages of plants. Therefore, we examined the induction of dehydrins in blueberry cell suspension cultures derived from leaf and stem tissue in response to low temperature, ABA, and polyethylene glycol (PEG) treatments (Parmentier-Line et al., 2002).

By one dimensional SDS-PAGE followed by immunoblotting, we compared the induction of dehydrins with cold, ABA, and PEG treatments in cell suspension cultures of two cultivars, 'Gulfcoast' and 'Tifblue', to that seen with low temperature and drought treatments of whole plants (Parmentier-Line et al., 2002). Unfortunately, the response was quite different. Cold treatment of cell suspension cultures resulted in an increase in the level of a 30 kDa dehydrin rather than the major 65, 60, and 14 kDa dehydrins seen in whole plants. A 65 kDa dehydrin was detected in cell suspension cultures, as in whole plants, but its level did not change with cold treatment. Certain concentrations of ABA (10^{-5} and 10^{-6} M) resulted in increases in the levels of both the 65 and 30 kDa dehydrins. On the other hand, PEG treatment to induce water stress resulted in little change in the 65 or 30 kDa dehydrins of cell suspension cultures, providing further support for the idea that the blueberry dehydrins may not respond to desiccation itself but to increases in ABA levels associated with drought stress in whole plants. Regardless of the interpretation, however, because of the different responses from that seen in whole plants, and the difficulties in maintaining the cultures, we have concluded that cell suspension cultures are not a good system for studying dehydrin expression in blueberry.

Expression of Molecular Chaperones in Response to Low Temperature Stress in Blueberry

Members of another group of proteins called molecular chaperones have been identified in some annual plants as being responsive to low temperature stress. The primary function of molecular chaperones is to assist in the transport, folding, and assembly of other proteins (Ellis, 1990; Gething and Sambrook, 1992). Indeed, a chaperone-like function has been suggested as one possible function of the dehydrin family of proteins (Close, 1996). Examples of cold-induced molecular chaperones that have been identified in other plants are the CAP 79 protein from spinach, which is a member of the HSP 70 family of proteins (Neven et al., 1992), and cyclophilin from maize (Marivet et al., 1992). We used antiserum raised against CAP 79 and a cyclophilin from *Arabidopsis* to determine whether immunologically-related proteins are induced in floral buds of blueberry during low temperature exposure (Muthalif and Rowland, 1995). CAP 79 antiserum reacted with a 70 kDa protein in blueberry floral buds; however, its abundance did not change with low temperature treatment. Cyclophilin antiserum reacted with a 17.5 kDa polypeptide, which did appear to increase slightly in abundance with low temperature exposure in floral buds of three different cultivars tested, 'Bluecrop' (1.9×), 'Tifblue' (1.3×), and 'Berkeley' (*V. corymbosum* cultivar) (2.8×).

ISOLATION AND DNA SEQUENCE ANALYSIS OF BLUEBERRY DEHYDRIN cDNAs

Purification and Sequencing of the 65, 60, and 14 kDa Dehydrins

The 65, 60, and 14 kDa dehydrins were gel-purified for sequencing following either isoelectric focusing in solution combined with one-dimensional SDS-PAGE (Figure 2) or one-dimensional SDS-PAGE alone of soluble proteins extracted from floral buds of blueberry plants that had received 600-900 chill units (Muthalif and Rowland, 1994a; Levi et al., 1999). N-terminal sequencing of the 60 and 14 kDa dehydrins revealed that both are blocked at the N-terminus. Therefore, in subsequent analyses, all the dehydrins were subjected to endoproteinase Lys-C digestion and selected peptides were sequenced. Table 1 summarizes the sequence data for all the peptides sequenced to date from the dehydrins.

Cloning a Full-Length cDNA That Encodes the 60 kDa Dehydrin

Peptide sequence information from the 65 and 60 kDa dehydrins of blueberry was used to synthesize degenerate primers for amplification of part of the gene(s) encoding these proteins (Levi et al., 1999). The amino acid sequences with the least codon degeneracy were chosen for synthesis of the degenerate primers. Because it was not known which peptide sequences were N-terminal and which were C-terminal, e.g., which of the primer sequences were 5′ and which were 3′, pairs of primers, assuming a 5′-3′ and 3′-5′ orientation, were used in polymerase chain reactions (PCRs). One pair of primers, based on amino acid se-

FIGURE 2. Comparison of profiles of soluble proteins, fractionated first by isoelectric focusing in solution and second by one-dimensional SDS-PAGE, from floral buds of *V. corymbosum* cultivar 'Berkeley' collected after 0 and 950 chill units. Isoelectric focusing was performed using 9 mg of proteins from each time point. Fractions 16-19 (from the basic end of the isoelectric focusing cell, pH 7.0-8.5) are shown. Because of the dramatic increase in the levels of the 65, 60, and 14 kDa dehydrins (indicated to the right of the gel), fractions from 950 chill unit buds were loaded at 1/5 (labeled as 950′-0.2x) and 1/10 (labeled as 950′-0.1x) the volume of the comparable 0 chill unit fractions. Proteins were separated through a 12.5% gel. It should be noted that isoelectric focusing varied slightly between the 0 and 950 chill unit samples. Proteins were shifted by one fraction toward the basic end in the 0 chill unit sample as compared with the 950 chill unit sample. Thus, fractions 17-19 of the 0 chill unit sample are comparable to fractions 16-18 of the 950 chill unit sample. Adapted from Muthalif and Rowland (1994b).

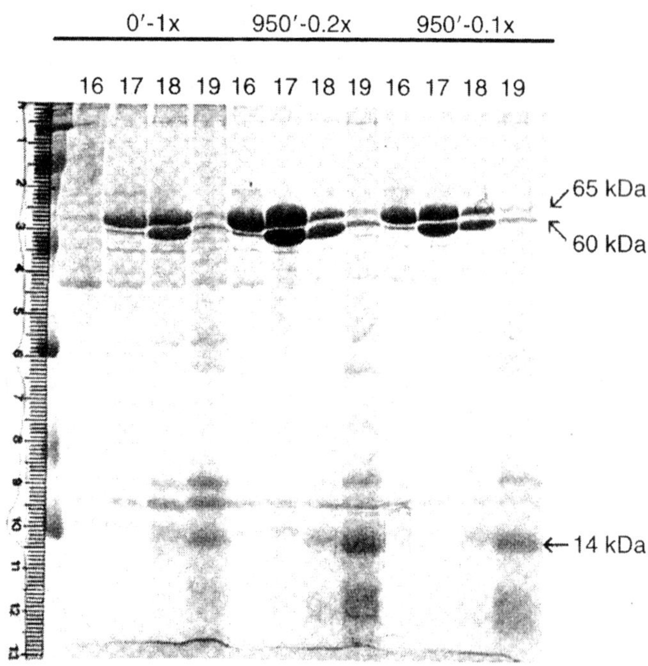

quences QDQLH and SADQNQ derived from the 65 kDa dehydrin, resulted in amplification of a 174 bp fragment. The fragment was cloned and sequenced. The presence of a K box (EGGGLADKVKDKIHG) within the sequence confirmed that part of a dehydrin gene had been cloned.

TABLE 1. Peptide sequences obtained after Lys-C digestion of the 65 and 60 kDa dehydrins of blueberry. The asterisk indicates a lower confidence secondary sequence. All others are high confidence primary sequences.

Dehydrin	Peptide Sequence
65 kDa	KQDQYRVG
	KEGGGLVDK
	KQDQLHGGYK
	KIHGGGG-SADQNQGGYK*
60 kDa	KEGGGLIYK
	KQDQLHGGYK
	KEGGGLMGELK
	KGGEQHQQQYNK
	KIHGGDGGSADQHQGIYGQDQQL–GY
14 kDa	K[S/A/I/K]GGGLMDK
	K[S/A/I/K]PGV[H/I][Y/G]GGAD[D/G]Q*

The 174 bp PCR fragment was used to screen a cDNA library prepared from RNA from dormant, cold-hardy floral buds of 'Bluecrop' (Levi et al., 1999). The buds for the cDNA library construction were collected from field plants having acquired approximately 600 chill units, since previously we showed that dehydrin levels are maximal by about 600-900 chill units (Muthalif and Rowland, 1994a). Hybridization with the 174 bp PCR fragment resulted in the isolation and purification of a clone with a 2.0 kb insert.

The 2.0 kb cDNA was sequenced and found to be full length. Inspection of the sequence confirmed that it encodes a member of the dehydrin family of proteins. Like dehydrins (Close, 1996), the deduced protein is hydrophilic, has a preponderance of glycine residues, is rich in polar and charged amino acids (such as glutamine, aspartic acid, lysine, tyrosine, histidine, glutamic acid, and arginine) and contains no phenylalanine or tryptophan. The deduced protein sequence contains five lysine-rich repeats or K boxes indicative of dehydrins. These sequences are contained within larger contiguous imperfect repeats composed of 48-62 amino acids (consensus sequence for blueberry cDNA QDQQLGGYRQDQRKEGGGLMDKVKDKIHG GGGGSADQHQGGY (K/G)QDQQ(H/L)GGYR). The cDNA does not contain a tract of serine residues (S segment) or the consensus amino acid sequence (V/T) DEYGNP (Y segment) present in some dehydrins (Close 1996). In addition, a computer search of the GenBank database revealed similarity to other dehydrins, the following five being the highest scoring matches: alfalfa cold acclimation protein, CAS15 (Monroy et al., 1993); *Pistacia* inflorescence bud protein, 32 kDa (accession Y07600); spinach cold acclimation protein, CAP85 (Neven et al., 1993); peach dehydrin, PCA60 (Artlip et al., 1997); and citrus cold-stress protein, COR19 (Cai et al., 1995).

A sequence identical to the 174 bp sequence (amplified from degenerate primers derived from peptide sequences from the 65 kDa dehydrin and used as a probe to isolate this cDNA) is not present within the cDNA sequence. However, very similar sequences are present as part of the five large imperfect repeats. The five high-confidence peptide sequences, ranging from 9 to 25 amino acids long (Table 1), obtained from the 60 kDa dehydrin exactly match sequences encoded within the cDNA clone. Amino acid composition of the 60 kDa dehydrin also agrees well with the expected amino acid composition based on the cDNA sequence. Based on this information, we concluded that the 2.0 kb dehydrin cDNA encodes the 60 kDa dehydrin, and the gene represented by this clone was named as *bbdhn1*, for blueberry dehydrin 1.

Glycosylation of Blueberry Dehydrins

Our conclusion that the 2.0 kb dehydrin cDNA encodes the 60 kDa dehydrin was made despite the fact that the DNA sequence and coupled *in vitro* transcription/translation reactions of the clone followed by SDS-PAGE indicated that it encodes a dehydrin with a native molecular mass of ~40 kDa instead of 60 kDa (Levi et al., 1999). The most obvious explanation for this discrepancy is that the 60 kDa dehydrin undergoes extensive post-translational modification such as heavy glycosylation resulting in a higher molecular weight than that predicted from the DNA sequence or *in vitro* transcription/translation reactions alone. Indeed, previously, we had suspected that the 65 and 60 kDa dehydrins of blueberry were glycosylated because of their negative reaction to silver staining (Muthalif and Rowland, 1994b). To test this idea, we used a commercially available glycoprotein detection system and showed that floral bud proteins of 65 and 60 kDa, which co-migrate with the 65 and 60 kDa dehydrins (detected with antidehydrin antibody) are glycosylated (Levi et al., 1999). To our knowledge, this is the first report of dehydrins being glycosylated. Whether other researchers have explored this possibility is unknown. The 2.0 kb dehydrin cDNA does not encode any of the Asn-XXX-Ser potential sites for N-linked glycosylation, but does encode six serine and three threonine residues, which could serve as potential sites of O-linked glycosylation.

Cloning Four Partial-Length Dehydrin cDNAs

The 2.0 kb blueberry dehydrin cDNA was then itself used as a probe to screen our cDNA library prepared from RNA from cold-hardened floral buds of 'Bluecrop'. In this screening, several positively hybridizing plaques were detected and, to date, four have been purified and completely sequenced. Analysis of the sequences confirm that they all are dehydrin cDNAs. All contain between two and five K boxes, all are hydrophilic, and all are rich in glycine and polar and charged amino acids such as glutamine. All are similar but unique-none of the sequences exactly match any of the other clones including *bbdhn1*. None, except for *bbdhn1*, are full length as their open reading frames do not begin with an ATG start codon. All four partial-length cDNAs have 3′-untranslated sequences which range in length from 196 to 224 bp, none of them being as long as the 714 bp 3′-untranslated sequence of *bbdhn1*. The four partial-length cDNA clones have inserts that are 0.6, 0.8, 0.9, and 1.2 kb long and the genes represented by these sequences were named *bbdhn2*, *bbdhn3*, *bbdhn4*, and *bbdhn5*, respectively. The cDNA sequences were entered into the GenBank and assigned accession numbers AF222738, AF222739, AF222740, and AF222741.

DNA and Predicted Protein Sequence Comparison of Blueberry Dehydrin cDNAs

To determine the degree of similarity among the cloned blueberry dehydrin cDNAs, DNA and predicted protein sequences were aligned using Vector NTI 5.0 software. The best alignment of the predicted protein sequences is shown in Figure 3. The alignments revealed that the five dehydrins are very similar at the DNA (data not shown) and protein levels. The sequences are more conserved at the 3′/carboxy end than at the 5′/amino end. In fact, all five cDNA clones have the same 61-amino

FIGURE 3. Comparison and best alignment of the predicted protein sequences of five dehydrin cDNAs isolated from blueberry. Sequences were compared using Vector NTI software. The numbers on top of the sequences represent total number of amino acid residues compared. The numbers that follow immediately after the cDNA name represent position of the amino acid relative to the beginning of the protein sequence. In addition, a consensus sequence, comprised of all amino acid residues found in the same positions of the alignment in a minimum of 3 of the 5 sequences, is shown below the sequences of the five clones. Amino acid residues of the individual clones that differ from the consensus sequence are indicated with bold lettering.

```
                    1                                                50
BBDHN1      (1)  --MAGIMNKIGQTLPCGGNKEEDKYKGGEQHQQQQYNKPGQHQGESRQDQ
BBDHN2      (1)  --------------------------------------------------
BBDHN3      (1)  --------------------------------------------------
BBDHN4      (1)  -------------RHEGGYKQD------------------QQLGGHRQDQ
BBDHN5      (1)  TNSDCETANGSLNQHQGGYKQDQLHGGYRQDQQLGGHEQDQQRGGYRQDQ
Consensus   (1)                  GG K                    Q  G  RQDQ

                 51                                              100
BBDHN1     (49)  RKEGGGLMGELKDKIHGGGVGTDGGGVGSVDPHGHQGGKYRHDQQHSEYV
BBDHN2      (1)  --------------------------------------------------
BBDHN3      (1)  -----RHEDKVKDKIHGGGVGSVDQHQGGYKQDQLHGG-YRQDQQLGGHG
BBDHN4     (20)  HNEG--LADKVKDKIHGGGVGSADQHQGGYKQDQLHGG-YRQDQQLGGHG
BBDHN5     (51)  HNEG--LADKVKDKIHGGGVGSVDQHQGGYKQDQLHGG-YRQDQQLGGHG
Consensus  (51)    EG  L DKVKDKIHGGGVGS DQHQGGYKQDQLHGG YRQDQQLGGHG

                 101                                             150
BBDHN1     (99)  QDQQPGGYRQDQRKEGGGLMDKVKDKIHGGGVGSADQHQDGYKQD-----
BBDHN2      (1)  ----------------------RHEKIHGGGVGSADQHQDGYKQD-----
BBDHN3     (45)  QDQQPGGYRQDQHKEGGGLIYKVKDKIHGGDGGSADQHQGIYGQD-----
BBDHN4     (67)  QDQQRGGYRQDQHNEG--LADKVKDKIHGGGVGSADQHQGGYKQD-----
BBDHN5     (98)  QDQQRGGYRQDQHNEG--LADKVKDKIHGGGVGSADQHQGGYKQDQLLGG
Consensus (101)  QDQQ GGYRQDQH EG  L DKVKDKIHGGGVGSADQHQGGYKQD

                 151                                             200
BBDHN1    (144)  ----QLHGGYKQDQQLGGYRQDQHKEGGGLIYKVKDKIHGGDGGSADQHQ
BBDHN2     (24)  ----QLHGGYKQDQQLGGHRQDQHNEG-----------------------
BBDHN3     (90)  ----QQLGGYRQDQQLGGYRQD--QQ------------------------
BBDHN4    (110)  ----QQPGGYGQDQQPGGYRQDQHKEG-----------------------
BBDHN5    (146)  YRQDQQPGGYGQDQQPGGYRQDQHKEG-----------------------
Consensus (151)      QQ GGY QDQQLGGYRQDQHKEG

                 201                                             250
BBDHN1    (190)  GIYGQDQQLGGYRQDQQLGGYRQDQQHGEYKQDQRKEGGGLMDKVKDKIP
BBDHN2     (47)  ----------------------------------------LADKVKDKIP
BBDHN3    (110)  --------------------------HGEYKQDQRKEGGWLMDKVKDKIP
BBDHN4    (133)  --------------------------------------GGLIYKVKDKIH
BBDHN5    (173)  --------------------------------------GGLIYKVKDKIH
Consensus (201)                                        GGLIDKVKDKIP

                 251                                             300
BBDHN1    (240)  GGNGGSAADQHQGVYGQDQQLGGYRQDQ----------------------
BBDHN2     (57)  GGNGGSAADQHQGVYGQDQQLGGYRQDQ----------------------
BBDHN3    (134)  GGNGGSAADQHQGVYGQDQQLGGYRQDQ----------------------
BBDHN4    (145)  GGDGG-SADQHQGIYGQDQQLGGYRQDQ----------------------
BBDHN5    (185)  GGDGG-SADQHQGIYGQDQQLGGYRQDQQLGGYRQDQQHGEYKQDQRKEG
Consensus (251)  GGNGGSAADQHQGVYGQDQQLGGYRQDQ

                 301                                             350
BBDHN1    (268)  ----------------------------------------QRGEYKQDQR
BBDHN2     (85)  ----------------------------------------QRGEYKQDQR
BBDHN3    (162)  ----------------------------------------QRGEYKQDQR
BBDHN4    (172)  ----------------------------------------QRGEYKQDQR
BBDHN5    (234)  GWLMDKVKDKIPGGNGGSAADQHQGVYGQDQQLGGYRQDQQRGEYKQDQR
Consensus (301)                                          QRGEYKQDQR
```

```
                       351                                 387
BBDHN1      (278)  KEGGGLMDKVKDTIHGGAGGGADKHRGEYKQDQYRGD
BBDHN2       (95)  KEGGGLMDKVKDTIHGGAGGGADKHRGEYKQDQYRGD
BBDHN3      (172)  KEGGGLMDKVKDTIHGGAGGGADKHRGEYKQDQYRGD
BBDHN4      (182)  KEGGGLMDKVKDTIHGGAGGGADKHRGEYKQDQYRGD
BBDHN5      (284)  KEGGGLMDKVKDTIHGGAGGGADKHRGEYKQDQYRGD
Consensus   (351)  KEGGGLMDKVKDTIHGGAGGGADKHRGEYKQDQYRGD
```

acid sequence at the carboxy end, suggesting that this sequence may serve an important function. As the sequences diverge toward the 5′/amino end, the dehydrins appear to differ from each other by a series of insertions/deletions and single base changes. It is possible that several of the cDNA clones represent alleles of each other rather than separate genes because 'Bluecrop', the cultivar from which these clones were isolated, is a tetraploid. A consensus K box for blueberry dehydrins, based on the sequences of all 20 blueberry K boxes that are encoded within the five dehydrin cDNAs isolated to date (shown in Figure 4), is comprised of 16 amino acid residues, EGGGLMDKVKDKIHGG. However, based on a closer examination of the K box sequences available so far, the blueberry dehydrin K boxes appear to fall into two basic types, one having the consensus sequence EGGGLMDKVKDKIHGG and the other having a more highly conserved sequence HNEGLADKVKDKIHGG. The dehydrin cDNAs analyzed so far appear to have either K boxes that are all of the EGGGLMDKVKDKIHGG type or a mixture of the EGGGLMDKVKDKIHGG and HNEGLADKVKDKIHGG types, with the EGGGLMDKVKDKIHGG types being found more toward the 3′/carboxy end. Of course, it is possible that other types will be found as more dehydrin cDNAs are isolated and sequenced.

Dehydrin Gene Copy Number in Blueberry

The number of dehydrin genes in blueberry has been estimated by hybridization of the *bbdhn1* probe to gel blots of genomic DNA of the original parent plants and a few F_1s of our diploid blueberry mapping populations (Levi et al., 1999). Genomic DNAs were digested with several different restriction enzymes, the sites of which were not present in the *bbdhn1* cDNA clone. Washes were carried out under either highly stringent or moderately stringent conditions. Each of the restriction enzyme digestions combined with highly stringent washes resulted in two or three strongly hybridizing fragments (not including allelic fragments, when possible to determine) and two to five weaker hybridizing fragments. Less stringent washes resulted in no increase in number of fragments hybridizing, but some of the weakly hybridizing fragments (from

FIGURE 4. The consensus blueberry dehydrin K segment along with a listing of the 20 individual K segments found in the five dehydrin cDNAs sequenced thus far. The number of times each amino acid in the consensus sequence occurs in the individual K segments is indicated in subscript.

Consensus	E_{14}	G_{14}	G_{14}	G_{17}	L_{19}	M_{10}	D_{15}	K_{19}	V_{19}	K_{20}	D_{20}	K_{15}	I_{20}	H_{16}	G_{20}	G_{20}
	H_5	N_5	E_5	W_2		A_5	Y_4	E_1	L_1			T_5		P_4		
						I_4	G_1									
BBDHN1	E	G	G	G	L	M	G	E	L	K	D	K	I	H	G	G
	E	G	G	G	L	M	D	K	V	K	D	K	I	H	G	G
	E	G	G	G	L	I	Y	K	V	K	D	K	I	H	G	G
	E	G	G	G	L	M	D	K	V	K	D	K	I	P	G	G
	E	G	G	G	L	M	D	K	V	K	D	T	I	H	G	G
BBDHN2	H	N	E	G	L	A	D	K	V	K	D	K	I	P	G	G
	E	G	G	G	L	M	D	K	V	K	D	T	I	H	G	G
BBDHN3							D	K	V	K	D	K	I	H	G	G
	E	G	G	G	L	I	Y	K	V	K	D	K	I	H	G	G
	E	G	G	W	L	M	D	K	V	K	D	K	I	P	G	G
	E	G	G	G	L	M	D	K	V	K	D	T	I	H	G	G
BBDHN4	H	N	E	G	L	A	D	K	V	K	D	K	I	H	G	G
	H	N	E	G	L	A	D	K	V	K	D	K	I	H	G	G
	E	G	G	G	L	I	Y	K	V	K	D	K	I	H	G	G
	E	G	G	G	L	M	D	K	V	K	D	T	I	H	G	G
BBDHN5	H	N	E	G	L	A	D	K	V	K	D	K	I	H	G	G
	H	N	E	G	L	A	D	K	V	K	D	K	I	H	G	G
	E	G	G	G	L	I	Y	K	V	K	D	K	I	H	G	G
	E	G	G	W	L	M	D	K	V	K	D	K	I	P	G	G
	E	G	G	G	L	M	D	K	V	K	D	T	I	H	G	G

stringent washes) hybridized more strongly under these conditions. These results, along with the protein and peptide sequence data, suggest that blueberry dehydrins are encoded by a multigene family, with about two to three genes with high homology to the *bbdhn1* cDNA clone and a few other less related genes.

Accumulation of Dehydrin Transcripts in Response to Low Temperature

The *bbdhn1* cDNA clone, which encodes the 60 kDa dehydrin, and the 174 bp PCR fragment that was amplified from degenerate primers based on peptide sequences from the 65 kDa dehydrin have both been used as probes on RNA blots to monitor the correlation in dehydrin mRNA levels in blueberry field plants with chill unit accumulation (Levi et al., 1999; Rowland et al., 1999). Both probes hybridized to two chilling-responsive messages of 2.0 and 0.5 kb (Figure 5), the 2.0 kb message being the same size as the *bbdhn1* cDNA. Consequently, it seems reasonable to conclude that the 2.0 kb size class of messages actually represents two messages, one encoding the 65 kDa dehydrin and one encoding the 60 kDa dehydrin, and that the 0.5 kb message encodes the 14 kDa dehydrin, as it is of an appropriate size to encode that dehydrin. RNA blots comparing dehydrin message levels in the more cold hardy cultivar 'Bluecrop' to that in the less hardy cultivar 'Tifblue' (Figure 5) revealed that, in both cultivars, the levels of both the 2.0 and 0.5 kb transcripts increased noticeably as early as 50 chill units, as compared with the 0 chill unit levels. However, the message levels reached maximum more quickly in the hardier cultivar 'Bluecrop' (by 300 chill

FIGURE 5. Northern blot of total RNA extracted from blueberry cultivars 'Tifblue' and 'Bluecrop' and hybridized with the 2.0 kb *bbdhn1* cDNA probe. RNA for northern was extracted from floral buds collected from field plants after different lengths of chilling (from 0 to 1500 chill units). Chill units are given above each lane. The probe hybridized to two messages of 2.0 and 0.5 kb, indicated by arrows to the right of the autoradiogram. Adapted from Rowland et al. (1999).

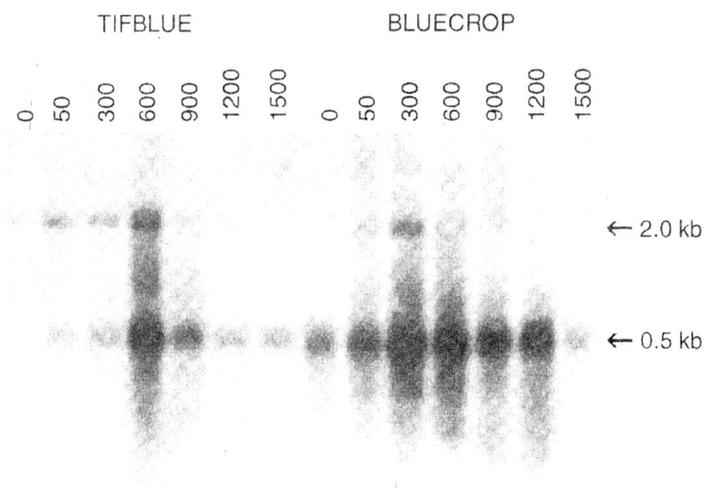

units) than in 'Tifblue' (by 600 chill units). In addition, the level of the 0.5 kb message remained higher for longer in 'Bluecrop' than in 'Tifblue', not declining dramatically until resumption of growth in the spring. The overall maximum level of the 2.0 kb message was about the same in both cultivars, whereas maximum level of the 0.5 kb message was about two fold higher in 'Bluecrop' than in 'Tifblue'.

In comparison to RNA accumulation, the 65, 60, and 14 kDa dehydrin proteins accumulated to higher levels quickly and remained at higher levels longer in 'Bluecrop' than in 'Tifblue' (Muthalif and Rowland, 1994a). The maximum level of all three proteins was higher in 'Bluecrop' than in 'Tifblue', although the largest difference seen was in the level of the 14 kDa dehydrin. The lag in reaching maximum protein levels (600-900 chill units) in 'Bluecrop' as compared to RNA levels (300 chill units) suggests that the dehydrin proteins are quite stable. This lag was not seen in 'Tifblue', where protein and RNA levels both peaked at about 600 chill units. Thus, the dehydrins of 'Bluecrop' may be more stable than those of 'Tifblue'. If dehydrins prove to play a causal role in determination of cold hardiness in blueberry, then the difference in hardiness between 'Bluecrop' and 'Tifblue' could be explained by a combination of earlier expression, overall higher expression (especially for the 14 kDa dehydrin), and greater stability of the dehydrins in 'Bluecrop' than in 'Tifblue'.

CONCLUSIONS

A recognized problem in the blueberry industry is the susceptibility of cultivars to freezing damage (Moore, 1993). Depending upon the geographical area, this type of damage may be caused by extreme low temperatures occurring during the winter while plants are still dormant, or by late winter or early spring frosts occurring while plants are beginning to resume growth. Therefore, a blueberry cultivar needs to have an appropriate chilling requirement and level of cold hardiness for the area in which it is intended to be grown. A cultivar with a chilling requirement that is too low for a particular area will resume growth too soon if exposed to fluctuating winter temperatures, making it susceptible to late frost damage. Conversely, a cultivar with a chilling requirement that is too high for a particular area may receive insufficient chilling resulting in delayed and erratic budbreak. Similarly, midwinter-freezing injury may occur if a cultivar is not cold hardy enough for certain growing areas.

Application of molecular genetic techniques to the study of cold hardiness and other dormancy-related traits should result in the identification of genes controlling these traits and to a better understanding of the role of these genes in control of cold hardiness, chilling requirement, etc. Once the genes are identified, theoretically, the level of cold hardiness or chilling requirement of a particular cultivar could be lowered or raised by over-expression or under-expression of the appropriate genes.

Our studies using blueberry floral buds have indicated that three major dehydrins of 65, 60, and 14 kDa increase with cold acclimation and decrease during deacclimation and resumption of growth. Furthermore, it has been shown that dehydrin expression in blueberry is closely associated with level of cold hardiness. More recently, the aim of our work has been to clone and sequence the cDNAs encoding the blueberry dehydrins, compare the sequences to each other and to dehydrins found in other plants, and further study the expression of the dehydrins at the RNA and protein levels in response to cold and drought stress in various organs of whole plants and in cell suspension cultures.

From proteins gels, peptide sequence data, and Southern blot data using one full-length 2.0 kb dehydrin cDNA as a probe, it appears that blueberry dehydrins are encoded by about 2-3 genes with high homology to the 2.0 kb cDNA, and a few other less related genes. In total, to date, one full length and four partial-length unique dehydrin cDNAs have been cloned and sequenced. To distinguish among them, they have been named, beginning with the full length clone, as *bbdhn1* (for *blueberry dehydrin 1*), *bbdhn2*, *bbdhn3*, *bbdhn4*, and *bbdhn5*. Based on peptide sequences encoded within the clone, we have concluded that the full-length *bbdhn1* cDNA encodes the 60 kDa dehydrin. Although dehydrin sequences are not colinear, e.g., short consensus sequences are present, but sequences outside the consensus are divergent in terms of length and sequence, a search of the GenBank database for sequences homologous to the *bbdhn1* cDNA has been performed. Highest scoring matches are to other dehydrins, alfalfa cold acclimation protein CAS15 and a 32 kDa *Pistacia* inflorescence bud protein. A comparison of the blueberry dehydrin clones reveal that there is very high homology both at the DNA and protein levels among the clones, particularly at the 3'/carboxy ends, and all contain multiple copies of the highly conserved K boxes. It is possible that several of the clones represent alleles of each other rather than separate genes.

RNA blots using RNA extracted from various organs, including floral buds, leaves, and stems, and the *bbdhn1* cDNA as a probe reveal homology to two chilling/cold-responsive messages of 2.0 and 0.5 kb. Our interpretation is that the 2.0 kb size class of messages actually represents two messages, one encoding the 65 kDa dehydrin and one encoding the 60 kDa dehydrin, and that the 0.5 kb message likely encodes the 14 kDa dehydrin. In addition, expression of the blueberry dehydrins in whole plants has been compared to that in cell suspension cultures. Cold and drought stress protein expression studies using whole plants indicate that blueberry dehydrins are induced by cold stress in all organs examined including floral buds, leaves, stems, and roots, and by drought stress in primarily stems. Furthermore, although dehydrin accumulation correlates positively with cold hardiness levels, it does not correspond precisely to the degree of drought tolerance or drought avoidance. For the most part, expression of the blueberry dehydrins in cell suspension cultures does not mimic that seen in whole plants. For example, cold treatment of cell suspension cultures results in an increase in the level of a 30 kDa cross-reacting protein rather than the major 65, 60, and 14 kDa dehydrins seen in whole plants. A 65 kDa dehydrin is detected in cell suspension cultures, as in whole plants, but its level does not change with cold treatment. Because of these differences and others, we have concluded that cell suspension cultures are not a good system for studying blueberry dehydrin expression.

Finally, our current efforts are focused in two areas. One area is to establish if the dehydrins, or other cold-responsive genes that we isolate, play a causal role in determination of cold hardiness in blueberry. To do this, collaborative efforts with other scientists are underway to use gene constructs of the *bbdhn1* cDNA in transformation experiments to determine their effect(s) on cold hardiness. Also, we are mapping the dehydrin genes to determine if any are associated with QTLs that control cold hardiness that we identify in our concurrent mapping work. The second area is to isolate and characterize other unique dehydrin cDNAs, as well as cDNA clones representing other cold-responsive messages from blueberry. To date, the only cold-responsive proteins that have been characterized in blueberry are the dehydrins and a cyclophilin. Because our own genetic studies with blueberry indicate that cold hardiness is a multigenic trait controlled largely by additive gene effects and, to a lesser extent, by dominance gene effects (Arora et al., 2000), we expect other genes, in addition to the dehydrins, to be involved in determination of cold hardiness.

REFERENCES

Arora, R., L.J. Rowland, J.S. Lehman, C.C. Lim, G.R. Panta and N. Vorsa. (2000). Genetic analysis of freezing tolerance in blueberry (*Vaccinium* section *Cyanococcus*). *Theor. Appl. Genet.* 100:690-696.

Arora, R., L.J. Rowland and G.R. Panta. (1997). Cold hardiness and dormancy transitions in blueberry and their association with accumulation of dehydrin-like proteins. *Physiol. Plant.* 101:8-16.

Arora, R. and M.E. Wisniewski. (1995). Ultrastructural and protein changes in cell suspension cultures of peach associated with low temperature-induced cold acclimation and abscisic acid treatment. *Plant Cell, Tissue and Organ Culture* 40:17-24.

Artlip, T.S., A.M. Callahan, C.L. Bassett and M.E. Wisniewski. (1997). Seasonal expression of dehydrin gene in sibling deciduous and evergreen peach (*Prunus persica* [L.] Batsch). *Plant Mol. Biol.* 33:61-70.

Cai, Q., G.A. Moore and C.L. Guy. (1995). An unusual group 2 LEA gene family in citrus responsive to low temperature. *Plant Mol. Biol.* 29:11-23.

Chen, T.H.H. and L.V. Gusta. (1983). Abscisic acid-induced freezing resistance in cultured plant cells. *Plant Physiol.* 73:71-75.

Close, T.J. (1996). Dehydrins: Emergence of a biochemical role of a family of plant desiccation proteins. *Physiol. Plant.* 97:795-803.

Close, T.J. (1997). Dehydrins: A commonality in the response of plants to dehydration and low temperature. *Physiol. Plant.* 100:291-296.

Downs, R.J. and H.A. Borthwick. (1957). Effects of photoperiod on the growth of trees. *Bot. Gaz.* 117:310-326.

Ellis, R.J. (1990). Molecular chaperones: The plant connection. *Science* 250:954-959.

Fuchigami, L.H., C.J. Weiser, K. Kobayashi, R. Timmis and L.V. Gusta. (1982). A degree growth stage (°GS) model and cold acclimation in temperate woody plants. Pp. 93-116. In: P.H. Li and A. Sakai (eds.), *Plant Cold Hardiness and Freezing Stress*. Academic Press, New York.

Gething, M.J. and J. Sambrook. (1992). Protein folding in the cell. *Nature* 355:33-45.

Hellergren, J. (1983). Cold acclimation of suspension cultures of *Pinus sylvestris* in response to light and temperature treatments. *Plant Physiol.* 72:992-995.

Lang, G.A. (1987). Dormancy: A new universal terminology. *HortScience* 22:817-820.

Lang, G.A., J.D. Early, G.C. Martin and R.L. Darnell. (1987). Endo-, para-, and ecodormancy: Physiological terminology and classification for dormancy research. *HortScience* 22:371-377.

Levi, A., G.R. Panta, C.M. Parmentier, M.M. Muthalif, R. Arora, S. Shanker and L.J. Rowland. (1999). Complementary DNA cloning, sequencing, and expression of a unique dehydrin from blueberry floral buds. *Physiol. Plant.* 107:98-109.

Marivet, J., P. Frendo and G. Burkard. (1992). Effects of abiotic stresses on cyclophilin gene expression in maize and bean and sequence analysis of bean cyclophilin cDNA. *Plant Sci.* 84:171-178.

Monroy, A.F., Y. Castonguay, S. Laberge, F. Sarhan, L.P. Vezina and R.S. Dhindsa. (1993). A new cold-induced alfalfa gene is associated with enhanced hardening at subzero temperature. *Plant Physiol.* 102:873-879.

Moore, J.N. (1993). The blueberry industry of North America. *Acta Hort.* 346:15-26.

Muthalif, M.M. and L.J. Rowland. (1994a). Identification of dehydrin-like proteins responsive to chilling in floral buds of blueberry (*Vaccinium*, section *Cyanococcus*). *Plant Physiol.* 104:1439-1447.

Muthalif, M.M. and L.J. Rowland. (1994b). Identification of chilling responsive proteins from floral buds of blueberry. *Plant Sci.* 101:41-49.

Muthalif, M.M. and L.J. Rowland. (1995). The search for chilling-responsive proteins in blueberry continues. *J. Small Fruit & Vitic.* 3:53-60.

Neven, L.G., D.W. Haskell, C.L. Guy, N. Denslow, P.A. Klein, L.G. Green and A. Silverman. (1992). Association of 70-kilodalton heat-shock cognate proteins with acclimation to cold. *Plant Physiol.* 99:1362-1369.

Neven, L.G., D.W. Haskell, A. Hofig, Q-B. Li and C.L. Guy. (1993). Characterization of a spinach gene responsive to low temperature and water stress. *Plant Mol. Biol.* 21:291-305.

Nissila, P.C. and L.H. Fuchigami. (1978). The relationship between vegetative maturity and the first stage of cold acclimation. *J. Amer. Soc. Hort. Sci.* 103:710-711.

Nitsch, J.P. (1957). Growth responses of woody plants to photoperiodic stimuli. *Proc. Amer. Soc. Hort. Sci.* 70:512-525.

Nooden, L.D. and J.A. Weber. (1978). Environmental and hormonal control of dormancy in terminal buds of plants. Pp. 222-268. In: M.E. Clutter (ed.), *Dormancy and Developmental Arrest.* Academic Press, New York.

Panta, G.R., M.W. Rieger and L.J. Rowland. (2001). Effect of cold and drought stress on blueberry dehydrin accumulation. *J. Hort. Sci. & Biotech.* 76:549-556.

Parmentier-Line, C.M., G.R. Panta and L.J. Rowland. (2002). Changes in dehydrin expression associated with cold, ABA and PEG treatments in blueberry cell cultures. *Plant Sci.* 162:273-282.

Perry, T.O. (1971). Dormancy of trees in winter. *Science* 171:29-36.

Rowland, L.J., E.L. Ogden, R. Arora, C.C. Lim, J.S. Lehman, A. Levi and G.R. Panta. (1999). Use of blueberry to study genetic control of chilling requirement and cold hardiness in woody perennials. *HortScience* 34:1185-1191.

Sakai, A. and W. Larcher. (1987). *Frost Survival of Plants.* Springer-Verlag, Berlin, p. 99.

Shinozaki, K. and K. Yamaguchi-Shinozaki. (2000). Molecular responses to dehydration and low temperature: Differences and cross-talk between two stress signaling pathways. *Current Opinion in Plant Biology* 3:217-223.

Sutinen, M.L., J.P. Palta and P.B. Reich. (1992). Seasonal differences in freezing stress resistance of needles of *Pinus nigra* and *Pinus resinosa*: Evaluation of the electrolyte leakage method. *Tree Physiol.* 11:241-254.

Thomashow, M.F. (1999). Plant cold acclimation: Freezing tolerance genes and regulatory mechanisms. *Ann. Rev. Plant Phys. Plant Mol. Biol.* 50:573-599.

Tremblay, M.F., P. Nadeau and M. Lalonde. (1992). Effect of ABA on freezing resistance of *Betula papyrifera* and *Alnus incana* woody plant cell suspensions. *Tree Physiol.* 10:317-326.

Vince-Prue, D. (1975). *Photoperiodism in Plants.* McGraw Hill, London. 443 p.

Wallner, S.J., M.T. Wu and S.J. Anderson-Krengel. (1986). Changes in extracellular polysaccharides during cold acclimation of cultured pear cells. *J. Amer. Soc. Hort. Sci.* 111:769-773.

Photoperiodic Regulation of Apical Growth Cessation in Northern Tree Species: The Role of Phytochrome and Gibberellin

Jorunn E. Olsen
John B. Jensen
Jörgen A. Mölmann
Arild Ernstsen
Olavi Junttila

SUMMARY. Tree species adapted to the climatic conditions of the northern boreal and subarctic vegetation zones have a capacity to develop a very high level of frost hardiness, even to survive the tempera-

Jorunn E. Olsen, John B. Jensen, Jörgen A. Mölmann and Arild Ernstsen are affiliated with the Department of Biology, University of Tromsö, N-9037 TROMSÖ, Norway.

Olavi Junttila is affiliated with the Department of Applied Biology, University of Helsinki, P.O. Box 27, FIN-00014 University of Helsinki, Finland.

Address correspondence to: Olavi Junttila, Department of Applied Biology, University of Helsinki, P.O. Box 27, FIN-00014 University of Helsinki, Finland (E-mail: olavi.junttila@helsinki.fi).

The invitation to contribute to this issue of the *Journal of Crop Improvement* is gratefully acknowledged.

Research of the authors was supported by grants from the Norwegian Research Council.

[Haworth co-indexing entry note]: "Photoperiodic Regulation of Apical Growth Cessation in Northern Tree Species: The Role of Phytochrome and Gibberellin." Olsen, Jorunn E. et al. Co-published simultaneously in *Journal of Crop Improvement* (Food Products Press, an imprint of The Haworth Press, Inc.) Vol. 10, No. 1/2 (#19/20), 2004, pp. 77-112; and: *Adaptations and Responses of Woody Plants to Environmental Stresses* (ed: Rajeev Arora) Food Products Press, an imprint of The Haworth Press, Inc., 2004, pp. 77-112. Single or multiple copies of this article are available for a fee from The Haworth Document Delivery Service [1-800-HAWORTH, 9:00 a.m. - 5:00 p.m. (EST). E-mail address: docdelivery@haworthpress.com].

http://www.haworthpress.com/web/JCRIP

Digital Object Identifer: 10.1300/J411v10n01_06

ture of liquid nitrogen in midwinter. Proper timing of hardening, as well as of dehardening, is crucial for winter survival of these species. In northern tree species, cessation of apical elongation growth and bud set is a prerequisite for developmental and metabolic processes leading to hardening, and this chain of events is induced by photoperiod. The northern tree species are closely adapted to the local light climate and display photoperiodic ecotypes. The critical photoperiod is under genetic control and increases with increasing latitude of origin of the ecotype. The photoperiod is probably perceived by the phytochrome system, but the role of other pigment systems, like cryptochrome, has not been studied in woody plants. Phytochrome genes have been cloned from both conifers and deciduous species, but so far we do not have any information about possible differences between photoperiodic ecotypes at the phytochrome level. Northern and southern ecotypes have different responses to red:far red ratios, which could indicate differences in composition of their phytochrome systems, for example, the proportions of phytochrome A and B. Both phytochrome A and phytochrome B can be involved in photoperiodic responses. Experiments with transgenic hybrid aspen suggest that responses to photoperiod could be affected by the amount of phytochrome A present in plants. In deciduous species, the plant hormone gibberellin A_1 (GA_1) can completely substitute for a long photoperiod, and short day induced cessation of growth is preceded by a significant reduction of GA_1 levels, particularly in the elongation zone. Photoperiodic control of GA metabolism is supported by several studies, but very little is known about the interaction between phytochrome and GA metabolism and/or responsiveness to GA_1. Although our knowledge is still very fragmentary, available results suggest that cessation of growth and initiation of hardening in trees can be controlled both through the phytochrome and the GA mediated systems. Research with tree species is a tedious and slow process, but with the emerging new methods and approaches, we may expect exciting new results in the near future.

KEYWORDS. Critical photoperiod, cold hardening, gibberellin, phytochrome, plant hormones

INTRODUCTION

The influence of photoperiod on the regulation of flowering in herbaceous plants and seasonal growth in woody plants has been well known.

since the discovery by Garner and Allard (1923) early in this century. Generally, photoperiodism is connected to the regulation of processes which are important for reproduction (flowering in seed plants) and survival (woody plants) of the plants (Nitsch, 1957). In both cases, the timing of these processes is of great importance and must be synchronized with the local climatic conditions. At a given locality, the photoperiod provides an accurate and reliable signalling system for timing of these developmental events. In northern tree species, cessation of apical elongation growth is a prerequisite for cold hardening and therefore photoperiod is a decisive factor for timing of development of cold hardiness in these species. Due to this, photoperiodism in northern tree species is a highly adaptive characteristic and the presence of photoperiodic ecotypes in trees related to latitudinal and altitudinal origin is well documented (for references, Junttila and Robberecht, 1998). Photoperiodism in such tree species provides an interesting system for studies of physiological and genetic aspects of climatic adaptation in plants. In addition, photoperiodic regulation of growth, cessation of growth and frost hardening has important practical aspects both in horticulture and in forestry.

Photoperiodic regulation of growth and hardening in trees has been widely studied at the physiological level in several northern tree species. Application of molecular genetic methods in studies of photoperiodism in herbaceous model plants, particularly in *Arabidopsis*, has resulted in very rapid progress indeed. The main focus in this research has been on the role of various light perception systems, phytochrome and cryptocrome in particular, in photoperiodic control of growth and flowering. With a complete genome sequence available, and increasing possibilities to utilize microarray technology for functional analysis of the whole genome, new dimensions to such developmental studies will be added. For obvious reasons, these studies have been much slower coming in trees, but the basic principles, as well as the new experimental methods, will be applicable to trees as well, and we may expect a rapid extension of our knowledge on photoperiodism in tree species in the near future.

The purpose of this paper is to give a short presentation of the basic aspects of photoperiodic regulation in northern tree species and to review the latest studies in the field. These studies have, to a great extent, focused on the perception of photoperiod and on the hormonal regulation of growth and growth cessation, and these topics will be discussed. Molecular aspects of bud development and bud dormancy have recently been reviewed (Rohde et al., 2000) and we will not cover dormancy as such in this presentation.

SHOOT ELONGATION, PHOTOPERIOD AND COLD HARDINESS

Tree species exhibit great variation in the pattern of shoot elongation, but two main patterns can be distinguished, indetermined or free growth, and determined or fixed growth (Kramer and Kozlowski, 1979). In a free growth pattern, organ formation or initiation of new nodes and internodes (stem units) takes place simultaneously with elongation of the internodes. In a fixed growth pattern, these two developmental processes are separated in time, all the stem units are present in the bud, and shoot elongation consists of elongation growth of the preformed internode initials. Organ formation and stem elongation take place in different parts of the apical meristem, organogenesis is located in the apical meristem proper and elongation in the subapical meristem (Sachs, 1965). Most of the northern hardwood species have a free growth pattern, while the northern conifers, with the exception of the first-year growth in some species, have a fixed growth pattern. Generally, a free growth pattern is a juvenile character and, even in free growing hardwood species, there is a tendency for increasing dominance of a fixed growth pattern with increasing maturity. Timing of growth cessation in species with a free growth pattern is controlled primarily by environmental factors, above all by the photoperiod. In trees with a completely fixed growth pattern, like Scots pine (*Pinus sylvestris* L.), the timing of growth cessation is affected only to a limited degree by environmental factors (Junttila, 1986). The physiological and molecular mechanisms underlying the fixed growth pattern have not been studied to any extent, but, may involve interaction or dominance between the various parts of the apical meristem. Consequently, studies on the photoperiodic regulation of apical growth cessation have concentrated on species with a free growth pattern.

The critical photoperiod is defined either as the longest photoperiod which induces growth cessation in a given plant population or clone, or as the shortest photoperiod sustaining continuous shoot elongation. The definition of a "critical night length" is also used (see, e.g., Clapham et al., 2001), based on the importance of the dark period. The term photoperiod is, however, well established in the literature dealing with photoperiodism, and we prefer to use it. Criteria for the critical photoperiod are not always clearly defined. In some cases, it refers to the photoperiod where all the plants stop growing, and, in other cases, to a photoperiod leading to bud set in 50% of the plants. Uniformity in the methods would facilitate comparisons between various studies, but the main picture emerging from the literature is rather unambiguous. Depending on the origin of the population, the critical photoperiod may vary from about 12 h to 22 h (Håbjörg, 1972; Heide, 1974; Håbjörg, 1978; Junttila, 1980; Howe et al., 1995). The suggestion that tree species contain photoperiodically different types was proposed by Sylven (1940) and the term photoperiodic ecotype was adopted to describe tree populations which, during evolution, have adapted to the photoclimate of their environment (Vaartaja, 1954; 1959). Such ecotypes have been demonstrated, among others, in *Acer platanoides*, *Alnus incana*, *Betula pubescens*, *Betula pendula*, *Coryllus avellana*, *Hippophae rhamnoides*, *Salix caprea*, *Salix pentandra*, *Populus* species, *Tilia cordata*, *Ulmus glabra*, *Picea abies* and *Pinus sylvestris* (for references, see Håbjörg, 1978). The critical photoperiod is normally determined experimentally under controlled conditions using low irradiance light to extend the main photosynthetic light period. Several studies have shown that the time course of growth cessation under natural light conditions can be estimated based on the critical photoperiod determined under controlled conditions (e.g., Heide, 1974; Junttila, 1980). This is shown for two ecotypes of Norway spruce at two latitudinal locations in Figure 1. At both locations (Steinkjer, 64°N, and Lankowitz, 47°N), the local ecotype (L1 and Lankowitz, respectively) receives the daylength signal for growth cessation early in August. If the southern ecotype (Lankowitz) is planted at a northern location (Steinkjer), it will receive this signal about a month later and will be vulnerable to frost injury. On the other hand, daylength at Lankowitz will never exceed the critical photoperiod for the northern ecotype, L1, and it will not be able to grown normally.

Contrasting ecotypes do not only differ in critical photoperiod but also in their requirements for irradiance and composition of the light used as extension light (Junttila and Kaurin, 1985; Clapham et al., 2001). Generally, the threshold level of light is higher for ecotypes from far north compared to those from southern latitudes, and the northern ecotypes appear to require more far red light than the southern ecotypes in order to give a long day response. However, these aspects need further studies. The critical photoperiod appears to be relatively constant over a range of temperatures (Heide, 1974), but low night temperatures (about 5°C) combined with high day temperatures (about 18°C) may override the photoperiod and induce bud set even under a 24-h photoperiod (Heide, 1974; Junttila, 1980). In many species stem elongation is strongly affected by thermoperiodicity and the combination of a high night temperature with a lower day temperature appears to reduce internode elongation (Erwin and Heins, 1995; Grindal et al., 1998).

FIGURE 1. Climate photothermographs for Steinkjer, Norway (64°N) and Lankowitz, Austria (47°04′N) together with critical "photothermas" for apical growth cessation of corresponding ecotypes (L1 and Lankowitz) of Norway spruce. Figures on the photothermographs stand for months of the year. (Heide 1974, by permission from Physiologia Plantarum.)

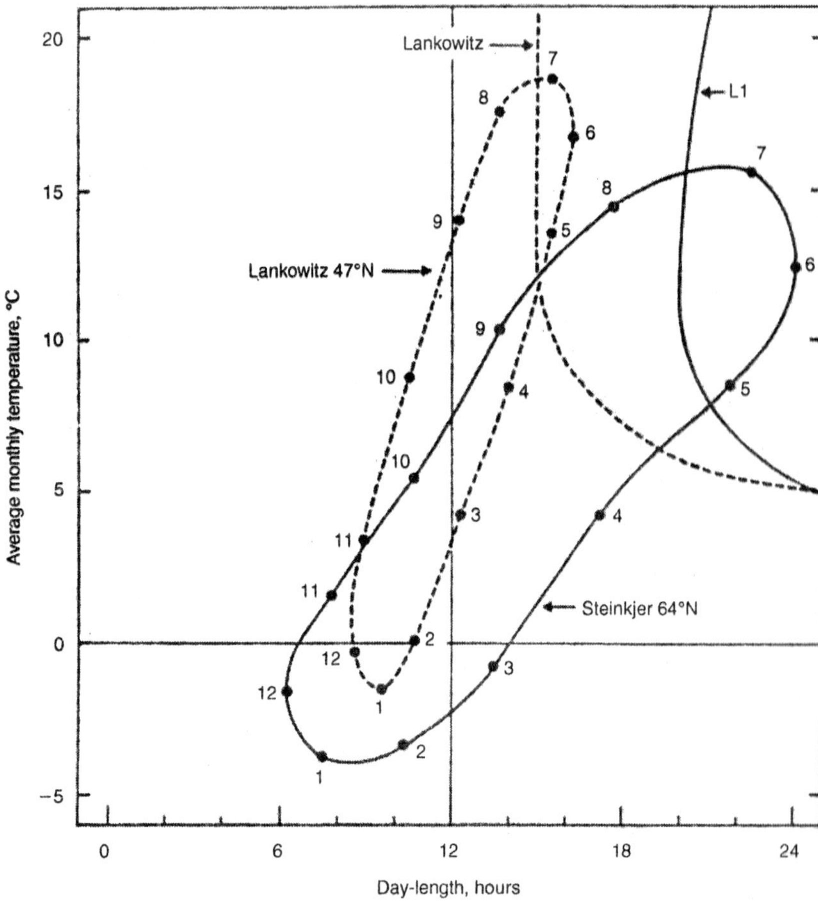

Regulation of gibberellin metabolism can be involved in these thermoperiodic responses (Grindal et al., 1998), and also interaction between thermoperiod and phytochrome B has been demonstrated (Grindal, personal communication). However, time course of SD-induced growth cessation in hybrid aspen was not significantly affected by thermoperiodicity (Junttila, unpublished results).

Within the northern temperate and boreal zones, the number of native tree species decreases significantly with increasing latitude. For example, in North America there are about 120 species in the zone no. 8 (minimum temperature −1.1 to −6.1°C) and only about 10 species in the zone no. 1 (minimum temperature below −45.6°C) (Bannister and Neuner, 2001). Generally, the frost hardiness of those species correlates with the minimum temperature of the area, as documented for conifer species by Bannister and Neuner (2001). However, the maximum hardiness of native species, observed in midwinter or induced artificially, greatly exceeds the annual minimum temperatures observed in the various zones. This difference appears to increase with increasing northern latitude (Bannister and Neuner, 2001), and, in a frost hardy stage, many of the extreme northern species are able to survive the temperature of liquid nitrogen. Thus, for such species, winter survival is not limited by their capacity for frost hardiness, but proper timing of hardening and dehardening. Although several phases of the annual cycle of hardiness are of importance, the timing of hardening is particularly important under northern conditions. This statement is supported by the existence of the significant genetic cline in critical photoperiod for growth cessation and hardening observed in many tree species. Such strong clines suggest that these traits are under relatively strong selection (Aitken and Hannerz, 2001). In Scots pine, which has a fixed shoot elongation while needle elongation is under photoperiodic control, the time course of cold hardening in latitudinal ecotypes changes in a similar clinal pattern as the regulation of growth cessation in species with a free growth pattern (Hurme et al., 1997).

Although the photoperiod may have a direct effect on cold hardening in woody plants, the main effect can be considered to be an indirect one through regulation of apical growth cessation. Cessation of elongation growth as an important prerequisite for cold hardening has been documented in numerous studies, recently by experiments with transgenic plants of hybrid aspen overexpressing the oat phytochrome A gene (Olsen et al., 1997b). These overexpressing lines did not set bud under short day conditions and were incapable of hardening irrespective of the photoperiod or the temperature at which they were exposed. Thus, short photoperiod alone, without affecting the elongation growth, was not able to initiate cold hardening in hybrid aspen. This also applies other tree species with a free growth pattern. However, recent studies have shown that short photoperiod induces frost hardening also in mature leaves of *Betula pendula* (Li et al., personal communication). This effect may be due to a direct photoperiodic modulation of metabolism in the leaf, but it may also be related to the photoperiodic regulation of stem growth.

PHYTOCHROME AND PHOTOPERIODIC RESPONSES IN TREES

Plants respond to light through at least three groups of photoreceptors, UVB-receptors, cryptochromes, and phytochromes, which absorb in the ultraviolet (UV), the blue/near UV, and the red/far-red ranges of light, respectively (Ahmad, 1999). Of the three classes of photoreceptors, the phytochromes are the best characterised. Native phytochromes are soluble homodimer proteins with a total molecular mass about 250 kDa. Each apoprotein subunit is covalently attached to a light-absorbing linear tetrapyrrole chromophore, phytochromobilin. The chromophore exists in two photoconvertible forms, the red light-absorbing Pr form ($\lambda max = 660$ nm) and the far-red light-absorbing Pfr form ($\lambda max = 730$ nm). Phytochromes are synthesised as Pr, which is converted in red light to Pfr (Braslavsky et al., 1997). The Pfr is generally considered to be the physiologically active form. The native Pr form of phytochromes is synthesised in the cytoplasm and upon conversion to the Pfr form it is imported into the nucleus where it is believed to be involved in gene regulation (Neff et al., 2000). Phytochromes are involved in the regulation of a large number of light-responsive genes and many photomorphogenic events, including seed germination, photoperiodic control of development, fruit ripening, stem elongation, and chloroplast development (Kendrick and Kronenberg, 1994; Smith, 1995; Jackson and Thomas, 1997).

Angiosperms have been found to contain a small number of phytochrome apoproteins (PHY) encoded by a small gene family. For example in *Arabidopsis thaliana,* five phytochrome genes (*PHYA* through *PHYE*) have been found (Sharrock and Quail, 1989; Clack et al., 1994). *PHYA*, *PHYB*, *PHYC*, and *PHYE* encode evolutionary divergent proteins sharing only 46-53% amino acid sequence identity (Clack et al., 1994). *PHYD* encodes an apoprotein (PHYD) that shares 80% amino acid sequence identity with PHYB and is a member of the same subgroup. Thus, the five members of the *Arabidopsis PHY* family fall into four subfamilies. Phylogenetic analyses, including sequences of *PHY* genes from many angiosperms and nonangiosperms support the existence of four subfamilies (Mathews et al., 1995; Mathews and Sharrock,

1997; Alba et al., 2000), although monocotyledonous plants may not contain *PHYE* homologues (Mathews and Sharrock, 1996).

Phytochrome in Tree Species

cDNAs encoding phytochromes have been cloned from several woody plants (Kolukisaoglu et al., 1995; Mathews et al., 1995; Eriksson and Moritz, 1997; Howe et al., 1998). However, full-length phytochrome cDNAs have only been cloned from a few tree species. Cottonwood (*Populus trichocarpa*) is the tree species in which the *PHY* family has been best characterised and the family has been found to contain one *PHYA* locus and two *PHYB* loci (*PHYB1* and *PHYB2*), but to lack members of the *PHYC* and *PHYE* subfamilies (Howe et al., 1998). PHYA in hybrid aspen (*Populus tremula x tremuloides*) has an amino acid sequence with high homology to *PHYA* of dicots, e.g., 80% identity and 90% similarity with PHYA of both *Arabidopsis thaliana* and *Cucurbita pepo* but less to monocots (Eriksson and Moritz, 1997). Southern blot analysis indicates that *PHYA* in hybrid aspen represents a single locus, as has also been found in *Populus trichocarpa* (Eriksson and Moritz, 1997; Howe et al., 1998). In the two coniferous species, Norway spruce (*Picea abies*) and Scots pine (*Pinus sylvestris*), two full length phytochrome genes have been isolated, called *PHYO* and *PHYP*, respectively (Clapham et al., 1999). In phylogenetic studies the *PHYP* from *Pinus* was found to be an early descendant of the line leading to angiosperm *PHYB*, *PHYD* and *PHYE*, whereas the *Picea PHYO* was found to be an early descendant of the line leading to angiosperm *PHYA* and *PHYC* (Schneider-Poetsch et al., 1998). Additional phylogenetic studies with a partial *Picea* phytochrome sequence have also indicated the presence of a sublineage of *PHYO*, called *PHYN* (Schneider-Poetsch et al., 1998).

PHYA is known to be light-labile and both the *PHYA* transcript and the protein level are rapidly reduced when plants are transferred from dark to light. On the other hand, PHYB through E are light-stable and their levels are not affected by light. Although *PHYO* in *Picea* is related to *PHYA*, the level of the *PHYO* transcript was affected little by light and was only down regulated by a factor of two (Clapham et al., 1999). This level of light regulation is in striking contrast with the situation in some angiosperm species where *PHYA* transcripts are decreased 10- to 100-fold after illumination (Quail et al., 1995). Burgin et al. (1999) studied light-mediated regulation of the protein levels of phytochrome in *Pinus elliotti* and *Pseudotsuga menziesii* using monoclonal antibodies against oat PHYA. They found that the detectable phytochrome pool decayed with a half-life of approximately 4 hours in both species. In comparison, a typical half-life in angiosperms is 1-2 hours (Clough and Vierstra, 1997). It has been suggested that this strong light-dependent regulation was acquired late in the phytochrome evolution, after gymnosperms separated from the angiosperms lineage (Clapham et al., 1999).

Photoperiodic Timekeeping and Phytochromes in Trees

The involvement of phytochromes in the regulation of photoperiodic responses has been well documented in *Arabidopsis*, tomato and a number of other herbaceous plants through analysis of mutants and transgenic plants (for reviews, see Smith, 1995; Quail, 1997; Casal et al., 1998; Neff et al., 2000). Various phytochromes appear to have partly different functions in plants, but they can also interact and may, in part, substitute for each other. The specific roles of the different phytochromes in the photoperiodic responses are as yet not well understood. Generally, however, phyA as well as phyB are thought to be involved in photoperiodism, and have been suggested to be the daylength receptors in long-day and short-day plants, respectively (Jackson and Thomas, 1997). In a recent study of flowering in *Arabidopsis*, a facultative long-day plant, Bagnall and King (2001) concluded that the roles of phytochromes in the regulation of flowering include an end-of-day phyB-type response, a minor phyA photoperiodic response, and a large indirect phyA effect involving photosynthesis.

Our knowledge of the involvement of phytochrome in the photoperiodic regulation of growth and cessation of growth in woody plants is mainly based on several physiological studies (D'Aoust and Hubac, 1986; Howe et al., 1996; further references in Thomas and Vince-Prue, 1997). Recently, however, transgene technology has been applied to address these issues in woody plants (Olsen et al., 1997b; Eriksson, 2000). Olsen et al. (1997b) studied the effects of overexpression of the oat *PHYA* gene in hybrid aspen. Transgenic lines with a strong expression of the oat *PHYA* gene (among others line no. 22) did not stop growing even under a photoperiod of 6 hours, whereas the wild type and transgenic control plants had a critical daylength of 15-16 h (Olsen et al., 1997b). When these plants were grown at controlled temperatures under natural light conditions at the latitude of about 70°N, the wild type stopped growing during the second half of September, while plants of the transgenic line 22 were still growing when the experiment was terminated, October 14 (Figure 2). Figure 2 also shows that temperature treatments used in this experiment had only a minor effect on the time

FIGURE 2. Cumulative elongation growth in wild type and phytochrome A overexpressing line (*PHYA* OE22) of hybrid aspen when grown under natural daylength at 69°39′N at various temperatures as indicated. For alternating temperature treatments (18/12 and 18/9°C), high and low temperature periods were 12 h. Photoperiod during the experimental period is also shown. The critical photoperiod for wild type was reached September 4-9.

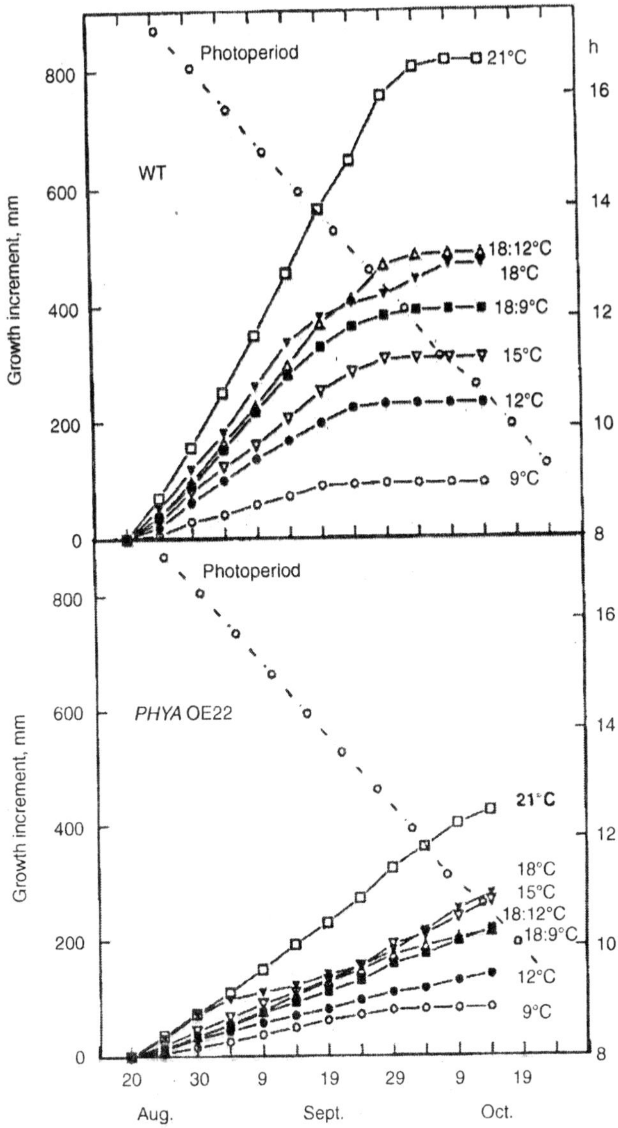

course of bud set in wild type hybrid aspen. However, effect of temperature on the rate of stem elongation during the linear phase of elongation growth, was stronger in the wild type than in line 22, indicating that overexpression of the oat phytochrome A also affected temperature response of hybrid aspen (Figure 3).

The results of Olsen et al. (1997b) could indicate that the response to photoperiod in hybrid aspen may be related to the amount of PHYA. Eriksson (2000) observed that transgenic lines with reduced levels of PHYA due to an introduced *PHYA*-antisense construct showed earlier bud formation in response to short days than the wild type, as could be expected on the basis of the results of Olsen et al. (1997). These studies indicate that phyA is involved in photoperiodic control of bud set and growth cessation in *Populus* (Olsen et al., 1997b; Eriksson, 2000). So far, no studies of transgenic plants expressing sense or antisense con-

FIGURE 3. Rate of elongation growth (mm d^{-1}) in wild type and and phytochrome A overexpressing line (*PHYA* OE22) of hybrid aspen as affected by temperature. Growth rates were calculated from the experiments shown in Figure 2 and show the maximum rates (Y) at the various temperatures (X). Wild type: Y = 1.26X − 7.49, r^2 = 0.90 (P = 0.0012); *PHYA* OE22: Y = 0.47X − 2.46, r^2 = 0.98 (P = 0.0001).

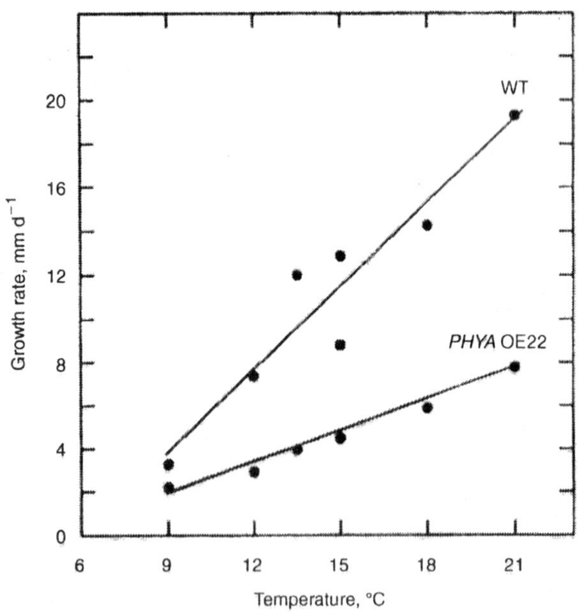

structs of *PHYB* have been reported for woody plants. Extension of such studies to include contrasting photoperiodic ecotypes would be highly interesting for further studies on the role of phytochrome in photoperiodic adaptation of woody plants.

Photoperiodism appears to be based on complicated mechanisms of timekeeping. Studies of regulation of flowering have resulted in the hypothesis that plants have two systems of daylength perception, one for night timekeeping (dark-dominant plants) and one for day timekeeping (light-dominant plants). Plants induced to flower under short daylengths (short day plants) utilize the system of night timekeeping and plants flowering in long daylengths (long day plants) rely on the system of day timekeeping (Thomas and Vince-Prue, 1997). In night timekeeping, the dark period is measured as the time from dusk to dawn. During this period plants are responsive to short periods of illumination (night-breaks), which give a long-day effect. The maximum sensitivity to night-breaks is time dependent indicating the involvement of a circadian clock in photoperiodism.

In tree species with an extensive north-south distribution, southern ecotypes possess characteristics of dark-dominant plants, while extreme northern ecotypes behave like light-dominant plants showing limited response to night-breaks and having a requirement for a low red:far red ratio during daylength extension (Junttila and Kaurin, 1985; Clapham et al., 1998; 2001). Responses to night-break treatments show a classical red-far red reversibility, as has also been demonstrated for *Populus* (Howe et al., 1996). This indicates an involvement of a light-stable phytochrome of type B, or similar one. Circadian rhythms in the effectiveness of night-break have been shown, among others, for a southern ecotype of Norway spruce (Clapham et al., 1998). In extreme northern ecotypes the photoperiodic response is like a far red high irradiance response, which is supposed to be mediated by phyA (Smith, 1995). Taken together, these studies indicate that photoperiodic responses in trees may be based on a combination of two main mechanisms, and that differences between ecotypes might reflect different inputs of the dark- and the light-dominant system. Thus, a phyB-based system appears to be dominating in southern ecotypes, while responses of northern ecotypes might be controlled by a phyA-based system. Also, in Norway spruce, the systems of night timekeeping and day timekeeping have been found to co-exist in intermediate populations in Sweden, between 64°N and 66°N (Clapham et al., 1998). In addition to the molecular characterization of phytochromes in various ecotypes under different growth conditions, as well as studies of transgenic plants, further investigations on the responses of contrasting ecotypes to various light treatments are needed to further elucidate the mechanisms underlying the differences in photoperiodic responses of tree ecotypes.

Also other pigments besides phytochromes, for example cryptochromes and other blue light receptors, are probably involved in the perception of daylength in *Arabidopsis* (Devlin and Kay, 2000). Blue light is known to reduce internode elongation in birch (*Betula pubescens*) and willow (*Salix pentandra*) (Junttila and Kaurin, 1985). Influence of blue light on daylength perception in tree species has not been studied at all. Such studies with contrasting ecotypes, as well as cloning and investigations of cryptochromes from trees, will be an interesting topic for future studies.

Detection of photoperiod is closely connected with the internal circadian rhythm, a basic character of most living organisms. Information on mechanisms of this biological clock is rapidly evolving and progress in this field has been frequently reviewed (e.g., McClung, 2000). Circadian rhythm also underlies photoperiodic responses in trees (Clapham et al., 2001), but so far we do not have any information at all on the role of circadian rhythm in photoperiodic adaptation.

PHOTOPERIODISM AND HORMONAL CONTROL OF GROWTH

Plant hormones are supposed to play a central role in the translation of environmental signals such as photoperiod, which is perceived by phytochrome and possibly cryptochromes, to endogenous signals that ultimately give rise to the manifestation of a growth process or developmental change. Of the main types of plant hormones, gibberellins (GA) appear to be of special importance with respect to the photoperiodic regulation of shoot elongation, and therefore a discussion of other plant hormones will not be covered here. Hormonal influence is dependent both on hormone metabolism and hormone signalling. Our emphasis here is on the hormone metabolism, mainly because of the available information and data. Almost nothing is known of hormonal signalling in woody plants.

Gibberellins and Shoot Elongation in Trees

Ever since Kurosawa in 1926 discovered the existence of gibberellin (GA), substantial amounts of evidence have accumulated showing that elongation growth of plants is controlled by GA. Numerous application experiments, analytical, genetic, and lately molecular studies all support this conclusion. GAs form a large group of more than 120 tetracyclic diterpene carboxylic acids that are found in plants, fungi, and bacteria. Only two or three of the known GAs are considered to possess hormone activity, whereas the others are metabolic precursors or inactivation products.

A great number of application experiments using various GAs and inhibitors of GA biosynthesis point to a major role of GA on shoot elongation in tree species. Numerous studies carried out in the 1950s and 1960s were reviewed by Westing (1959), Melchior and Knapp (1962), and later studies by Junttila (1991). The most pronounced responses have usually been reported for species with a free growth pattern. Some examples of application studies are discussed here. In *S. pentandra*, several application experiments support the conclusion that GAs, and particularly GA_1, play a prominent role in the control of shoot elongation (Junttila, 1976; Junttila and Jensen, 1988; Junttila et al., 1991; 1992; Junttila, 1993a; 1993b). An important role of GA_1 in elongation growth has also been observed in application studies with GAs and GA biosynthesis inhibitors in *Populus tremula x tremuloides* (unpublished results). In dwarfed oat *PHYA* overexpressing hybrid aspen plants, levels of GAs were reduced as compared with the wild type, but normal elongation growth was restored for all transgenic lines by exogenous GA_1.

Although plants with a free growth pattern usually have shown the most prominent responses, there are also examples of plants with a fixed growth pattern responding to exogenous GAs (Junttila, 1991). Results from early experiments with exogenous GA_3 in conifers were commonly negative (Westing, 1959), but numerous later studies have demonstrated positive effects of GAs on shoot elongation in Pinaceae, Cupressaceae, and Taxodiaceae (for references, see Junttila, 1991). In addition to GA_3, exogenous $GA_{4/7}$ has shown stimulating activity on longitudinal growth in numerous studies, like in *Pinus elliotii* (Pharis and Kuo, 1977), *Pinus radiata* (Pharis, 1976), and *Pinus sylvestris* (Wang et al., 1992; 1997). In *Picea mariana* $GA_{4/7}$ enhanced elongation growth in trees from slow-growing families, but not in trees of fast-growing families (Pharis et al., 1991). The effect of GAs on growth in conifers appears to be dependent on the method of application, whereas the timing of application appears to be important for species with a fixed growth pattern. In *Pseudotsuga menziesii,* exogenous $GA_{4/7}$ enhanced shoot elongation when applied before bud swelling, but not after flushing (Ross, 1983; Owens et al., 1985). In *Picea abies*, exogenous GA_1 or GA_4 stimulated shoot elongation (Moritz, 1995).

The rather complex biosynthesis of GAs (for reviews, see Graebe, 1987; Hedden and Kamiya, 1997; Hedden, 1999; Yamaguchi and Kamiya, 2000; Hedden and Phillips, 2001) can be regarded as consisting of two main parts. The early pathway, in which *trans*-geranylgeranyl-diphosphate is converted to GA_{12}-aldehyde, is common to all systems that have been studied (Hedden and Phillips, 2001). According to our knowledge, no study of this part of the GA biosynthetic pathway in woody species has been published.

In the later pathways GA_{12}-aldehyde is metabolized to active GA hormone. In vegetative tissues of most plants this is predominantly accomplished by the early 13-hydroxylation pathway (GA_{12}-GA_{53}-GA_{44}-GA_{19}-GA_{20}-GA_1), in which GA_{12} is hydroxylated at C-atom no. 13 and GA_1 is known to be the active hormone. An alternative route is the non-13-hydroxylation pathway (GA_{12}-GA_{15}-GA_{24}-GA_9-GA_4-GA_1), in which biosynthesis of GA_4 occurs (Graebe, 1987; Hedden and Kamiya, 1997). In angiosperm tree species, the early 13-hydroxylation pathway is the major route of GA biosynthesis (e.g., Davies et al., 1985; Koshioka et al., 1985; Poling and Maier, 1988; Rood et al., 1988; Steffens and Hedden, 1992; Olsen et al., 1994; Zanewich and Rood, 1994; Hasan et al., 1994; Ridoutt et al., 1995), although there are also some examples of the presence of some non-13-hydroxylated GAs in some species (Junttila et al., 1988; Hasan et al., 1994; Ridoutt et al., 1995). In conifers, the non-13-hydroxylation pathway is dominating in vegetative tissues (e.g., Odén et al., 1982; 1987; Moritz et al., 1989a; Moritz, 1995; Wang et al., 1996; Doumas et al., 1992), although some 13-hydroxylated GAs also occur (Odén et al., 1987; Moritz et al., 1989a; Moritz, 1995). The non-13-hydroxylation pathway can alternatively be called the late 13-hydroxylation pathway, which points to the late conversion of GA_4 to GA_1 by hydroxylation at C-atom no. 13 (Moritz et al., 1989a; Odén et al. 1995; Wang et al., 1996).

It is widely recognized that the 3β-hydroxylated GA_1 is the primary active GA controlling shoot elongation in most species, and that the majority of other GAs exhibiting bioactivity do so as a consequence of their metabolism to GA_1. This has been shown also in a number of angiosperm tree species, like *Salix pentandra* (Junttila et al., 1991; 1992; Junttila, 1993a), birch (*Betula pubescens*), and alder (*Alnus glutinosa*) (Junttila, 1993b). In these studies, the 3β-hydroxylated GA_1 was able to restore growth in plants treated with prohexadione, an inhibitor of the

3β-hydroxylase responsible for the formation of GA_1 from GA_{20}. Commonly also GA_4 is considered to have intrinsic hormone activity and to be the most important GA hormone in those plants where it is the dominating 3β-hydroxylated GA, like in *Arabidopsis* (Hedden, 1999). In *Salix*, GA_4 has been tentatively identified as a minor GA (Junttila et al., 1988). Exogenous GA_4 exhibits biological activity in *S. pentandra*, but has been shown to be metabolised to GA_1 (Junttila, 1993a). Although GA_4 is the most abundant 3β-hydroxylated GA in coniferous species, it has been suggested that GA_1 is the active GA in shoot elongation even in conifers. In conifers GA_4 is readily converted to GA_1, and its activity in application studies thus appears to be due to its conversion to GA_1 (Moritz et al., 1989a; Odén et al., 1995; Wang et al., 1996). In a number of species, e.g., *Picea abies* (Odén et al., 1987), *Picea sitchensis* (Moritz et al., 1989b), *Psudotsuga menziesii, Cupressus arizonica*, and *Juniperus scopulorum* (Pharis and Kuo, 1977), the presence of GA_3, another 3β-hydroxylated GA usually considered to have intrinsic hormone activity, has also been reported. However, Moritz (1995) found that growth cessation in *Picea abies* was not correlated with a decrease in GA_3, but with lowered amounts of GA_1 and partly GA_4. Taken together, these data might suggest that GA_1 is the active hormone in control of shoot extension growth in woody angiosperms as well as in conifers.

Correlation between endogenous levels of gibberellins, as analysed by combined gas chromatography-mass spectrometry (GC-MS), and active shoot elongation has been shown in a number of tree studies. In *S. pentandra*, there are close correlations between growth, cell divisions, and levels of GA_1 as well as other GAs, in zones of elongating shoot tips (Olsen et al., 1995a; 1995c; 1997a; Hansen et al., 1999). In *Populus* (Bate et al., 1988; Olsen et al., 1997b), and *Eucalyptus globulus* (Hasan et al., 1994), correlation between high levels of GA_1 and other GAs and shoot elongation has been shown. Dwarf growth of *Populus tremula x tremuloides* overexpressing the oat *PHYA* correlated with lowered GA levels relative to the wild type (Olsen et al., 1997b). Dwarfism was completely suspended by a far red end of day treatment, and this correlated with restoration of GA_1 contents to similar levels as in the wild type (Olsen and Junttila, 2002). Also, in alder and birch rough estimates of relative levels of GAs have indicated high amounts of GA_1 and GA_{20} in flushing buds (Zanewich and Rood, 1994). In *Picea abies*, a positive correlation between growth and levels of GA_1, as well as GA_4 and GA_9 has been reported (Moritz, 1995). In fast-growing families of *Picea mariana*, higher GA levels were found than in slow-growing families

(Pharis et al., 1992). In *Pinus sylvestris*, inhibition of shoot elongation by prohexadione application was correlated with reduced contents of GA_1, GA_4, and GA_3 (Wang et al., 1995), and exogenous $GA_{4/7}$ stimulated longitudinal growth and increased the levels of GA_4, GA_7, and GA_9 (Wang et al., 1992).

At the molecular level, data are as yet scarce for woody species, but a correlation between high expression of a key regulatory gene in the GA biosynthetic pathway, GA20-oxidase (GA20ox), and internode extension has been reported for *Populus tremula x tremuloides* (Eriksson et al., 2000). Higher expression of the gene was observed in elongating internodes than in mature internodes, and cessation of apical growth was correlated with decreased expression. The multifunctional GA20ox is responsible for forming the immediate precursor of GA_1 or GA_4, i.e., GA_{20} or GA_9, respectively, in an oxidation sequence involving several steps. Moreover, ectopic expression of the *Arabidopsis thaliana* GA20ox1 in hybrid aspen increased the height growth rate and biomass, and the plants contained higher levels of GA_1 and GA_4 in internode tissue than the wild type control (Eriksson et al., 2000). However, there was no direct relationship between degree of stimulated elongation growth and absolute GA_1 levels in the different transgenic lines. Eriksson (2000) has also reported on hybrid aspen plants with an antisense construct of a segment of the gene encoding the GA20ox. However, contrary to what might be expected, elongation growth of these plants did not differ from the wild type, and there was no significant difference in GA levels in elongating internodes. Similar studies with other genes of the GA biosynthetic pathway and other species, included conifers, would be of great interest.

General Regulation of Gibberellin Biosynthesis

Although very little data on this is available for woody species, substantial evidence has accumulated that GA biosynthesis is negatively regulated by GA activity in various herbaceous species (Yamaguchi and Kamiya, 2000). GA20ox and GA3ox, which is responsible for the formation of GA_1 from GA_{20} and GA_4 from GA_9, are both families of enzymes that are subjected to feedback-regulation, i.e., the mRNA levels of the genes encoding them are down regulated by GAs (Yamaguchi and Kamiya, 2000). The only such information available for a woody tree species points to a negative feedback control of GA20ox at the expression level in hybrid aspen (Eriksson, 2000). The deactivation of GAs, known to be catalyzed by GA2ox, is up regulated by GA activity and is thus subjected to feedforward-regulation. However, there is no evidence for a feedback-regulation of expression of the genes early in the GA biosynthetic pathway encoding *ent*-copalyl diphosphate synthetase, *ent*-kaurene synthase and *ent*-kaurene oxidase (Yamaguchi and Kamiya, 2000). Thus, the level of bioactive GA is transcriptionally controlled both at the level of biosynthesis and deactivation. It has been reported that mRNA levels do not always correlate with enzyme activities, which points also to post-transcriptional regulation (Lange et al., 1997). To obtain an understanding of regulation of GA biosynthesis by endogenous as well as environmental conditions in woody species, studies of genes and enzymes of the pathways are strongly needed.

Regulation of biologically active GA biosynthesis might also occur by the formation and hydrolysis of GA-conjugates (Sembdner et al., 1991). Two main types of GA conjugates exist in plants, the GA glucosyl esters and the GA-glucosides. To the best of our knowledge, only one GA-conjugate has been conclusively identified in tree species, i.e., GA_9-glucosyl ester in shoots of *Picea sitchensis* (Lorenzi et al., 1976; Moritz, 1992) and *Pseudotsuga menziesii* (Doumas et al., 1992). However, nothing is known about its regulation and function in these species.

Photoperiodic Regulation of GA Biosynthesis

In order to investigate the involvement of different plant growth regulators in photoperiodic control of shoot elongation, many experiments have been performed in which growth regulators were applied under short day (SD) conditions. Although correlations have been found between high endogenous levels of IAA and cytokinins and active elongation growth, and low levels and lack of growth and dormancy, exogenous auxins and cytokinins have hitherto not been shown to be able to substitute for the long day (LD) requirement for shoot elongation (e.g., Junttila, 1976; Olsen et al., 1997a; Thomas and Vince-Prue, 1997). ABA has commonly been thought to be responsible for inducing and maintaining dormancy, but single applications commonly do not induce growth cessation, and ABA contents of leaves and buds commonly do not increase after transfer to SD conditions. However, there are also several examples to the contrary (Thomas and Vince-Prue, 1997). Thus, there is not unequivocal evidence for a role of these hormones in photoperiodic control of shoot elongation, as there is for GA, and so the possible roles of IAA, cytokinins and ABA will not be further discussed here. Recently described interactions between IAA and GA metabolism (Ross et al., 2000) warrant, however, further investigations on hormone interactions also with respect to photoperiodic regulation. Endogenous levels of GA_1 and IAA are often found to change in a similar pattern under various photoperiodic treatments (see, e.g., Olsen et al., 1997a; 1997b).

In several woody species, cessation of apical elongation growth under SD conditions has been shown to be prevented by application of certain GAs, such as GA_3 that was used in numerous early studies (Nitsch, 1957; Wareing and Saunders, 1971; Junttila, 1976; Junttila, 1991; Thomas and Vince-Prue, 1997). Also, certain exogenous GAs can induce new growth when applied in the initial period after growth cessation has occurred. On the basis of such experiments, GA is thought to play an important role in photoperiodic control of shoot elongation in woody species. The greatest responses have typically been shown in plants exhibiting a free growth pattern (Thomas and Vince-Prue, 1997).

Several studies have focused also on endogenous GA as important mediators of photoperiodic growth control. Early studies applying bioassays for recording endogenous GA-like substances usually showed reduced GA activity under SD conditions (Wareing and Saunders, 1971; Junttila, 1982; Dunberg and Odén, 1983; Lavender and Silim, 1987; Junttila, 1991). With the development of sensitive and selective physiochemical methods for GA analyses, it has become clear that bioassays do not give conclusive information on the identity of substances exhibiting activity. Both uptake, transport and metabolism as well as inhibitors present, can complicate assessment of the assay and may give unreliable estimates of GA levels. Up to now the involvement of endogenous GAs has been studied with reliable analytical methods to a limited extent only. However, a number of thoroughly studied species provides evidence for an important role of GA in photoperiodism in woody plants.

In *Salix pentandra*, the most thoroughly studied woody angiosperm in this respect, levels of most GAs including the active GA_1, decreased rapidly under short photoperiod (Olsen et al., 1995a; 1995c; 1997a). In a northern ecotype from 69°N, having a critical photoperiod of about 22 h, the levels of GA_1 were 30-40% lower in young stem and leaf tissues after only 2 SD cycles (Olsen et al., 1995a). This was well before growth cessation, which in this ecotype occurs after about 8-10 days under12-h photoperiod. Under prolonged SD conditions, GA_1 levels decreased further by more than 80% after 30 SD cycles (Olsen et al., 1995a). Furthermore, using a microscale method for analysing GAs in 5 mm long stem segments of individual plants (Moritz and Olsen, 1995), it was shown that active shoot elongation correlated with a gradient of

GA_1 in shoot apices (Olsen et al., 1995c). The highest GA_1 contents were found 5-15 mm below the apical tip. After 5 SD the gradient had disappeared and the decrease in GA_1 content was about 50% in the segments that had the highest levels under LD (Olsen et al., 1995c). Upon transfer from SD to growth-inducing LD conditions after 15 SD, a gradient in GA_1 was again established within a few days (Olsen et al., 1997a). Bud break and initiation of growth correlated with a rapid, transient increase in GA_1 in defined zones of shoot tips up to levels twice those in LD-grown control seedlings at day 7 after transfer. Thereafter, the GA_1 levels rapidly decreased to levels similar to the control plants (Olsen et al., 1997a). The precursors of GA_1, i.e., GA_{53}, GA_{19}, and GA_{20} as well as the inactivation products GA_8 and GA_{29}, also declined under SD (Olsen et al., 1995a; 1995c; 1997a). Furthermore, in several application experiments, exogenous GAs (GA_{20}, GA_1, GA_4, GA_9, and GA_3) have been shown to substitute for the LD requirement for growth (Junttila, 1976; Junttila and Jensen, 1988; Junttila et al., 1992; Junttila, 1993a).

In a study of *Populus tremula x tremuloides*, a decrease in GA levels under SD was also observed (Olsen et al., 1997b). The clone used had a critical photoperiod of 15 h and ceased growth after 3-4 weeks under a 10 h photoperiod. At that time levels of GA_1 and other GAs had diminished by more than 50% in shoot apices as compared with levels under LD. Also, exogenous GA_{20} and GA_1 were able to prevent growth cessation and induce new growth under SD (unpublished results). Plants of hybrid aspen ectopically expressing the oat *PHYA* were impaired in their photoperiodic growth control, and did not cease growth under SD. Also no change in GA levels under SD was observed (Olsen et al., 1997b). Thus, this provides evidence for an interaction between phytochrome and GA as an important regulatory system in photoperiodic control of shoot elongation. Furthermore, Eriksson et al. (2000) reported on construction of hybrid aspen plants with ectopic expression of GA20ox. These plants had elevated GA levels and showed a delay in growth cessation under SD. Hybrid aspen plants with an introduced antisense construct of a fragment of this gene showed no significant increase in GA levels, but nevertheless ceased shoot elongation before the wild type (Eriksson, 2000).

In *Picea abies,* lowered levels of GA_1, GA_9, and partly GA_4, were found under SD conditions (Moritz, 1995). In addition, exogenous GA_1 and GA_4 were able to substitute for LD under the SD conditions used (Moritz, 1995). However, it appears that the comparison of GA contents under LD and SD was not completely relevant, as SD treated plants were exposed to 460 μmol m^{-2} s^{-1} photosynthetic active irradiation for 10 h, whereas LD exposed plants were given this light intensity for 24 h, instead of a more comparable low-intensity daylength extension for 14 h. Therefore, how much of the relatively small differences in GA levels observed could be ascribed to the great difference in photosynthetic active irradiation, and how much to a photoperiodic control of GA metabolism is unknown.

In addition to photoperiodic control of GA levels, there is also evidence for a modulation of sensitivity for GA. When plants have attained full dormancy, no response to exogenous GA is usually observed, as sensitivity to GA is greatly decreased (Vegis, 1964; Lavender and Silim, 1987). In *Salix pentandra*, the ability of GAs to induce new growth diminished with increasing numbers of SD cycles (Junttila and Jensen, 1988), indicating a decrease in sensitivity to GA. Transgenic hybrid aspen overexpressing the oat *PHYA* that did not stop growing under SD, but they contained similar GA levels as wild type plants in which growth was arrested (Olsen et al., 1997b). Thus, lack of response to SD could be caused by a reduced sensitivity to GA in the transgenic plants.

Possible Steps of Gibberellin Biosynthesis Under Photoperiodic Control

In *Salix pentandra,* shoot elongation in resting plants was restored by exogenous GA_{20} and GA_1 (Junttila and Jensen, 1988). Earlier precursors of the GA-biosynthetic pathway, GA_{19}, GA_{44}, and GA_{53} were not able to substitute for the LD requirement for growth. This is suggestive of a photoperiodic control or block of GA biosynthesis at the level of the GA20ox, i.e., on the conversion of GA_{19} to GA_{20}. The fact that there was no increase in amount of GA_1 in internodes or leaves after application of GA_{19} under SD, confirms the existence of such a block in the GA biosynthetic pathway (unpublished results). This hypothesis has been tested by application of labelled GA_{19} with subsequent monitoring of their metabolism under SD and LD. No effect of daylength on the metabolism of $[^3H_2]GA_{20}$ and $[^3H_2]GA_1$ was observed (Rood and Junttila, 1988; Olsen et al., 1995b). Also, using 16,17-$[^3H_2]$-dihydro-GA_{19} to mimic GA_{19} in *S. pentandra*, no such daylength control was detected (Olsen et al., 1995b). However, this could be explained by the fact that dihydro-GAs have very low biological activity as compared with GAs (Olsen and Junttila, 1997). Later $[^{14}C]GA_{19}$ was used in a similar study of *Populus tremula x tremuloides*, but neither did these experiments reveal a photoperiodic effect of its conversion to [^{14}C]GA_{20}. However, using GA_{19} labelled with a stable isotope in application experiments, followed by GC-MS analysis, a reduced conversion of GA_{19} to GA_{20} under SD as compared with LD was observed (unpublished data). Thus, this experiment supports the conclusion that conversion of GA_{19} to GA_{20} is a target of photoperiodic control. No accumulation of endogenous GA_{19} was observed under SD, which could have been expected, but there was a delay in decline of its levels compared with GA_{20} and GA_1 (Olsen et al., 1995a). Evidence for a photoperiodic control point at the metabolism of GA_{19} to GA_{20} is also provided by studies on a native GA20ox of hybrid aspen. It was shown that cessation of apical growth was correlated with reduction in expression of the gene in expanding leaves and roots compared with a higher expression in elongating plants under LD conditions (Eriksson, 2000).

It should also be noted that photoperiod might exert its control at earlier steps in the GA biosynthesis pathway, like on the levels of *ent*-kaurene or GA_{53}. This has been shown for some herbaceous rosette plants in which the bolting process is under photoperiodic control (Gilmour et al., 1986; Zeevaart and Gage, 1993). So far there is no experimental evidence for SD-control at these steps in trees, but this should certainly be investigated.

To obtain a further understanding of photoperiod and other environmental cues affecting the biosynthesis and metabolism of GAs, studies of genes encoding for the enzymes of the pathways are strongly needed.

Mode of Action of Gibberellin in Photoperiodic Control of Shoot Elongation

Since stem elongation is a consequence of the two basal growth processes cell division and cell extension, it is reasonable to believe that GA, as a mediator of photoperiodic growth control, plays a role in control of either or both of these processes. In herbaceous species, there are several reports on GAs affecting cell elongation in stems (Jones, 1983), probably by influencing microtubule orientation (Katsumi and Ishida, 1991). There are also several reports on GA involvement in cell division in herbaceous species (Sachs, 1965; Sauter et al., 1995; Sauter, 1997). In *Pseudotsuga*, application of $GA_{4/7}$ prior to bud swelling under LD conditions enhanced stem elongation, whereas after flushing no effect of exogenous GA was found (Owens et al., 1985). According to these authors, frequent cell divisions take place during initiation of shoot growth, whereas shoot elongation after flushing is characterized mainly by cell expansion. Thus, it appears that the enhanced elongation as a consequence of early GA application was due to a stimulation of cell division. However, the authors also argue for an effect of GA on cell elongation at this early stage (Owens et al., 1985). In *S. pentandra*, no effect of GA_1 on cell extension under short photoperiod was found. In a northern ecotype, appearance of new internodes ceased within a period of 10-15 days under short photoperiod but no effect of the concomitantly decreasing GA_1 levels was found on elongation of cells in internodes already initiated at the start of the SD treatment (Olsen et al., 1995c). This does not exclude the possibility that GA_1 might play a role in regulating cell extension at early stages of cell development, and that there might be a threshold length beyond which cell elongation is no longer affected by GA_1. However, there is evidence of a role for GA in photoperiodic control of cell division in *S. pentandra*. Cessation and resumption of cell elongation (bud set and bud break) correlated with cessation and resumption of cell division in the subapical region of shoot tips, along with reduced and increased GA_1 levels, respectively (Hansen et al., 1999, Olsen et al., 1995b; 1995c; 1997a). Also, exogenous GA_1 applied after SD-induced growth cessation had occurred greatly stimulated subapical cell divisions (Hansen et al., 1999). Furthermore, in dwarfed *Populus tremula x tremuloides* overexpressing the oat *PHYA*, there was a correlation between reduction in cell numbers and lowered levels of GA_1, whereas no difference in cell length relative to the wild type was found (Olsen et al., 1997b).

Rohde et al. (1997) has reported on high promoter activity of the cell-cycle regulating gene cyclin 1 in elongating transgenic *Populus tremula x alba* plants with an introduced promoter-GUS fusion (the promoter region of the cyclin 1 from *Arabidopsis* and β-glucuronidase). In dormant buds, expression levels of the construct were low or absent. No data for the effect of GAs on the expression of this gene construct was presented, but in light of the correlation between active shoot extension and high GA_1 levels under LD in *Populus tremula x tremuloides* (Olsen et al., 1997a) and the related species *Salix pentandra* (Olsen et al., 1995a; 1995c; 1997b), the study by Rohde et al. (1997) supports to the view that GAs are involved in photoperiodic control of cell division. Direct molecular evidence for this has only been reported for herbaceous species, e.g., for rice, where GA stimulates cell division by promoting cell histone H1 kinase activity and the expression of cdc2 and cyclin genes during the induction of rapid growth (Sauter et al., 1995; Sauter 1997). Also, in light grown pea, Yang et al. (1996) have proposed that

GA acts in control of cell division. It could be argued that the observed effects of exogenous and endogenous GA_1 on cell division are indirect as GA might control cell extension in early stages of cell development, and attainment of a minimum critical cell size appears to be an essential determinant for cell division (John et al., 1993). Thus, the processes of cell enlargement and cell division seem closely inter-related. A complete understanding of the action of GAs on these basic growth responses is still lacking in woody species and more studies are needed to elucidate GA control at different stages of cell development relative to photoperiodism. Molecular studies of the effect of GA on the cell division cycle in trees would be of great interest.

Generally, striking morphological and developmental changes are involved in the photoperiodically regulated transformation of the active vegetative meristem to a terminal, resting bud. It has been reported that symplastic fields in the tunica of the apical meristem of *Betula pubescens* are important in coordinating activity of the meristem (Rinne and van der Schoot, 1998). These authors also observed a breakdown of these fields under SD, implicating cessation of the communication between apical cells and morphogenetic inactivation. The cell isolation was due to the activation of glucan synthase complexes resulting in callose plugs formed on all plasmodesmata. It would be very interesting to investigate the role of GA in regulation of communication between cells in the apical meristem. Organogenesis in the apex continues within the bud even after cessation of shoot elongation. With respect to the cessation of apical elongation growth, the possible role of cell-to-cell communication as a regulatory mechanism involved in the control of bud scale differentiation and communication between the apical and the subapical meristem should be investigated.

Another challenge will be to identify the developmental switch, or switches, that determine the initiation of bud formation, controlling the formation of bud scales. Weigel and Nilsson (1995) report on a gene (*LEAFY*) that is sufficient for switching vegetative lateral meristem to inflorescent meristems in *Arabidopsis* and *Populus tremula x tremuloides*. In *Arabidopsis*, GAs were shown to promote flowering by activating the *LEAFY* promoter (Blázquez et al., 1998). Similar studies focusing on the change from the active vegetative meristem to a resting terminal bud would be of great interest. It can be hypothesized that one or more similar switches as *LEAFY* could play a role in this process, and that GA might play a similar role in activating such switches. As mentioned above, molecular studies on the nature of GA action on the basic growth processes will be a prerequisite for obtaining a detailed understanding of the role of GAs in photoperiodic growth control.

CONCLUDING COMMENTS AND FUTURE ASPECTS

The geographic areas where northern tree species have evolved are not known. If the main direction of spreading of these species has been from the south to the north, the southern, SD type of response must have been the earliest response, followed by an increase in the critical daylength as a result of an adaptation during the distribution to the more northern locations. Thus, such tree species may provide an interesting system for further studies of the physiological and genetic mechanisms for adaptation. The traditional understanding of photoperiodism involves perception of the daylength signal in the leaves, and translocation of the stimuli to the apex where developmental reprogramming will take place (Figure 4). In spite of comprehensive studies with experimental systems for flower induction, the character of the putative daylength signal from leaves to the apex is not known. The site of daylength perception in trees is not completely settled, but some earlier studies indicate that, in some cases, even the apical parts of a tree may perceive the signal (Wareing, 1954). There are, however, no recent studies on this question, and modern methods, including laser and LED technology, could provide new methods to study the localisation of daylength perception in trees. It is known, for example, that perception of end-of-day far red irradiation, which stimulates internode elongation, is perceived by the internodes (e.g., Rousseaux et al., 1997). Also, studies with transgenic plants, modified with phytochrome genes regulated by tissue specific promoters could be another interesting approach for such studies.

The putative photoperiodic signal results in a reprogramming of the meristematic activities in the apex, including cessation of cell division in the subapical meristem and, in many cases, temporal differentiation of bud scales or abscission of the apex. Neither the signal nor the type of correlative interactions between bud scale differentiation, or apical abscission and subapical cell division, are known in any detail. Coordination of activities in the apical meristem may involve the regulation of symplastic connections through the plasmodesmata, but so far we do not have any information about such regulation between the cells in the apical meristem proper and those in the subapical meristem. Regulation

FIGURE 4. Schematic model for physiological aspects of the photoperiodic regulation of shoot elongation in deciduous tree species with a free growth pattern. Photoperiod, mediated by the phytochrome systems, is suggested to control gibberellin biosynthesis. Current evidence indicates that this control could be through regulation of the multifunctional C20 oxidase and short photoperiod blocks the GA biosynthesis. The early 13-hydroxylation pathway, most common in woody plants, is illustrated in the figure. In addition, GA-independent pathways can be expected to be involved, but they are not well-studied in woody species. (1) Traditional models of photoperiodic responses are based on perception of daylength signal in leaves. Possibility for perception of this signal in apical regions of woody plants needs further studies. (2) Leaf perception requires translocation of the stimuli to the apical parts. Nothing is known about the putative SD-induced translocation signal in woody plants. GA levels in the elongating apical stem tissue are reduced by SD, and if these GA levels are based on metabolism in situ, a putative signal could be an inhibitor of GA biosynthesis. Sucrose is known to be involved in regulation of flowering, but the role of sucrose in regulation of bud set is not known. (3) The subapical meristem is one probable site of action for GA_1. Cell division in the subapical meristem appears to be blocked by the absence of GA_1, resulting in a cessation of elongation growth. (4) Cessation of the elongation growth is accompanied by initiation of the apical bud, marked by differentiation of bud scales. Interdependence between these two developmental processes (bud scale initiation, cessation of cell division in the subapical meristem) is not known. Particularly, in mature trees of many northern deciduous species, instead of forming an apical bud, the whole meristematic region of the stem apex is aborted. Physiological and molecular aspects of this process are not known.

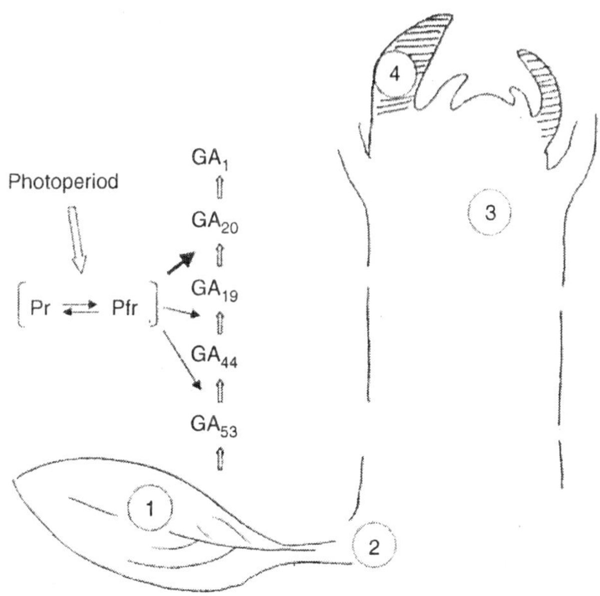

of cell-to-cell communication may be more connected to the development of dormancy than to the initial processes of growth cessation.

Elucidation of the genetic and physiological basis of adaptation to photoperiod remains a challenging task. This adaptation is expressed as a change of the critical daylength, change in the rate of response to daylength, and change in the response to spectral composition of light. The critical photoperiod is a quantitative trait, but the genes encoding this character have not been identified. Phytochromes are involved both as a sensory system and as a regulatory system. Present observations of contrasting ecotypes and research with transgenic lines could indicate both quantitative and qualitative differences between the ecotypes. Extensive genetic analyses are needed to investigate these possibilities. Studies of photoperiodism in trees have focused exclusively on the role of phytochrome. As cryptochromes, mediating some of the blue light responses, also appears to be involved in photoperiodism, there is an obvious need for the analysis of blue light effects, occurrence and regulation of blue light receptor genes in tree species.

As presented in this review, we have a substantial amount of data indicating interactions between the phytochrome systems(s) and plant hormones, in particular gibbberellins. So far, we do not have any conclusive data on differences in hormonal metabolisms between contrasting ecotypes; neither do we have any information on the molecular nature of the possible interaction between phytochrome and plant hormones. Signal transduction pathways for phytochrome and various plant hormones are currently being studies very intensively in *Arabidopsis*, and those studies will, without doubt, lead to interesting studies, as well as the genetic manipulation, of daylength responses in trees. Functional studies, both at genome and protein level, are expected to bring rapid progress in our understanding of various regulatory mechanisms in plants, including tree species. Such functional studies may also be of great importance for further analysis of differences between various ecotypes. In spite of these possibilities, extensive use of sophisticated transgenic plants, for example with inducible and tissue specific promoters, are also needed. Several common tree species are now amenable for transformation either by *Agrobacterium* or biolistic methods (Walter et al., 1998; Nilsson et al., 1996). With respect to the great variation in patterns of shoot growth found in tree species, studies of the involved mechanisms should include an increasing range of species. For example, conifers like *Juniperus*, with a large geographic area of distribution and without formation of an apical bud, could provide highly interesting systems for studies of photoperiodic regulation of growth cessation and frost hardening. In general, lack of suitable developmental mutants of tree species greatly hampers research progress in such species. However, development of new technologies, and accelerated cloning of plant genes, will, hopefully, ensure significant progress in studies of photoperiodism in trees during the years to come.

REFERENCES

Ahmad, M. (1999). Seeing the world in red and blue: insight into plant vision and photoreceptors. *Current Opinion in Plant Biology* 2: 230-235.

Aitken, S.N. and M. Hannerz. (2001). Genecology and gene resource management strategies for conifer cold hardiness. Pp. 23-54. In: F.J. Bigras and S.J. Colombo (eds.), *Conifer Cold Hardiness*. Kluwer Academic Publ., Dordrecht.

Alba, R., P.M. Kelmenson, M.-M., Cordonnier, and L.H. Pratt. (2000). The phytochrome gene family in tomato and the rapid differential evolution of this family in angiosperms. *Mol. Biol. Evol.* 17: 362-373.

Bagnall, D.J. and R.W. King. (2001). Phytochrome, photosynthesis and flowering of *Arabidopsis thaliana*: Photophysiological studies using mutants and transgenic lines. *Aust. J. Plant Physiol.* 28: 401-408.

Bate, N.J., S.B. Rood, and T.J. Blake. (1988). Gibberellins and heterosis in poplar. *Can J. Bot.* 66: 1148-1152.

Bannister, P. and G. Neuner. (2001). Frost resistance and distribution of conifers. Pp. 3-22. In: F.J. Bigras and S.J. Colombo (eds.), *Conifer Cold Hardiness*. Kluwer Academic Publ., Dordrecht.

Blazquez, M.A., R. Green, O. Nilsson, M.R. Sussman, and D. Weigel. (1998). Gibberellins promote flowering of Arabidopsis by activating the *LEAFY* promoter. *Plant Cell* 10: 791-800.

Braslavsky, S.E., W. Gärtner and K. Schaffner. (1997). Phytochrome photoconversion. *Plant, Cell and Environment* 20: 700-706.

Burgin, M.J., J.J. Casal, G.C. Whitelam, and R.A. Sanchez. (1999). A light-regulated pool of phytochrome and rudimentary high-irradiance under far-red light in *Pinus elliottii* and *Pseudotsuga menziesii*. *J. Exp. Bot.* 50: 831-836.

Casal, J.J., R.A. Sanchez, and J.F. Botto. (1998). Modes of action of phytochromes. *J. Exp. Bot.* 49: 127-138.

Clack, T.S., Mathews, and R.A. Sharrock. (1994). The photochrome apoprotein family in *Arabidopsis* is encoded by five genes: the sequences and expression of *PHYD* and *PHYE*. *Plant Mol. Biol.* 25: 413-427.

Clapham, D., I. Dormling, I. Ekberg, G. Eriksson, M. Qamaruddin, and D. Vince-Prue. (1998). Latitudinal cline of requirement for far-red light for the photoperiodic control of bud set and extension growth in *Picea abies* (Norway spruce). *Physiol. Plant.* 102: 71-78.

Clapham, D.H., H.Ü. Kolukisaoglu, C.-T. Larsson, M. Qamaruddin, C. Wiegmann-Eirund, H.A.W. Schneider-Poetsch, and S. von Arnold. (1999). Phytochrome types in *Picea* and *Pinus*. Expression patterns of *PHYA*-related types. *Plant Mol. Biol.* 40: 669-678.

Clapham, D., I. Ekberg, A.C.H. Little, and O. Savolainen. (2001). Molecular biology of conifer frost tolerance and potential application to tree breeding. Pp. 187-220. In: F.J. Bigras and S.J. Colombo (eds.) *Conifer Cold Hardiness*. Kluwer Academic Publ. Dordrecht.

Clough, R.C., and Vierstra, R.D. (1997). Phytochrome degradation. *Plant, Cell and Environment* 20: 713-721.

D'Aoust, A.L. and C. Hubac. (1986). Phytochrome action and frost hardening in black spruce seedlings. *Physiol. Plant.* 67: 141-144.

Davies, J.K., E. Jensen, O. Junttila, L. Rivier, and A. Crozier. (1985). Identification of endogenous gibberellins from *Salix pentandra. Plant Physiol.* 78: 473-476.

Devlin, P.F. and S.A. Kay. (2000). Chryptochromes are required for phytochrome signaling to the circadian clock but not for rhythmicity. *The Plant Cell* 12: 2499-2509.

Doumas, P., N. Imbault, T. Moritz, and P.C. Odén. (1992). Detection and identification of gibberellins in Douglas fir (*Pseudotsuga menziesii*) shoots. *Physiol. Plant.* 85: 489-494.

Dunberg, A. and P.C. Odén. 1983. Gibberellin and conifers. Pp 221-295. In: A. Crozier (ed.), *The Biochemistry and Physiology of Gibberellin*. Vol 2. Praeger, New York.

Eriksson, M.E. (2000). The role of phytochrome A and gibberellins in growth under long and short day conditions. Studies in hybrid aspen. *Acta Universitatis Agriculturae Sueciae, Silvestria* 164. ISBN 91-576-5898-6.

Eriksson, M.E. and T. Moritz. (1997). Isolation of a cDNA encoding a phytochrome A (accession no. AJ001318) from *Populus tremula x tremuloides. Plant Physiol.* 115: 1731.

Eriksson, M.E., M. Israelsson, O. Olsson, and T. Moritz. (2000). Increased gibberellin biosynthesis in transgenic trees promotes growth, biomass production and xylem fiber length. *Nat. Biotechnol.* 18: 784-788.

Erwin, J.E. and R.D. Heins. (1995). Thermomorphogenic responses in stem and leaf development. *HortSci.* 30: 940-949.

Gilmour, S.J., J.A.D. Zeevart, L. Schwenen, and J.E. Graebe. (1986). Gibberellin metabolism in cell-free extracts from spinach leaves in relation to photoperiod. *Plant Physiol.* 82: 190-195.

Graebe, J.E. (1987). Gibberellin biosynthesis and control. *Annu. Rev. Plant Physiol.* 38: 419-465.

Grindal, G., A. Ernstsen, J.B. Reid, O, Junttila, B. Lindgård, and R. Moe. (1998). Endogenous gibberellin A_1 levels control thermoperiodic stem elongation in *Pisum sativum. Physiol. Plant.* 102: 523-531.

Garner, W.W. and H.A. Allard. (1923). Further studies in photoperiodism, the response of the plant to the relative length of day and night. *J. Agric. Res.* 23: 871-920.

Hansen, E., J.E. Olsen, and O. Junttila. (1999). Gibberellins and subapical cell divisions in relation to bud set and bud break in *Salix pentandra. J. Plant Growth Regul.* 18: 167-170.

Hasan, O., B.G. Ridoutt, J.J. Ross, N.W. Davies, and J.B. Reid. (1994). Identification and quantification of endogenous gibberellins in apical buds and the cambial region of *Eucalyptus. Physiol. Plant.* 90: 475-480.

Hedden, P. (1999). Recent advances in gibberellin biosynthesis. *J. Exp. Bot.* 50: 553-563.

Hedden, P. and Y. Kamiya. (1997). Gibberellin biosynthesis: enzymes, genes and their regulation. *Annu. Rev. Plant Physiol. Plant Mol. Biol.* 48: 431-460.

Hedden, P. and A.L. Phillips. (2001). Gibberellin metabolism: new insights revealed by the genes. *Trends in Plant Science* 5: 523-530.

Heide, O.M. (1974). Growth and dormancy in Norway spruce ecotypes (*Picea abies*) I. Interaction of photoperiod and temperature. *Physiol. Plant.* 30: 1-12.

Howe, G.T., W.P. Hackett, G.R. Furnier, and R.E. Klevorn. (1995). Photoperiodic responses of a northern and southern ecotype of black cottonwood. *Physiol. Plant.* 93: 695-708.

Howe, G.T., G. Gardner, W.P. Hackett, and G.R. Furnier. (1996). Phytochrome control of short-day-induced bud set in black cottonwood. *Physiol. Plant.* 97: 95-103.

Howe, G.T., P.A. Bucciaglia, W.P. Hackett, G.R. Furnier, M.-M. Cordonnier-Pratt, and G. Gardner. (1998). Evidence that the phytochrome gene family in black cottonwood has one *PHYA* locus and two *PHYB* loci but lacks members of the *PHYC/F* and *PHYE* subfamilies. *Mol. Biol. Evol.* 15: 160-175.

Hurme, P., T. Repo, O. Savolainen, and T. Pääkkönen. (1997). Climatic adaptation of bud set and frost hardiness in Scots pine (*Pinus sylvestris*). *Can. J. For. Res.* 27: 716-723.

Håbjørg, A. (1972). Effects of photoperiod and temperature on growth and development of three latitudinal and three altitudinal populations of *Betula pubescens* Ehrh. *Medd. Nor. Landbrukshøgsk.* 51: 1-27.

Håbjörg, A. (1978). Photoperiodic ecotypes in Scandinavian trees and shrubs. *Medd. Nor. Landbrukshøgsk.* 57: 1-20.

Jackson, S., and B. Thomas. (1997). Photoreceptors and signals in the photoperiodic control of development. *Plant, Cell and Environment* 20: 790-795.

John, P.C.L., K. Zhang, and C. Dong. (1993). A $p34^{cdc2}$-based cell cycle: Its significance in monocotyledonous, dicotyledonous and unicellular plants. Pp 9-34. In: J.C. Ormrod and D. Francis (eds.), *Molecular and Cell Biology of the Plant Cell Cycle*. Kluwer Academic Publishers, Dordrecht.

Jones, R.L. (1983). The role of gibberellins in plant cell elongation. *CRC. Crit. Rev. Plant Sci.* 1: 23-47.

Junttila, O. (1976). Apical growth cessation and shoot tip abscission in *Salix*. *Physiol. Plant.* 38: 278-286.

Junttila, O. (1980). Effect of photoperiod and temperature on apical growth cessation in two ecotypes of *Salix* and *Betula*. *Physiol. Plant.* 48: 347-352.

Junttila, O. (1982). Gibberellin-like activity in shoots of *Salix pentandra* as related to the elongation growth. *Can. J. Bot.* 60: 1231-1234.

Junttila, O. (1986). Effects of temperature on shoot growth in northern provenances of *Pinus sylvestris*. *Tree Physiol.* 1: 185-192.

Junttila, O. (1991). Gibberellins and the regulation of shoot elongation in woody plants. Pp. 199-210. In: N. Takahashi, B.O. Phinney, and J. MacMillan (eds.), *Gibberellins*. Springer Verlag, New York. ISBN 0-387-97259-5.

Junttila, O. (1993a). Exogenously applied GA_4 is converted to GA_1 in seedlings of *Salix pentandra*. *J. Plant Growth Regul.* 12: 35-39.

Junttila, O. (1993b). Interaction of growth retardants, daylength, and gibberellins A_{19}, A_{20} and A_1 on shoot elongation in birch and alder. *Physiol. Plant.* 12: 123-127.

Junttila, O. and Å. Kaurin. (1985). Climatic control of apical growth cessation in latitudinal ecotypes of *Salix pentandra* L. Pp. 83-91. In: Å. Kaurin, O. Junttila, and J. Nilsen, (eds.), *Plant Production in the North*, Norwegian University Press, Oslo.

Junttila, O., H. Abe, and R.P. Pharis. (1988). Endogenous gibberellin in elongating shoots of clones of *Salix dasyclados* and *S. viminalis*. *Plant Physiol.* 87: 781-784.

Junttila, O. and E. Jensen (1988). Gibberellins and photoperiodic control of shoot elongation in *Salix*. *Physiol. Plant.* 74: 371-376.

Junttilla, O., E. Jensen, and A. Ernstsen. (1991). Effects of prohexadione (BX-112) and gibberellins on shoot growth in seedlings of *Salix pentandra*. *Physiol. Plant.* 83: 17-21.

Junttila, O., E. Jensen, D.W. Pearce, and R.P. Pharis. (1992). Stimulation of shoot elongation in *Salix pentandra* by gibberellin A_9; activity appears to be dependent upon hydroxylation to GA_1 via GA_{20}. *Physiol. Plant.* 84: 113-120.

Junttila, O. and R. Robberecht. (1998). Ecological aspects of cold-adapted plants with a special emphasis on environmental control of cold hardening and dehardening. Pp. 61-77. In: R. Margesin and F. Schinner (eds.), *Cold Adapted Organisms–Ecophysiology, Enzymology, Molecular Biology*. Springer-Verlag, Berlin.

Katsumi, M. and K. Ishida. (1991). The gibberellin control of cell elongation. Pp. 211-219 In: N. Takahashi, B.O. Phinney, and J. MacMillan (eds.), *Gibberellins*. Springer Verlag, New York. ISBN 0-387-97259-5.

Kendrick, R.E., and G.H.M. Kronenberg. (eds). (1994). *Photomorphogenesis in Plants*. 2. Kluwer Academic, Dordrecht. 828 p.

Kolukisaoglu, H.A., S. Marx, C. Wiegmann, S. Hanelt, and H.A.W. Schneider-Poetsch. (1995). Divergence of the phytochrome gene family predates angiosperm evolution and suggests that *Selaginella* and *Equisetum* arose prior to *Psilotum*. *J. Mol. Biol.* 41: 329-337.

Koshioka, M., J.S. Taylor, G.R. Edwards, et al. (1985). Identification of gibberellin A_{19} and A_{20} in vegetative apple tissue. *Agric. Biol. Chem.* 49: 1223-1226.

Kramer, P.J. and T.T. Kozlowski. (1979). *Physiology of Woody Plants*. Academic Press, New York.

Lange, T., S. Robatzek, and A. Frisse. (1997). Cloning and expression of a gibberellin 2-beta,3-beta-hydroxylase cDNA from pumpkin endosperm. *Plant Cell* 9: 1459-1467.

Lavender, D.P. and S.N. Silim. (1987). The role of plant growth regulators in forest trees. *Plant Growth Regul.* 6: 171-191.

Lorenzi, R., R. Horgan, and J.K. Heald. (1976). Gibberellin A_9 glucosyl ester in needles of *Picea sitchensis*. *Phytochemistry* 92: 789-790.

Mathews, S., M. Lavin, and R.A. Sharrock. (1995). Evolution of the phytochrome gene family and its utility for phylogenetic analysis of angiosperms. *Ann. Mol. Bot. Gard.* 82: 296-321.

Mathews, S. and R.A. Sharrock. (1996). The phytochrome gene family in grasses (Poaceae): a phylogeny and evidence that grasses have a subset of the loci found in dicot angiosperms. *Mol. Biol. Evol.* 13: 1141-1150.

Mathews, S. and R.A. Sharrock. (1997). Phytochrome gene diversity. *Plant, Cell and Environment* 20: 666-671.

McClung, C.R. (2000). Circadian rhythms in plants: a millennial view. *Physiol. Plant.* 109: 359-371.

Melchior, G.H. and R. Knapp. (1962). Gibberellin-wirkungen am Bäumen. *Silvae Genet.* 11: 29-39.

Moritz, T. (1992). GA-conjugates he use of combined capillary liquid chromatography/mass spectrometry for the identification of a gibberellin glucosyl conjugate. *Phytochemical Analysis* 3: 32-37.

Moritz; T. (1995). Biological activity, identification and quantification of gibberellins in seedlings of Norway spruce (*Picea abies*) grown under different photoperiods. *Physiol. Plant.* 95: 67-72.

Moritz, T., J.J. Phillipson, and P.C. Odén. (1989a). Metabolism of tritiated and deuterated gibberellins A_1, A_4 and A_9 in Sitka spruce (*Picea sitchensis*) shoots during the period of cone-bud differentiation. *Physiol. Plant.* 77: 39-45.

Moritz, T., J.J. Philipson, and P.C. Odén. (1989b). Detection and quantification of gibberellins in Sitka spruce (*Picea sitchensis*) of different ages and coning ability by bioassay, radioimmunoassay and gas chromatography-mass spectrometry. *Physiol. Plant* 75: 325-332.

Moritz, T. and J.E. Olsen. (1995). Comparison between high resolution, selected reaction monitoring and four-sector tandem mass spectrometry in quantitative analysis of gibberellins in milligram amounts of plant tissue. *Anal. Chem.* 67: 1711-1716.

Neff, M.M., Fankhauser, C., and Chory, J. (2000). Light: an indicator of time and place. *Genes and Development* 14: 257-271.

Nilsson, O., Little, C.H., Sandberg, G., and O. Olsson. (1996). Expression of two heterologous promoters, *Agrobacterium rhizogenes rolC* and cauliflower mosaic virus 35S, in the stem of transgenic hybrid aspen plants during the annual cycle of growth and dormancy. *Plant Mol. Biol.* 31: 887-895.

Nitsch, J.P. (1957). Photoperiodism in woody plants. *Proc. Amer. Soc. Hort. Sci.* 70, 526-544.

Odén, P.C., B. Andersson, and R. Gref. (1982). Identification of gibberellin A_9 in extracts of Norway spruce [*Picea abies* (L.) Karst.] by combined gas chromatography-mass spectrometry. *J. Chromatography* 247: 133-140.

Odén, P.C., L. Schwenen, and J.E. Graebe. (1987). Identification of gibberellins in Norway spruce [*Picea abies* (L.) Karst.] by combined gas chromatography-mass spectrometry. *Plant Physiol.* 84: 516-519.

Odén, P.C., Q. Wang, K. Högberg, and M. Werner. (1995). Transport and metabolism of gibberellins in relation to flower bud differentiation in Norway spruce (*Picea abies*). *Tree Physiol.* 15: 451-456.

Olsen, J.E., T. Moritz, E. Jensen, and O. Junttila. (1994). Comparison of endogenous gibberellins in roots and shoots of elongating *Salix pentandra* seedlings. *Physiol. Plant* 90: 378-381.

Olsen, J.E., E. Jensen, O. Junttila, and T. Moritz. (1995a). Photoperiodic control of endogenous gibberellins in *Salix pentandra. Physiol. Plant.* 93: 639-644.

Olsen, J.E., T. Moritz, E. Jensen, and O. Junttila. (1995b). Lack of effect of photoperiod on metabolism of exogenous GA_{19} and GA_1 in *Salix pentandra* seedlings. *Physiol. Plant.* 94: 522-528.

Olsen, J.E., O. Junttila, and T. Moritz. (1995c). A localised decrease of GA_1 in the elongation zone of *Salix pentandra* seedlings precedes cessation of shoot elongation induced by short photoperiod. *Physiol. Plant.* 95: 627-632.

Olsen, J.E. and O. Junttila. (1997). Growth-promoting activity of gibberellins on shoot elongation in *Salix pentandra* is reduced by 16, 17-dihydro-derivatisation. *Physiol. Plant.* 99: 63-66.

Olsen, J.E., O. Junttila, and T. Moritz. (1997a). Long day-induced bud break in *Salix pentandra* is associated with transiently elevated levels of GA_1 and gradual increase in indole-3-acetic acid. *Plant and Cell Physiol.* 38: 536-540.

Olsen, J.E., O. Junttila, J. Nilsen, M. Eriksson, I. Martinussen, O. Olsson, G. Sandberg, and T. Moritz. (1997b). Ectopic expression of phytochrome A in hybrid aspen changes critical daylength for growth and prevents cold acclimation. *Plant Journal* 12: 1339-1350.

Owens, J.N., J.E. Webber, and S.D. Ross. (1985). Interaction between gibberellin $A_{4/7}$ and root-pruning on the reproduction and vegetative processes in Douglas-fir. III. Effects on anatomy of shoot elongation and terminal bud development. *Can. J. For. Res.* 15: 354-364.

Pharis, R.P. (1976). Probable roles of plant hormones in regulating shoot elongation, diameter growth and crown form of coniferous trees. Pp. 291-306. In: M.G.R. Cannell, F.T. Last (eds.), *Tree Physiology and Yield Improvement*. Academic press, London.

Pharis, R.P. and C.G. Kuo. (1977). Physiology of gibberellins in conifers. *Can. J. For. Res.* 7: 299-325.

Pharis, R.P., F.C. Yeh, and B.P. Danick. (1991). Superiod growth potential in trees. What is its basis, and can it be tested for at an early age? *Can. J. For. Res.* 21: 368-374.

Pharis, R.P., R. Zhang, I. Jiang, B.P. Dancik, and F.C. Yeh. (1992). Differential efficacy of gibberellins in flowering and vegetative shoot growth, including heterosis and inherently rapid growth. Pp. 13-27. In: C.M. Karssen, L.C. Van Loon, and D. Vreugdenhil (eds.), *Progress in Plant Growth Regulation*, Kluwer Academic Publishers, Dordrecht.

Poling, S.M. and V.P. Maier. (1988). Identification of endogenous gibberellins in naval orange shoots. *Plant Physiol.* 88: 639-642.

Quail, P., Boylan, M., Parks, B.M., Short, T.W., Xu, Y., and Wagner, D. (1995). Phytochromes: Photosensory perception and signal transduction. *Science* 268: 675-680.

Quail, P.H. (1997). An emerging molecular map of the phytochromes. *Plant, Cell and Environment* 20: 657-665.

Ridoutt, B.G., R.P. Pharis, and R. Sands (1995). Identification and quantification of cambial region hormones of *Eucalyptus globulus*. *Plant Cell Physiol.* 36: 1143-1147.

Rinne, P. and C. van der Shoot. (1998). Symplastic fields in the tunica of the shoot apical meristem coordinate morphogenetic events. *Development* 125: 1477-1485.

Rohde, A., M. Van Montagu, D. Inzé, and W. Boerjan. (1997). Factors regulating the expression of cell cycle genes in individual buds of *Populus*. *Planta* 201: 43-52.

Rohde, A., G.T. Howe, J.E. Olsen, T. Moritz, M. van Montagu, O. Junttila, and W. Boerjan. (2000). Molecular aspects of bud dormancy in trees. Pp. 89-134. In: S.M. Jain and S.C. Minocha (eds.), *Molecular Biology of Woody Plants*. Vol. 1. Kluwer Academic publishers.

Rood, S.B. and O. Junttila. (1988). Lack of effect of photoperiod on the metabolism of gibberellin A_{20} in *Salix pentandra*. *Physiol. Plant.* 75: 506-510.

Rood, S.B., N.J. Bate, L.N. Mander, and R.P. Pharis. (1988). Identification of gibberellins A_1 and A_{19} from *Populus balsamifera* x *P. deltoides*. *Phytochemistry* 27: 11-14.

Ross, J.J., D.P. O'Neill, J.J. Smith, L.H.J. Kerckhoffs, and R.C. Elliott. (2000). Evidence that auxin promotes gibberellin A_1 biosynthesis in pea. *Plant J.* 21: 547-552.

Ross, S.D. (1983). Enhancement of shoot elongation in Douglas fir by gibberellin $A_{4/7}$ and its relation to the hormonal production of flowering. *Can. J. For. Res.* 13: 986-994.

Rousseaux, M.C., C.L. Ballare, E.T. Jordan, and R.D. Viestra. (1997). Directed overexpression of *PHYA* locally suppresses stem elongation and leaf senescence responses to far-red radiation. *Plant, Cell and Environment* 20: 1551-1558.

Sachs, R.M. (1965). Stem elongation. *Annu. Rev. Plant Physiol.* 16: 73-96.

Sauter, M. (1997). Differential expression of a CAK (*cdc2*-activating kinase)-like protein kinase, cyclins and *cdc2* genes from rice during cell cycle and in response to gibberellin. *Plant J.* 11: 181-190.

Sauter, M., S.L. Mekhedov, and H. Kende. (1995). Gibberellin promotes histone H1 kinase activity and the expression of *cdc2* and cyclin genes during the induction of rapid growth in deepwater rice internodes. *Plant Journal* 7: 623-632.

Schneider-Poetsch, H.A.W., Ü. Kolukisaoglu, D.H. Clapham, J. Hughes, and T. Lamparter. (1998). Non-angiosperm phytochromes and the evolution of vascular plants. *Physiol. Plant.* 102: 612-622.

Sembdner, G., W. Schliemann, and G. Schneider. (1991). Biochemical and physiological aspects of gibberellin conjugation. In: Gibberellins. N. Takahashi, B.O. Phinney, J. MacMillan (eds.) pp. 249-263. Springer-Verlag. New York. ISBN 0-387-97259-5.

Sharrock, R.A., and Quail, P.H. (1989). Novel phytochrome sequences in *Arabidopsis thaliana*: structure, evolution, and differential expression of a plant regulatory photoreceptor family. *Genes and Development* 3: 1745-1757.

Smith, H. (1995). Physiological and ecological function within the phytochrome family. *Annual Rev. Plant Physiol. Plant Mol. Biol.* 46:, 289-315.

Steffens, G.L. and P. Hedden. (1992). Effect of temperature regimes on gibberellin levels in thermosensitive dwarf apple trees. *Physiol. Plant.* 86: 539-543.

Sylven, N. (1940). Lang-och kortdagstyper av de svenska skogsträden. Long day and short day types of Swedish forest trees. *Svensk Papperstidn.* 43: 317-324, 332-342, 350-354.

Thomas, B. and D. Vince-Prue. (1997). *Photoperiodism in Plants*. Academic Press. San Diego. ISBN 0-12-688490-0. 428 p.

Vaartaja, O. (1954). Photoperiodic ecotypes of trees. *Can. J. Bot.* 32: 392-399.

Vaartaja, O. (1959). Evidence of photoperiodic ecotypes in trees. *Ecol. Monographs* 29: 91-111.

Vegis, A. (1964). Dormancy in higher plants. *Annu. Rev. Plant Physiol.* 15: 185-224.

Walter, C., Grace, L.J., Wagner, A., White, D., and Walden, A.R. (1998). Stable transformation and regeneration of transgenic plants of *Pinus radiata* D. Don. *Plant Cell Reports* 17: 460-468.

Wang, Q., C.H.A. Little, C. Sheng, P.C. Odén, and R.P. Pharis. (1992). Effect of exogenous gibberellin $A_{4/7}$ on tracheid production, longitudinal growth and the levels of indole-3-acetic acid and gibberellins A_4, A_7 and A_9 in the terminal shoot of *Pinus sylvestris* seedlings. *Physiol. Plant.* 86: 202-208.

Wang, Q., C.H.A. Little, T. Moritz, and P.C. Odén. (1995). Effect of prohexadione on cambial and longitudinal growth and the levels of endogenous gibberellins A_1, A_3, A_4 and A_9 and indole-3-acetic acid in *Pinus sylvestris* shoots. *J. Plant Growth Regul.* 14: 175-181.

Wang, Q., C.H.A. Little. T. Moritz, and P.C. Odén. (1996). Identification of endogenous gibberellins, and metabolism of tritiated and deuterated GA_4, GA_9 and GA_{20}, in Scots pine (*Pinus sylvestris*) shoots. *Physiol. Plant.* 97: 764-771.

Wang, Q., C.H.A. Little, and P.C. Odén. (1997). Control of longitudinal and cambial growth by gibberellins and indole-3-acetic acid in current-year shoots *of Pinus sylvestris. Tree Physiol.* 17: 715-721.

Wareing, P.F. (1954). Growth studies in woody species. VI. The locus of photoperiodic perception in relation to dormancy. *Physiol. Plant.* 7: 261-277.

Wareing, P.F. and P.F. Saunders. (1971). Hormones and dormancy. *Annu. Rev. Plant Physiol.* 22: 261-288.

Weigel, D. and O. Nilsson. (1995). A developmental switch sufficient for flower initiation in diverse plants. *Nature* 377: 495-500.

Westing, A.H. (1959). Effect of gibberellin and conifers: generally negative. *J. For.* 57: 120-122.

Yang, T., P.J. Davies, and J.B. Reid. (1996). Genetic dissection of the relative roles of auxin and gibberellin in the regulation of stem elongation in intact light-grown peas. *Plant Physiol.* 110: 1029-1034.

Yamaguchi, S. and Y. Kamiya. (2000). Gibberellin biosynthesis: Its regulation by endogenous and environmental signals. *Plant Cell Physiol.* 41: 251-257.

Zanewich, K.P. and S.B. Rood (1994). Endogenous gibberellins in flushing buds of three deciduous trees: alder, aspen and birch. *J. Plant Growth Regul.* 13: 159-162.

Zeevaart, J.A.D. and D.A. Gage. (1993). *Ent*-Kaurene biosynthesis is enhanced by long-day plants *Spinacia oleracea* L. and *Agrostemma githago* L. *Plant Physiol.* 101: 25-29.

Cell-Cell Communication as a Key Factor in Dormancy Cycling

Päivi L. H. Rinne
Christiaan van der Schoot

SUMMARY. Dormancy in perennial plants is the phenomenon that growth and development are temporarily suspended. The fact that development is orchestrated by the growing point or, more precisely, by the shoot apical meristem, suggests that dormancy is also a property of the growing point. Understanding dormancy in terms of suspended development requires the identification of events that are essential in development, but lacking in dormancy. Essential in the developmentally active meristem is that the behaviour of individual cells is integrated by the continuous exchange of correlative signals. Certain signals cross the plasma membrane and the cell walls, whereas other signals traffic via plasmodesmata, which are abundant in all cell walls. Signalling via both pathways is essential for meristem functioning. Recent findings show that during the onset of photoperiodically induced dormancy in birch the meristem cells become uncoupled from their cooperatively generated

---

Päivi L. H. Rinne is Senior Scientist, and Christiaan van der Schoot is Professor, Plant Developmental Physiology, Department of Biology and Nature Conservation, Agricultural University of Norway, N-1432 Ås, Norway.

Address correspondence to: Päivi L. H. Rinne, Plant Developmental Physiology, Department of Biology and Nature Conservation, Agricultural University of Norway, N-1432 Ås, Norway (E-mail: Paivi.Rinne@ibn.nlh.no).

This research is supported by the Finnish Academy, Helsinki, Finland (Grant 51350).

[Haworth co-indexing entry note]: "Cell-Cell Communication as a Key Factor in Dormancy Cycling." Rinne, Päivi L. H., and Christiann van der Schoot. Co-published simultaneously in *Journal of Crop Improvement* (Food Products Press, an imprint of The Haworth Press, Inc.) Vol. 10, No. 1/2 (#19/20), 2004, pp. 113-156; and: *Adaptations and Responses of Woody Plants to Environmental Stresses* (ed: Rajeev Arora) Food Products Press, an imprint of The Haworth Press, Inc., 2004, pp. 113-156. Single or multiple copies of this article are available for a fee from The Haworth Document Delivery Service [1-800-HAWORTH, 9:00 a.m. - 5:00 p.m. (EST). E-mail address: docdelivery@haworthpress.com].

http://www.haworthpress.com/web/JCRIP

Digital Object Identifer: 10.1300/J411v10n01_07

signal network by the closing of plasmodesmata and the sealing of cell walls. A hermetic closing of plasmodesmal entrances, via the production of 1,3-β-D-glucan-containing sphincter-complexes, eventually locks the meristem into a state of endo-dormancy. During chilling-induced release from endo-dormancy the sphincter-complexes are digested, most likely by cytoplasmic and spherosome-delivered 1,3-β-D-glucanase. This restores the symplasmic organization of the meristem and prepares it for future proliferation. We therefore suggest that the shuttling between proliferative and dormant states, known as 'dormancy cycling,' can be explained in terms of transitions in the intrinsic communication network of the meristem. Comparative evidence, using different plant systems, suggests that reversible blockage of cell-cell communication might be central to dormancy cycling in general. Based on this new paradigm we propose a network-model of dormancy cycling that depicts the meristem as transitioning through three subsequent states of cellular communication, with environmental factors triggering the transition from one state to the next. The characterization of endo-dormancy by sphincter-complexes may facilitate future efforts to diagnose and manipulate endo-dormancy by molecular technologies, which would greatly benefit horticulture and agriculture.

KEYWORDS. *Betula*, callose, cell-cell communication, dormancy, 1,3-β-D-glucan, 1,3-β-D-glucanase, plasmodesmata, plasmodesmal sphincter, *Populus*, shoot apical meristem, *Solanum tuberosum*

INTRODUCTION

Dormancy and Phenology

Dormancy is defined phenologically as the state in which visible growth and development are suspended (Lang, 1987). Since it is the growing point, or more specifically the shoot apical meristem that orchestrates development, dormancy can be regarded as the developmentally inactive state of the growing point (e.g., Romberger, 1963). Growing points can become inactive for different reasons, however, and three main categories of dormancy are distinguished, known as para-, eco- and endo-dormancy (Lang, 1987). In para-dormancy the immediate cause of the suppression originates within the plant, but outside (para =

other than) the dormant structure (Lang, 1987). For example, axillary shoot meristems are para-dormant when they are suppressed by the proliferating shoot apical meristem. This situation, also widely known as apical dominance, applies to many short-lived plants and the juvenile stages of some trees (Cline, 2000). In mature perennial plants and some annuals, the adjacent tissues can somehow suppress the axillary meristems (Pétel and Gendraud, 1996; Napoli, Beveridge, and Snowden, 1999; Sussex and Kerk, 2001). In eco-dormancy the suspension of development is simply due to environmental conditions that do not favor growth (Lang, 1987), whereas in endo-dormancy the cause originates within the meristem or is selectively perceived by the meristem (Lang, 1987). However, also in the case of endo-dormancy the initial cause comes from the environment in the form of alterations in photo- and thermoperiod.

It is important to note that the above definitions of dormancy classify dormancy phenomena on the basis of the origin of the inducing signals. Although there is no doubt that there are different dormant states, the above definitions do not address the actual state that a meristem assumes (Rappaport and Wolf, 1969; Saure, 1985; Dennis, 1994; Seeley, 1994). Commonly, the meristem state is implicitly deduced from the context in which the meristem becomes arrested. This lack of attention for meristem states is at least partly due to the fact that these definitions conceptualize dormancy as a state that is exclusively defined by the signals that lead to it. This focus on input signals, with the implicit understanding that the nature of the signals is more important than the nature of the responding structure, has prevented the pinpointing of state-specific features. The focus on input signals has not changed by the introduction of molecular approaches, which have been used exclusively within the perspective of classic hormone research (discussed by, e.g., Lang, 1994). This led to further refinement and detailing, and facilitated the 'uncovering of more steps in the transduction pathways from the environmental signal to horticultural response' (Lang, 1994), but not to a characterization of meristem states.

The novel molecular tools will not help much if we cannot more precisely pinpoint when the various dormant states are present. As pointed out by Saure (1985), much of the confusion that exists about dormancy is due to the comparison of data from plants that were identically treated while the meristems might have been in different states. To minimize such confusion the state of the meristem should be verified, at least phenologically, by bud burst experiments (e.g., Lavander, 1991). Such experiments show that in para- and eco-dormancy the meristems are poised for proliferation and removal of the shoot apical meristem or the return of favorable growth conditions, respectively, suffice to release the developmental potential of these meristems. In endo-dormancy, in contrast, the meristem is unable to proliferate even under conditions that normally favor development. Endo-dormancy, therefore, is the only meristem state that is intrinsically dormant (see Emilsson, 1949; Samish, 1954; Dennis, 1994; Seeley, 1994; Rinne, Kaikuranta, and van der Schoot, 2001).

The universal occurrence of endo-dormancy in genetically diverse plants-deciduous trees and other perennials as well as tubers and seeds-suggests that the intrinsic mechanisms that lock the meristem in this state are evolutionarily conserved. This does not imply that the signaling paths that connect the environment to the meristem-intrinsic mechanisms are also conserved, but rather that variable signal cascades target a common meristem-intrinsic mechanism. For example, in trees like *Betula pubescence* and *Acer pseudoplantanus* a photoperiodically generated signal cascade is required to suspend development, while temperature- and draught-triggered signaling may be used in addition in other tree species (Kawase, 1961; Vegis, 1964; Hallé, Oldeman, and Tomlinson, 1978; Dennis, 1994; Rinne, Welling, and Kaikuranta, 1998). Developmentally regulated pathways might co-operate with the above signalling pathways. This might be the case in potato where para-dormant buds enter endo-dormancy during tuber development (van Es and Hartmans, 1987). In conclusion, the signalling cascades connecting the perception of external cues to the intrinsic developmental lock are tailored to the needs of each plant species or eco-type, adapted to the environment where it originated. In characteriszing the dormant state the crucial question, therefore, is not so much what environmentally induced signal cascades induce endo-dormancy but, rather, what inbuilt response mechanisms lock the meristem into a developmentally inactive state.

Dormancy and Developmental Biology

From the perspective of developmental biology the phenological definition of dormancy as 'the temporary suspension of visible growth' (reviewed by Lang, 1987) can be reformulated as 'the absence of the principles that underlie development.' Since suspension of development takes place regardless of the genetic make-up of the species, the developmental principles that are suspended must also be very general. Recent investigations have underlined repeatedly that at the shoot apical meristem the most important developmental principle is that of cell-cell communication (e.g., van der Schoot, 1996; Laufs, Jonak and Traas, 1998; Laufs et al., 1998; Rinne and van der Schoot, 1998; Crawford and Zambryski, 1999; van der Schoot and Rinne, 1999a, b; Gisel et al., 1999; Meyerowitz, 1999; Fletcher and Meyerowitz, 2000; Rinne, Kaikuranta, and van der Schoot, 2001). The position dependent behavior of individual cells as well as their cooperation within a meristem (Sachs, 1991; Steeves and Sussex, 1989) would not be possible without cell-cell communication. Communication is the basis for integrating cellular and supracellular processes. Cellular processes sustain the physiological and genetic identity of individual cells, whereas supracellular processes subordinate cellular activities to that of the meristem as a whole (van der Schoot, 1996; Rinne and van der Schoot, 1998; van der Schoot and Rinne, 1999a, b). Supracellularity is sustained by information send via both extracellular and symplasmic pathways, both of which are essential for development.

We propose that endo-dormancy, as an interruption of development, can be depicted as the state of the meristem in which cell-cell communication is suspended by an innate mechanism. To substantiate this view we will first briefly examine the general organization of the duplex meristem, and the internal pathways that are available for developmental signalling between meristem cells. Such information is also important in order to understand how environmental signals reach the meristem cells. Secondly, we will investigate the nature of the innate mechanism that suspends the communication during the development of endo-dormancy, and the way it is removed during the release from endo-dormancy. The generality of this mechanism is examined by comparing different plant systems such as deciduous trees and potato tubers (Figure 1).

Understanding how plants regulate their growth and activity cycles is of great importance, not only scientifically, but also from a practical point of view since the onset of endo-dormancy occurs in tandem with a subsequent initiation of tolerance to freezing and other stresses (Weiser, 1970; Sakai and Larcher, 1987; Rinne, Welling, and Kaikuranta, 1998). Therefore, the potential importance of these findings for agricultural and horticultural practice will be discussed briefly. The results may provide both a conceptual framework and a starting point for the future control of plant dormancy phenomena by molecular technologies.

FIGURE 1. Phytochrome-mediated signalling paths trigger the development of endo-dormancy under shortening photoperiod, as illustrated by two kinds of systems. **(A)** Deciduous trees, birch (*Betula pubescens*) and poplar (*Populus tremula*). **(B)** Tuber bearing plants, potato (*Solanum tuberosum*). In both systems light perception and time-measurement take place in the leaf blades (green arrow), from where a putative signal (red arrows) is sent out, via the petioles, towards the axillary and shoot apical meristems. Meristems might, in addition, show a direct response to photoperiod. The leaf-derived signal traffics, at least partly, via symplasmic pathways. The perception of the signals by the meristems is followed by the gradual development of endo-dormancy, a process that involves changes in cell-cell communication and gene expression. The blue arrows in the diagrams represent 'apical dominance signals,' that are send out from the proliferating shoot apical meristems (or stolon, i.e., underground horizontal shoot) under long photoperiod to keep the nearby axillary buds in a state of para-dormancy. In potato, the development of endo-dormancy arrests the stolon apical meristem and redirects imported photosynthates to support tuber growth (orange arrows).

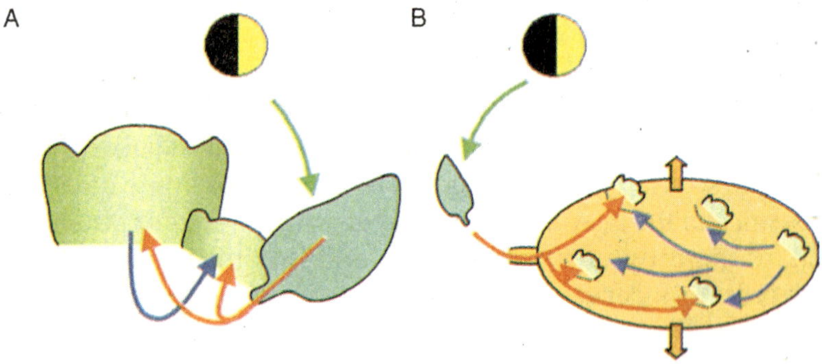

CELL-CELL COMMUNICATION DURING DEVELOPMENT

Patterns at the Shoot Apical Meristem

Development is orchestrated by a small group of cells that is seated at the shoot apex. This cell group, the shoot apical meristem, is committed to the production of leaf primordia and stem tissues. The cell-cell signalling events that underlie these developmental processes take place within a cellular matrix with a specific organization. In angiosperms, this matrix has a so-called duplex organization (Newman, 1965), that is, the meristem consists of two superimposed developmental zones, a tunica and a subtending corpus (Figure 2A). Both the tunica and corpus arise from the division activities of centrally located initial cells and

FIGURE 2. Duplex meristem organization in Angiosperms. **(A)** Two different cell division activities built the meristem. In the tunica cell divisions, as a rule, are confined to the anticlinal plane (red arrow), resulting in the formation of a centrifugally expanding cell sheet. In the underlying corpus cells divide in all planes (blue arrows), which results in a three-dimensionally growing cell body. **(B)** Clonal analysis revealed the presence of distinct and superimposed sets of initial cells (red and blue dots). The plane in which the initial cells and their derivatives divide is coordinated differently in the tunica and the corpus. **(C)** Two distinct cytohistological zones subdivide the meristem. Cells in the center stain differently (lighter color), show different organelle distributions and cell cycling activities compared to cells in the periphery of the meristem. **(D)** Symplasmically, the meristem is composed of at least four dynamically separated areas: the central tunica (red), the central corpus (yellow), the peripheral tunica (green), and the peripheral corpus (brown). The symplasmic subdivision of the meristem is in agreement with both the tunica-corpus model and the cytohistological zonation model.

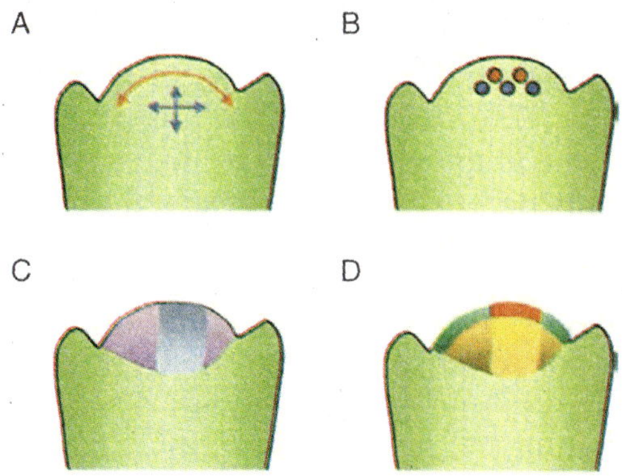

their recent derivatives (Figure 2B). The tunica, as a rule, restricts its cell divisions to the anticlinal plane, perpendicular to the meristem surface, and expands centrifugally. At the periphery of this expanding sheet the surplus of tunica cells is released to stem tissues and leaf primordia, where they differentiate into epidermal cells. In contrast, the corpus is composed of cells that divide in all planes without any obvious rules, and the divisions might just serve to fill space (Figure 2A). In terms of cell division patterns, the shoot apical meristem can thus be pictured as a unit composed of two parts, in which a cell sheet covers a central three-dimensional body of cells (Steeves and Sussex, 1989; van der Schoot and Rinne, 1999a, b).

The cells of the tunica and the corpus not only differ in the orientation of their cell divisions, but also physiologically and genetically (Sawhney, Rennie, and Steeves, 1981; Steeves and Sussex, 1989; Medford, 1992). For example, the 'house-hold' gene *lipid transfer protein* (LTP) is expressed solely in the tunica of the tomato and tobacco shoot apical meristem (Fleming et al., 1992, 1993). Similarly, in *Arabidopsis* the homeobox gene *ATML1* is expressed in the tunica (Lu et al., 1996). In contrast, the *knotted-1* gene, encoding a transcription factor, is restricted to the corpus in maize (Jackson, Veit, and Hake, 1994). Likewise, the expression of *NTH1* and *NTH15*, knotted 1-type homeobox genes, is restricted to the corpus in tobacco (Nishimura et al., 1999). These and other examples suggest that the tunica and corpus serve partly distinct tasks during development (reviewed in, e.g., Medford, 1992; Evans and Barton, 1997; Fletcher and Meyerowitz, 2000). When cells of the tunica occasionally insert cells into the corpus they take on the physiological and genetic identity of the corpus cells, and their fate is 'reprogrammed.' The fact that in angiosperms the meristem maintains such clear distinction between the tunica and the corpus implies that the overall organization is essential for meristem function.

In addition to the longitudinal division into tunica and corpus, the meristem is subdivided radially into a central and peripheral 'cytohistological zone' (reviewed in Steeves and Sussex, 1989). The central zone includes the central parts of the tunica and corpus, while the remaining parts belong to the peripheral zone (Figure 2C). The cytohistological zonation pattern originally referred to the different staining properties of these meristem zones, as well as to differences in cell cycling, organelle distribution and metabolic activity (e.g., Sawhney, Rennie, and Steeves, 1981; Berggren, 1984; Sussex, 1989). The cytohistological zonation model (Steeves and Sussex, 1989) reflects the existence of different functional areas in the meristem. The peripheral area is reserved for leaf primordia formation while the central area is kept free from this activity. At the periphery of the meristem, where cells have to be recruited into the primordia and the stem, cell cycling is higher and metabolism more intense (Sussex, 1989; Medford, 1992; Rinne and van der Schoot, 1998; van der Schoot and Rinne, 1999a, b). Not only the tunica and the corpus show differences in gene expression, also the central and peripheral zones show differences. For example, the gene *NTH20*, a knotted1-type homeobox gene, is only expressed in the peripheral zone of tobacco (Nishimura et al., 1999). On the other hand, the expression of the MADS box genes *SAMADS A* and *SAMADS B* is restricted to the central zone in *Sinapis alba* (Menzel, Apel and Melzer, 1996).

Projection of the cytohistological zonation model upon the tunica-corpus model suggests that at least four developmental zones can be distinguished in the meristem (van der Schoot and Rinne, 1999b). Recent research has indeed shown that expression of genes that are crucial in the overall coordination of the meristem is in accordance with both subdivisions in the meristem, longitudinal as well as radial. For example, the *Arabidopsis CLV3* is expressed in the tunica and *CLV1* in the corpus but only in their central parts (Clark, Running, and Meyerowitz, 1995; Clark, Williams, and Meyerowitz, 1997). The coordination of these gene expression patterns in time and space might require surveillance by supracellular signal networks (van der Schoot and Rinne, 1999a, b).

Pathways for Signal Exchange

The Extracellular Matrix

Part of the signals involved in the coordination of cellular activities at the shoot apical meristem operates via the 'extracellular matrix' (Roberts, 1994). In case of freely mobile signals the extracellular matrix might exert little control over the direction of signal flow. Certain signals simply might be produced in one cell, then pass the plasma membrane of the producer as well as the receiver cells and evoke a response. Alternatively, excreted signals might evoke a response after crossing the extracellular matrix and binding to a plasma membrane receptor of a neighboring cell. These signalling mechanisms are useful in creating patterns only if cells have a different capacity to produce them, and/or to respond to them. An important example of a receptor mediated signalling system in the shoot apical meristem that meets these requirements is the above mentioned CLV1-3 system in *Arabidopsis* (Fletcher and Meyerowitz, 2000). The central cells of the tunica in the vegetative as well as the floral shoot meristem produce the protein CLV3 that is excreted to the extracellular space (Clark, Running, and Meyerowitz, 1995). It probably binds to the putative receptor kinase CLV1, situated in the plasma membrane of the subjacent central corpus cells, which is believed to influence cell proliferation in the central corpus (Clark, Williams, and Meyerowitz, 1997). In turn, the central corpus cells convey a positive signal to the central zone in the tunica to co-ordinate the size of the two gene expression domains (Fletcher and Meyerowitz, 2000). The discovery of this system confirmed the functional subdivision of the meristem into distinct cell groups in tunica and corpus, as well as their cooperation in maintaining the meristem.

The Symplasmic Space

Another part of the correlative signals traffics from cell to cell via plasmodesmata, cytoplasmic bridges, hundreds of which connect individual plant cells through pores in the walls (Figure 3). Through the

FIGURE 3. Plasmodesmata in the shoot apical meristem. **(A)** Model of a generalized plasmodesma. Protein particles in the plasma membrane (green dots) are connected via proteinacious spokes (—) to proteins embedded in the appressed endoplasmic reticulum (red dots). Microinjected Lucifer Yellow CH (yellow ovals) diffuses through the subcanals between the spokes to the adjacent cell. cw, cell wall; ER, endoplasmic reticulum; AER, appressed ER; pm, plasma membrane, SEL, size exclusion limit. **(B)** Plasmodesma in the actively proliferating shoot apical meristem of birch, chemically fixed in the presence of 1% tannic acid. The plasmodesmata (arrow head) in the meristem are linear withouth constrictions or sphincters. Bar = 100 nm. cw, cell wall. **(C)** Iontophoretic microinjection (me) of Lucifer Yellow CH in any cell (1) of a proliferating meristem results in immediate diffusion to neighbouring cells (2), demonstrating that the plasmodesmata are in an open state (except at the boudaries). For the sake of simplicity only one plasmodesma and one neighboring cell is drawn.

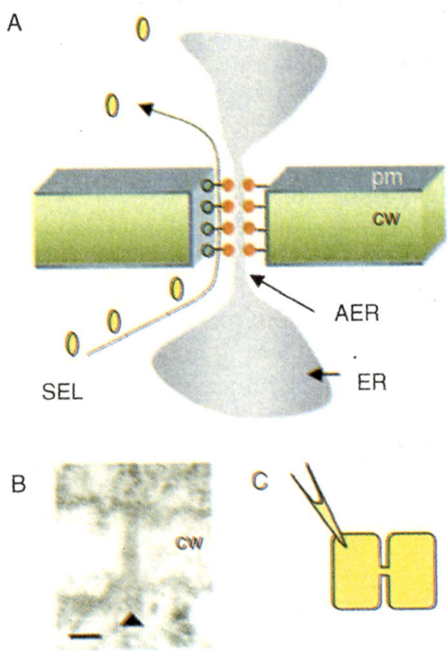

plasmodesmal lumen the endomembrane system is continuous from cell to cell in the form of appressed endoplasmic reticulum (AER), also called desmotubule due to its cylindrical form. The AER is anchored to the plasma membrane via proteinaceous spokes, giving rise to "subcanals" (e.g., Lucas, Ding, and van der Schoot, 1993). The size and properties of the subcanals determine the size class of molecules that can diffuse cytoplasmically from cell to cell, hence the term size exclusion limit (SEL). The AER itself can function as a pathway for the transfer of lipid molecules (Grabski, de Feijter, and Schindler, 1993).

The plasmodesmata in the shoot apical meristem originate in two fundamentally different ways (Cooke, Tilney, and Tilney, 1996; van der Schoot, 1996; Bergmans et al., 1997). During cell division plasmodesmata arise by entrapment of endoplasmic reticulum strings in the forming cell plate (Hepler, 1982). Ontogenetically, therefore, these primary plasmodesmata are continuities rather than interconnections (van der Schoot and Rinne, 1999a). In contrast, neighbouring cells that are separated by non-division walls do not share primary plasmodesmata. Instead, they become interconnected 'secondarily' by *de novo* produced plasmodesmata that penetrate the non-division walls. These 'secondary plasmodesmata' are ontogenetically spoken 'inter-connections' because they become inserted in pre-existing walls. Formation of secondary plasmodesmata is important as it allows the interconnection of cells from the different lineages or clonal groups within the meristem. For example, the cells of the clonally distinct tunica and corpus are interconnected by secondary plasmodesmata (van der Schoot, 1996). Within the tunica and the corpus cell lineage branches also become interconnected sideways via secondary plasmodesmata. In conclusion, the formation of secondary plasmodesmata ensures that in principle all cells of a meristem are integrated into a symplasmic network (van der Schoot and Rinne, 1999a, b).

Signal Exchange via Plasmodesmata

Passive Transport

Regardless whether plasmodesmata are primary or secondary in nature, a universal feature is that they permit diffusing signals to pass from cell to cell in their open or conductive state. As diffusion is independent of plasmodesmata type, it is also independent of lineage. Therefore, diffusion through the plasmodesmal canals can be a unifying force. Morphogenetic signalling via diffusion through these canals would require that the connected cells differ in the production or release of morphogenes, because otherwise chemical equilibrium would soon make such signalling ineffective.

Transmission electron micrographs of the shoot apical meristem in angiosperms consistently show that plasmodesmata connect all the cells of the meristem (e.g., Bergmans et al., 1997; Ormenese et al., 2000). This does not, however, provide information on their functional state: whether they are open or closed, what their SEL is in case they are open, and whether or not they posses a supramolecular complex that subjects them to regulation. Recent dye-coupling experiments have shown that in the vegetative shoot apical meristem most of the plasmodesmata are in an open state (Rinne and van der Schoot, 1998). At the boundary between the central and peripheral zones of the meristem plasmodesmata are closed, however, as microinjected dye-molecules, like lucifer yellow CH (LYCH), do not cross the plasmodesmata at this boundary. Similarly, LYCH does not move from the tunica to the corpus, indicating that the plasmodesmata that connect them are also constricted or closed. These radial and vertical barriers for the movement of symplasmic tracers show that the meristem possesses symplasmic fields that correspond to the structural subdivision of the meristem (Figure 2D). They reflect the division into tunica and corpus, as well as the subdivision into cytohistological zones. The discovery of these symplasmic barriers has revealed how the meristem might function as a unit, while simultaneously keeping certain tasks localised at distinct areas of the meristem (Rinne and van der Schoot, 1998; Gisel et al., 1999; van der Schoot and Rinne, 1999a, b).

The positionally determined constriction of the plasmodesmal canals between the symplasmic fields of the tunica, and between the tunica and the corpus, might be regulated via an actin-myosin based mechanism, proposed for plasmodesmata in mature tissues (White et al., 1994; Overall and Blackman, 1996). Disruption of the actin filaments by profilin and cytochalasin D results in an increase of the plasmodesmal SEL in mesophyll, indicating that actin is involved in keeping the functional diameter of the plasmodesmata at a certain value (Ding, Kwon, and Warnberg, 1996). This requires metabolic energy as application of anoxia or azide, which depletes ATP-pools by preventing ATP production, also yields an increased SEL (Cleland, Fujiwara, and Lucas, 1994). Alternatively, plasmodesmal constriction might be achieved by the calcium-binding contractile phosphoprotein centrin that has been localized at plasmodesmata (Blackman, Harper, and Overall, 1999). Each of these mechanisms, or a combination of them, might hypothetically function within the meristem to narrow down the plasmodesmal canals at the field boundaries far below its normal SEL. The fact that the boundaries remain in the same location, while the apical meristem proliferates, indicates that the positional closing of plasmodesmata is under dynamic regulation of the tunica, the corpus, or both, dependent on the location of the boundary (see further in Rinne and van der Schoot, 1998; van der Schoot and Rinne, 1999a, b; Crawford and Zambryski, 1999; Ormenese et al., 2002).

Symplasmic fields might be crucial for the functioning of the meristem for several reasons. Firstly, the cells of the symplasmic fields share small metabolites and signalling molecules, which might metabolically stabilize them. Secondly, they might function as 'morphogenetic fields' that coordinate morphogenesis in a gradient-dependent way as in animal systems (reviewed by Müller, 1997). Thirdly, a symplasmic field could facilitate a fast and unified response of the participating cells, even if only a single cell member is directly exposed to the external signal. Such a situation could be exemplified by the above-mentioned CLV1-3-system. Binding of the extracellular protein CLV3 to the receptor kinase CLV1, located in the plasma membrane of the central corpus cells (Fletcher and Meyerowitz, 2000), would lead to a unified response in the central corpus if one of the downstream components in the signal cascade could traffic through plasmodesmata to equally 'inform' the other cells of the field that are less well exposed to the ligand.

Active Transport

During the last decade evidence has accumulated that certain macromolecules might move through plasmodesmata (Ding, 1997; Ghoshroy et al., 1997). For example, plant viruses appeared to spread as RNA from cell to cell via plasmodesmata, a process that is aided by virally encoded movement proteins (MP) (see, e.g., Pennazio, Roggero, and Conti 1999). In plants infected with plant viruses, and in transgenic plants expressing their MP, the MP can be immunolocalized within the plasmodesmata (e.g., Tomenius, Clapham, and Meshi, 1987; Atkins et al., 1991; Storms et al., 1998). MPs possess a targeting sequence for plasmodesmata, and they might simultaneously function as chaperones for the viral nucleic acids, as shown by *in vitro* experiments (Citovsky et al., 1992; Ghoshroy et al., 1997; Crawford and Zambryski, 1999). After binding of the MP to the macromolecular assembly at the plasmodesmal entrance the canal dilates or gates temporarily, and the nucleic acid, singly or complexed with the MP, moves to the adjacent cell in a process that might require metabolic energy (reviewed in, e.g., Lucas, Ding, and van der Schoot, 1993; Ding, 1997; Ghoshroy et al., 1997).

The consideration that plant viruses might have evolved by employing plant-endogenous systems for cell-to-cell trafficking led to the idea that the trafficking of plant-endogenous proteins and RNAs might serve the regulation and coordination of non-cell-autonomous gene expression during plant development (Lucas, Ding, and van der Schoot, 1993). Evidence has now emerged that plasmodesmata, indeed, can traffic certain plant-endogenous macromolecules, proteins as well as RNAs (reviewed in Ding, 1997; Ghoshroy et al., 1997; van Bel, Günther, and van Kesteren, 1999). In the maize meristem, the homeodomain protein KNOTTED-1 is found in all areas of the meristem (except for the leaf buttress), even though its mRNA cannot be localized in the tunica. It has been suggested, therefore, that the protein traffics to the tunica via the plasmodesmata that interconnect tunica and corpus (Jackson, Veit, and Hake, 1994; Lucas et al., 1995).

In *Anthirrhinum* several cases of non-cell autonomous gene expression have been described, implying that factors that directly influence gene expression move to other cells. For example, selective, transposon induced expression of *deficiens* (*def*) and *globosa* (*glo*) in the lower meristem layers lead to downstream effects in the tunica, although the reverse is not found (Perbal et al., 1996). GLO might traffic through secondary plasmodesmata from the corpus to the tunica, where lateral spread is restricted. It was suggested that the latter was due to the inability of GLO to traffic through primary plasmodesmata (Perbal et al., 1996). In *Arabidopsis*, selective expression of *floricaula* (*flo*) in the tunica causes downstream effects in the corpus (Hantke, Carpenter, and Coen, 1995; Carpenter and Coen, 1995; Furner et al., 1996). These and other cases cases of non-autonomous behaviour clearly show that certain 'morphogenetic factors' can move selectively within the meristem. Recently, mosaic studies with mutants have provided evidence that the transcription factor LEAFY can move non-selectively within the meristem, thereby restoring the wild type phenotype (Sessions et al., 2000). Although these observations do not prove that such trafficking is a regular event in the non-regenerating and non-engineered meristem, they do demonstrate that the meristem at least has this important option for maintaining its organisational integrity.

Macromolecular movement across symplasmic field boundaries, such as that between tunica and corpus, must involve the opening of narrowed or closed plasmodesmata, a process called 'gating.' The precise mechanisms of gating are not entirely clear, but some models have been proposed based on studies with mesophyll and trichomes (see Ding, 1997; Goshroy et al., 1997; McLean, Hempel, and Zambryski, 1997; Waigman et al., 1997; Schultz, 1999). In general, the SEL may or may not increase during gating. Macromolecular trafficking between the corpus and the tunica probably does not involve an increase in SEL, as microinjected dyes remain confined to their symplasmic field. This conclusion seems contradictory to the observation that KNOTTED1, which probably moves from the corpus to the tunica, increases the plasmodesmal SEL in mesophyll of source leaves (Lucas et al., 1995). Considering that plasmodesmata appear to have very different characteristics even among leaves at different developmental stages and in different conditions (Imlau, Truernit, and Sauer, 1999; Oparka et al., 1999; Crawford and Zambryski, 2001), it is plausible that plasmodesmata in morphogenetic areas like the shoot apical meristem possess unique regulatory properties. Mesophyll might thus not be a good model tissue to study possible functions of plasmodesmata in meristematic areas.

LACK OF COMMUNICATION DURING ENDO-DORMANCY

Shutting off Symplasmic Pathways

Like discussed in the introduction, endo-dormancy can be conceptualized as the interruption of developmental signaling at the shoot apical meristem by an innate mechanism. Traditionally, the investigation of endo-dormancy in perennial plants focussed on its timing, and on the putative factors that induce dormancy. The preoccupation with inducing signals and the nature and the transmission of these signals through the shoot system (e.g., Rappaport and Wolf, 1969) might have precluded a focus on the state the meristem adopts in endo-dormancy (Seeley, 1994; van der Schoot, 1996). The state itself, which cannot be understood unambigiously from the inducing signals, should be the primary target of investigation because it determines the survival of the plant. We have proposed that the state of dormancy can be investigated from the perspective of cell-cell communication, and that endo-dormancy is the temporary blockage of developmental signaling in the meristem. Therefore, the crucial point determining the state of the meristem is whether the cells of the shoot apical meristem are 'online' or 'offline.' In support of this we found for birch (Rinne and van der Schoot 1998; Rinne, Kaikuranta, and van der Schoot, 2001) and poplar (unpublished results) that during the development of endo-dormancy microinjected fluorescent dyes like Lucifer Yellow LH (457 Da; Stewart, 1981) did no longer pass from cell to cell, demonstrating that all plasmodesmata were narrowed or closed. In potato, a similar situation was present, although the isolation was not complete; small irregular patches of cells could be coupled due to disrupted plasmodesmata (van der Schoot, 1996).

The obstruction of signalling channels during endo-dormancy cannot be due to the narrowing or closing of plasmodesmata via the activation of an actin-myosin/centrin based mechanisms, as proposed for the creation of symplasmic fields (described above). This is obvious from the fact that the above mechanisms need a continuous input of energy, which is not readily available in endo-dormant systems. Endo-dormancy requires a semi-permanent closing of the signal pathways, by a mechanism that does not need maintenance and energy input after it is installed, but yet, it should be easily removable when growth is renewed. Such a reversible mechanism could utilize 1,3-β-D-glucan (callose).

Deposition of 1,3-β-D-Glucan

Winter Phloem and Traumatized Cells

A classic example of physical blockage of symplasmic pathways is 'winter phloem,' where the sieve plate pores of individual sieve tubes become sealed by the deposition of 1,3-β-D-glucan, resulting from the activation of 1,3-β-D-glucan synthase (Esau, 1977). The deposition of 1,3-β-D-glucan by this enzyme can be so extensive that not only the pores, but also the entire sieve plate becomes covered. The enzyme 1,3-β-D-glucan synthase is also involved in the differentiation of the phloem system. During sieve tube development a cylindrical collar of 1,3-β-D-glucan is first deposited around each plasmodesma. After the sieve plate has expanded 1,3-β-D-glucan is gradually removed, resulting in the formation of mature sieve plate pores (Esau and Thorsch, 1985; Lucas, Ding, and van der Schoot, 1993). In this process, 1,3-β-D-glucan functions as a 'temporary and mouldable wall material' (Kauss, 1996).

1,3-β-D-glucan can also be deposited as a layer between the plasma membrane and the cell wall. This event is commonly viewed as a wound response, providing protection, and preventing the loss of cytoplasmic components. The deposition of 1,3-β-D-glucan is due to the activation of glucan synthase II complexes (1,3-β-D-glucan synthase) that are associated with the extracellular surface of the plasma membrane (Olesen, 1980 and references therein). This enzyme is normally involved in cellulose synthesis, but it can rapidly switch to the synthesis of the 1, 3-β-D-glucan after plasma membrane perturbation (Delmer, 1987). Perturbation leads first to sudden changes in the transmembrane potential, the opening of mechanosensitive ion channels (e.g., Delmer, 1987; Kauss, 1985; Garrill, Findlay, and Tyerman, 1996), the massive influx of Ca^{2+}, and the production of 1,3-β-D-glucan (Kauss, 1985, 1996; Hayashi et al., 1987).

Shoot Apical Meristems

As sieve plate pores arise from plasmodesmata, it is not surprising that 1,3-β-D-glucan deposition can also occur around plasmodesmata in general (Northcote, Davey, and Lay, 1989; Lucas, Ding, and van der Schoot, 1993). Immunolocalization studies have, indeed, shown the presence of 1,3-β-D-glucan synthase at mature plasmodesmata (Northcote, Davey, and Lay, 1989; Delmer et al., 1993). Although only few studies have investigated 1,3-β-D-glucan deposition at shoot apical meristems, they demonstrate that it can be present at specific locations. For example, in the active meristem 1,3-β-D-glucan is transiently present at the cell plates in dividing cells (Rinne, Kaikuranta, and van der Schoot, 2001), like shown earlier for dividing protoplasts (Kakimoto and Shibaoka, 1992). In the active meristem tiny amounts of 1,3-β-D-glucan were present along the plasmodesmal canal. These tiny amounts of 1,3-β-D-glucan do not seal of the plasmodesmata as demonstrated by microinjection of fluorescent dyes. In contrast, in the endo-dormant meristem substantial deposits of 1,3-β-D-glucan were present specifically around the neck region of the plasmodesmata (Figure 4A), and microinjections demonstrated that this physically blocked all cell-cell communication (Rinne and van der Schoot, 1998; Rinne, Kaikuranta, and van der Schoot, 2001). Plasmodesmata have been characterized as short-distance analogues of the phloem, or mini phloem systems (Olesen and Robards, 1990) but particularly the plasmodesmata in the over-wintering shoot apical meristem of perennials show phloem-like behaviour as in winter they are sealed by 1,3-β-D-glucan.

In contrast to the unselective deposition of 1,3-β-D-glucan over the entire sieve plate in winter phloem, the deposits in the endo-dormant

FIGURE 4. Plasmodesmata in endo-dormant meristems after immunolabelling **(A)**, and after staining with tannic acid **(B-D)**. The scheme at the right hand side indicates whether or not microinjected Lucifer Yellow CH diffused from any injected cell (1) via plasmodesmata to the adjacent neighboring cells (2). For the sake of simplicity only one plasmodesma and one neighbouring cell is drawn. **(A)** Immunolabelling with anti-1,3-β-D-glucan antibody. Plasmodesmata (arrow heads) in the endo-dormant meristem of birch show the presence of 1,3-β-D-glucan in the areas of the external sphincter ring and the plasmodesmal canal. **(B)** Endo-dormant meristems of birch, fixed with osmium tetroxide and tannic acid, display proteinacious external sphincter rings and plasmodesmal plugs at locations that correspond to the areas labelled with antibodies against 1,3-β-D-glucan (compare with 'A'). Plasmodesmata in the endo-dormant meristems of poplar **(C)** and potato **(D)** possess similar prominant sphincter-complexes. The sphincters are most pronounced in poplar, whereas in potato they are less well defined. In all cases, however, their presence categorically blocks cell-cell communication via plasmodesmata. Bar = 100 nm. cw, cell wall; s, sphincter ring.

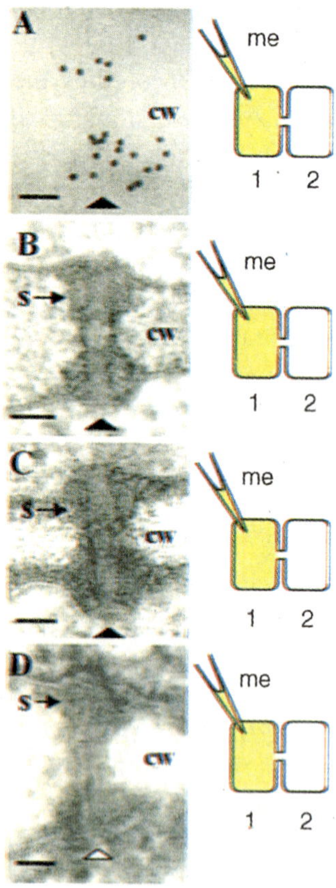

meristem were present only around the plasmodesmal entrances. Tannic acid staining revealed that 1,3-β-D-glucan was localized at a ring-like wall space that contained proteins (Figure 4B). The co-localization of proteins and 1,3-β-D-glucan suggests that one of the proteins could be the enzyme 1,3-β-D-glucan synthase. As the 1,3-β-D-glucan synthase is likely to be an integral membrane protein (Lucas, Ding, and van der Schoot, 1993), the 'lumen' of the ring-like space, somewhat distant from the plasma membrane, might contain other proteins. In addition, various polysaccharides might be present as 1,3-β-D-glucan typically occurs in an intimate mixture with other materials (Turner, Wells, and Roberts, 1994; Kauss, 1996). In the literature, comparable 1,3-β-D-glucan-containing structures at the plasmodesmal entrances have been interpreted as 'sphincters' (Olesen and Robards, 1990; Lucas, Ding, and van der Schoot, 1993).

Models of Plasmodesmal Sphincters

A Flux Regulating External Sphincter

Plasmodesmal sphincters are depicted as extracellular ring-like structures, located between the plasmodesmal neck region (orifice) and the cell wall (Olesen, 1979, 1980; Delmer et al., 1993; Badelt et al., 1994). Such sphincters should not be confused with the unrelated sphincter of grasses, which consist of an internal plasmodesmal diaphragm (reviewed by Beebe and Russin, 1999). The problems in defining clear roles and mechanisms for sphincter functioning arise from the fact that plasmodesmal dimensions are too small to be resolved by traditional TEM technology (e.g., Thompson and Platt-Aloia, 1985). Sphincters are not consistently found in particular tissues or cell types. It has been suggested, therefore, that conspicuous sphincters develop only at a certain stage of cellular differentiation, or that they might not be present in all types of tissue (Olesen and Robards, 1990). Often, it is also necessary to add tannic acid to the fixative to make the sphincters clearly visible (Robards, 1976; Olesen, 1979; Olesen and Robards, 1990; Mollenhauer and Morré, 1987; Badelt et al., 1994). In cases where sphincters are observed without addition of tannic acid (Olesen, 1979) naturally occurring tannic acids might be present. Some authors have suggested that the 1,3-β-D-glucan that is found in the sphincters is an artefact (Radford, Vesk, and Overall, 1998) or even that the sphincters themselves are artefacts (Ding, 1999). Ding (1999), for example, pointed out that tannic acid precipitation and poor chemical fixation might have created the structures, presented as sphincters, although he left open the possibility that in some species real sphincters may be present.

With these cautionary notes in mind we will examine a plausible mechanisms of sphincter functioning. The sphincter model of Olesen and Robards (1990) proposes that the sphincters operate in the fine-regulation of plasmodesmata as follows (Figure 5A). Signal transduction processes–such as propagated changes in the membrane potential or other local perturbations result in activation of Ca^{2+} channels at the plasmodesmal neck region, promoting the net influx of Ca^{2+}. In the model, the transient increase of $[Ca^{2+}]$ at the plasmodesmal entrance is followed by the activation of 1,3-β-D-glucan synthase and the deposition 1,3-β-D-glucan at the extracellular side of the plasma membrane. Recently, however, it has been demonstrated that local fluctuations in cytosolic $[Ca^{2+}]$ only transiently alter the conductive state of plasmodesmata, possibly via the modulation of contractile plasmodesmal proteins (Tucker and Boss, 1996; Holdaway-Clarke et al., 2000). These transient changes of plasmodesmal conductance do not involve deposition of 1,3-β-D-glucan. Therefore, the activation of 1,3-β-D-glucan synthase might require the additional action of the activator β-furfuryl-β-glucoside from the vacuole, like suggested earlier (Ohana et al., 1991, 1992; Ohana, Benziman, and Delmer, 1993; Lucas, Ding, and van der Schoot, 1993). The activity of 1,3-β-D-glucan synthase gradually fills the ring-like space at the plasmodesmal entrance with 1,3-β-D-glucan. The subsequent binding of water by 1,3-β-D-glucan (Stone and Clarke, 1992) will result in expansion of the ring. Because the outer part of the ring is against the cell wall it can expand only inwardly, thereby constricting the plasmodesmal canal (Olesen and Robards, 1990). 2-Deoxy-D-glucose, which inhibits 1,3-β-D-glucan synthesis, prevents the deposition of 1,3-β-D-glucan and the constriction of the plasmodesmal canal (Olesen and Robards, 1990; Radford, Vesk, and Overall, 1998). This demonstrates that 1,3-β-D-glucan can, indeed, alter the effective size of the plasmodesmal canal.

Possible Ca^{2+} sources include the cell walls, the sphincter ring area, and also the endoplasmic reticulum (ER) strands that are continuous with the AER in the plasmodesmata. Olesen and Robards (1990) preferred the first possibility, but it seems plausible that the ER sacculi that are close enough to the plasmodesmal entrance could even more precisely regulate the local concentration of Ca^{2+} (e.g., Schultz, 1999).

FIGURE 5. Models of 1,3-β-D-glucan-based plasmodesmal sphincters. A single plasmodesmal entrance is depicted in a longitudinal section. **(A)** Sphincter model according to Olesen and Roberts (1990). A local perturbation (red arrow on the left) induces opening of stress-sensitive ion channels in the membrane, resulting in the influx of Ca^{2+}. The transient increase in cytoplasmic free Ca^{2+} activates the enzyme 1,3-β-D-glucan synthase (blue dot) that is situated at the plasma membrane, and converts UDP-glucose to 1,3-β-D-glucan (yellow), which is deposited into the extracellular sphincter ring. Absorbtion of water by 1,3-β-D-glucan results in the swelling and inward expansion of the ring, thereby constricting the plasmodesmal canal at the neck region (modified after Olesen and Robards, 1990). **(B)** Sphincter-complex model according to Rinne et al. (2001). The triggering events are similar to 'A', except that β-furfuryl-β-glucoside functions as a co-activator of 1,3-β-D-glucan synthase. Free Ca^{2+} might be released into the cytoplasm from the cell walls, local ER-sacculi, or both. 1,3-β-D-glucan (yellow) is not only deposited in the extracellular ring but also in the plasmodesmal canal. As a result the inwardly expanding extracellular sphincter ring closes on an intracellular sphincter plug. Since the sphincter ring cannot expand into the canal lumen the excess 1,3-β-D-glucan is protruding into the cytoplasm, as visible in Figure 4. The red ovals are proteinaceous spokes. AER, Apressed Endoplasmic Reticulum; cw, cell wall; ER, endoplasmic reticulum; UDP, uridine diphosphate; UDPg, UDPglucose; βF-βG, β-furfuryl-β-glucoside.

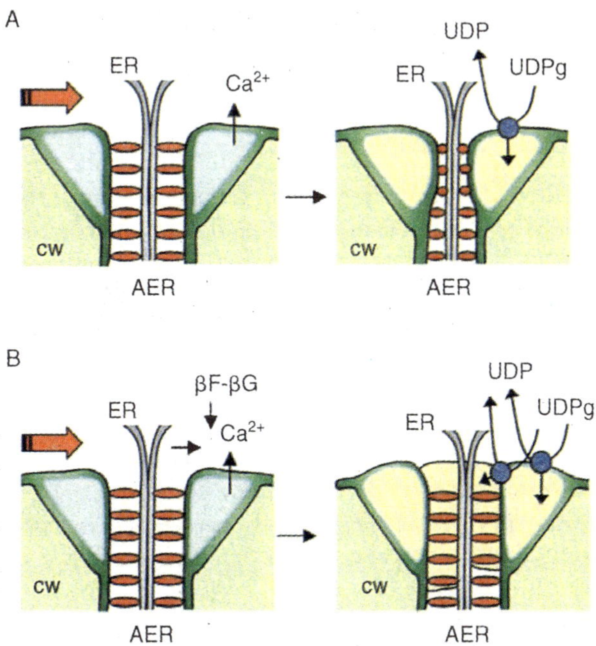

An Endo-Dormancy Sphincter-Complex

The sphincters we observed in the endo-dormant shoot apical meristem of birch, poplar and potato differed from the above model (Olesen and Robards, 1990) in several ways. Firstly, the sphincters were present at virtually all plasmodesmata of all cells. Secondly, their presence appeared to be under developmental control as they were entirely lacking from actively growing meristems and appeared only when endo-dormancy was established (Rinne and van der Schoot, 1998; Rinne, Kaikuranta, and van der Schoot, 2001). This state-dependent presence makes it unlikely that they are fixation artefacts, a conclusion which is supported by the fact that the sphincters co-existed with cellular features that characterize proper tissue fixation, such as a smooth and tightly appressed plasma membrane, and well-preserved microtubuli (Ding, 1999). Thirdly, there were distinct structural differences between the sphincters in the endo-dormant meristem and those depicted in the model of Olesen and Robards (1990).

Ultrastructural observations revealed that the sphincters in the dormant shoot apical meristems of birch, populus and potato did not constrict the plasmodesmal canal at the neck region, whereas in the model of Olesen and Robards (1990) the constriction of the neck is the central feature. The neck region of the canal in birch, populus and potato was filled with an electron dense, proteinaceous 'plug,' and immunolocalization in birch showed that the plug contained 1,3-β-D-glucan (Figure 4). This indicates that the AER and its associated macromolecular assembly are embedded in a 1,3-β-D-glucan-containing matrix that obstructs the sub-channels between the assembly and the plasma membrane tube (Figure 5B). This is significant, as it implies that communication via the plasmodesmata is shut off entirely during endo-dormancy. In contrast, the sphincter of the Olesen and Robards model (1990) might retain some permeability, as the plasmodesmal canal cannot be sealed off completely due to the lack of filling material (see Overall, 1999; Schultz, 1999).

Microinjection of LYCH in single cells of fully endo-dormant meristems demonstrated not only that cells were uncoupled, but also that the uncoupling was very persistent. As fully endo-dormant meristems are in a state of dehydration, impalement of the microelectrode led to an immediate absorption of LYCH-solution from the microelectrode tip. In many cases, this resulted in strong swelling of the impaled cells. The cells retained the swollen state at least until the next day, as assessed from their fluorescent outlines (Rinne and van der Schoot, 1998). This demonstrated that the plasmodesmal size exclusion limit (SEL) not only dropped below the value for LYCH (457 Da) but that the plasmodesmata were hermetically closed, not allowing even the passage of water to adjacent hygroscopic cells. Water did not by-pass the plasmodesmata by moving through the plasma membrane and the walls, indicating that cells walls were waterproofed. The difficulties to stain cell walls of the endo-dormant meristem with water-soluble dyes might visualize this (Rinne, Kaikuranta, and van der Schoot, 2001). Cell walls could be made water-repellent by impregnation with suberin-like substances (unpublished results). In case of endo-dormant floral buds extracellular phenolic compounds have been suggested to protect cells by blocking water movement and ice nucleation (see in Wisniewski et al., 1996b). Esterification of pectin epitopes may have a similar function (Wisniewski et al., 1996b).

In conclusion, during the onset of endo-dormancy 1,3-β-D-glucan is deposited, both as an external sphincter ring around the plasmodesmal entrances and as an internal plasmodesmal plug. The simultaneous formation of a ring and a plug not only stabilizes the plasmodesmal neck region and the internal AER-complex but it might also provide a hermetic seal as the external sphincter-ring expands inwardly, thereby closing on the internal canal-plug (Figure 5). Whether the formation of sphincter-complexes requires the presence of different forms of 1,3-β-D-glucan synthases at distinct locations in the plasma membrane, or possibly even the AER, remains to be established. This new 'bottle-neck-cork' model of a plasmodesmal sphincter, i.e., a 'sphincter-complex,' might be common in perennials, where the meristem first develops endo-dormancy, and subsequently pre-adapts its uncoupled cells to harsh winter conditions. In birch, the uncoupled meristem cells dehydrate, while simultaneously producing various protectants and dehydrins (Welling, Kaikuranta, and Rinne, 1997; Rinne, Welling and Kaikuranta, 1998). The production of dehydrins is a general feature of dehydrating tissues (e.g., Ingram and Bartels, 1996), and it has been well established for overwintering flower buds (Muthalif and Rowland, 1994) and bark tissues (Arora, Wisniewski, and Scorza, 1992; Arora and Wisniewski, 1994; Wisniewski et al., 1996a).

Sphincter-Complexes and CER

In the endo-dormant meristems of potato tubers plasmodesmal sphincter-complexes occurred together with so-called 'concentric configurations of endoplasmic reticulum' (CER) (Figure 6A). These CERs, first described for potato by Shih and Rappaport (1971), consist of concentrically wrapped layers of rough endoplasmic reticulum that might be up to 5 μm in diameter (van der Schoot, 1996). The function of CER has remained unknown, but they were proposed to reflect the low metabolic state of the cells (Shih and Rappaport, 1971). The sphincters in the potato meristem were initially regarded as devices that could constrict the plasmodesmata (van der Schoot, 1994, 1996), like in the model of Olesen and Robards (1990). Ultrastructural observation showed, how-

FIGURE 6. Sphincters and concentric configuration of endoplasmic reticulum (CER) in the endo-dormant meristem of potato (*Solanum tuberosum* cv. Desiree). To improve the contrast of membranes and sphincter-complexes 1% tannic acid was used in the fixtion medium. **(A)** Prominent CER in a corpus cell. Darkly stained sphincter-complexes are visible at the plasmodesmata (arrow head). A few spherosomes are visible (arrows), one in the center of the CER. Bar = 2 μm. **(B)** Cross section through the sphincter area of two adjacent plasmodesmata showing a closed (lower one) and an empty (upper one) plasmodesma. An electron-dense sphincter ring (S) surrounds the entrance of both plasmodesmata, although more pronounced in the closed one. In closed plasmodesmata the lumen between the appressed endoplasmic reticulum (AER) and the plasma membrane tube is filled with electron-dense material. In empty plasmodesmata the AER is missing and the lumen is devoid of any filling substance (arrow). Note that in case of the upper plasmodesma particles (arrow) can be discerned in the sphincter ring (S). Open triangle, plasma membrane tube. Bar = 50 nm.

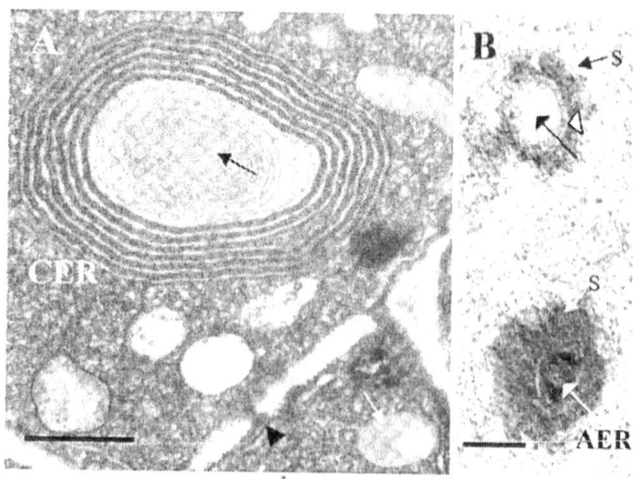

ever, that the plasmodesmal canal inside the sphincter ring was not constricted (Figure 4D), a situation similar to that in birch and poplar. In addition, in a small number of plasmodesmata the AER was no longer visible in the plasmodesmal canal, probably due to withdrawal during the formation of CER, resulting in enlarged plasmodesmal canals (Figure 6B). The appearance of both enlarged and closed plasmodesmal cannels in endo-dormant meristems explained the irregular dye-coupling patterns obtained after microinjection of LYCH and LYCH-10 kDa dextran (van der Schoot, 1996). Although in birch and poplar all plasmodesmata were closed by sphincter-complexes, it can be concluded that the presence of sphincter-complexes and empty plasmodesmata in the endo-dormant potato meristem led to the same end result, the obstruction of the patterns of communication that are necessary to sustain development.

RESTORATION OF COMMUNICATION PATHWAYS

The Effects of Chilling

It is well known that the endo-dormant bud regains the potential for renewed growth and development only after it has received an adequate exposure to cold (Coville, 1920; Nooden and Weber, 1978; Powell, 1987; Heide, 1993). Selective chilling of various plant parts has demonstrated that the stimulatory effect of cold on dormancy release cannot be imported from the rest of the plant (Coville, 1920; Metzger, 1996). It is the endo-dormant bud itself that must be exposed to cold. That cold acts directly on the meristem in promoting the release from endo-dormancy has been confirmed by experiments with isolated endo-dormant meristems (Metzger, 1988; Purvis, 1940, 1961). This makes sense in the light of our findings that the endo-dormant bud is in a state of cellular disconnection, with no pathways available for the import of signals from the rest of the plant (Rinne, Kaikuranta, and van der Schoot, 2001).

As cold releases the meristem from endo-dormancy it was expected that it promotes the partial or entire restoration of the communication pathways within the shoot apical meristem. Ultrastructural observations showed that the sphincter-complexes in the birch meristem had disappeared after a 6-week cold period (Figure 7A), which equals the length of the cold period that is required to release meristems from endo-dormancy. Microinjection experiments showed that after a 6-week cold

FIGURE 7. Ultrastructural details of birch (*Betula pubescens* Ehrh.) meristems, released from endo-dormancy by chilling. **(A)** Meristems were fixed with 1% tannic acid in the fixation medium. The sphincter-complexes of the endo-dormant state are no longer present at the plasmodesmata (arrowheads). The drawing shows diffusion of microinjected dyes from any injected cell (1) to the neighboring cells (2) through restored plasmodesmal canals (except for the plasmodesmata at field boundaries). For the sake of simplicity, only one plasmodesma and one neighboring cell is drawn. Bar = 200 nm. cw, cell wall; v, vacuole-like structure (spherosome). **(B)** Endo-dormant and chilled meristems contain large amount of small vacuole-like structures (spherosomes), but only in chilled meristems they align the cell wall (arrows) and make connections with plasmodesmata. Protein bodies (p) surround the nucleus, and contain electron-dense protein deposits. Bar = 5 μm. **(C)** Immunolocalization studies showed that the peripheral parts of the vacuole-like structures (spherosomes) contained 1,3-β-D glucanase. Bar = 200 nm. cw, cell wall.

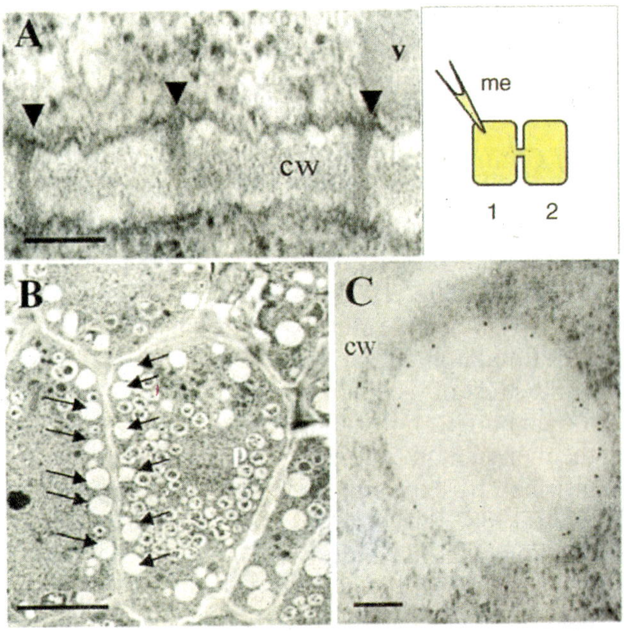

period the plasmodesmata had regained their capacity to transport microinjected fluorescent dyes, and the meristem displayed dye-coupling patterns characteristic of the active meristem (Rinne, Kaikuranta, and van der Schoot, 2001). These eco-dormant meristems are ready to proliferate as soon as the external conditions improve.

The Role of 1,3-β-D-Glucanase

In diffuse porous trees, where the phloem is functioning over several seasons, the 1,3-β-D-glucan of winter phloem is removed in spring by the action of 1,3-β-D-glucanase (Krabel et al., 1993). As the major component of the sphincter-complex is 1,3-β-D-glucan, plasmodesmal restoration of the meristem, similarly, implicates the local action of 1,3-β-D-glucanases. A first indication that the enzyme 1,3-β-D-glucanase was responsible for the hydrolysis of the 1,3-β-D-glucan deposits at the sphincter complexes came from the analyses of 1,3-β-D-glucanase expression patterns during the cold period of 2°C. Expression of 1,3-β-D-glucanase appeared to be up regulated with a maximum after 6 weeks of chilling (Rinne, Kaikuranta, and van der Schoot, 2001). The removal of intra- and extracellular 1,3-β-D-glucan deposits at the plasmodesmata requires the action of 1,3-β-D-glucanase inside and outside the plasma membrane. Immunolocalisation showed that, after adequate chilling, 1,3-β-D-glucanase was present free in the cortical cytoplasm and in numerous small vacuole-like structures that lined up at the plasma membrane and the plasmodesmata (Figure 7B,C). Interestingly, these 1,3-β-D-glucanase-containing vacuoles were present already in endo-dormant meristems before they received cold, but they were dispersed in the cytoplasm. Ultrastructural observations showed that exposure to cold had displaced the 1,3-β-D-glucanase-containing structures in the vicinity of the plasma membrane. After a cold period, sufficient to break endo-dormancy, a substantial number of these vacuole-like structures appeared in contact with the plasma membrane and plasmodesmata (Rinne, Kaikuranta, and van der Schoot, 2001). The free 1,3-β-D-glucanase that occurred at the end of the chilling period in the peripheral cytoplasm probably represented *de novo* synthesized enzyme. It is tempting to speculate that the two different locations of 1,3-β-D-glucanases–in small vacuoles-like structures and free in the cytoplasm–reflect two distinct routes for bringing together the enzyme and its substrate in the extracellular sphincter ring and in the intracellular plug, respectively.

The production of these 1,3-β-D-glucanase containing structures at an early phase of dormancy, for later use during endo-dormancy release, is a surprising example of pro-activity. The preproduction and subsequent compartmentalisation of 1,3-β-D-glucanase solves the strategic problem of how to produce sufficient amounts of enzyme for symplasmic restoration during endo-dormancy when metabolism is severely slowed down. This implies that the vacuole-like structures could have a prominent role in dormancy cycling and, therefore, we will examine in greater detail how these structures are formed and how they might deliver their enzymes to the plasmodesmal sphincters.

The Putative Role of Spherosomes

The vacuole-like structures were uniform in size–approximately 0.5-2.0 μm in diameter in both poplar and birch (Figure 8)–lacked a clear unit-membrane, and in view of their low electron-density contained neutral lipids. Together, these features identified the vacuole-like structures as oil bodies or spherosomes. Spherosomes were also present in the endo-dormant meristems of potato tubers, although clearly in lower numbers than in birch and poplar, and occasionally they occupied the core of a CER (Figure 6). Spherosomes are believed to arise from specific microdomains of the ER by the deposition of triacylglycerides and/or sterol esters inside the ER lipid layer (Frandsen, Mundy, and Tzen, 2001; Murphy, 2001). This results in the formation of small vesicles that possess a half-unit membrane and, occasionally, the remains of the inner membrane leaflet (Matile, 1975; Figure 8A,B). Spherosomes are commonly thought to function in the storage of lipids in oleineous seeds and fruits, but recent evidence suggests that spherosomes might be a more common cell constituent than previously thought, and have functions beyond the passive storage of triacylglycerols. Spherosomes are likely to be involved in lipid metabolism, lipid trafficking, and membrane recycling in most cell types of plants, animals as well as micro-organisms (Murphy, 2001).

The sudden increase in the number of spherosomes in the meristems of birch seedlings, newly exposed to short photoperiod, might relate to the arrest of plasma membrane proliferation, as demonstrated for yeast cells in their stationary phase (see Murphy, 2001). The sterol esters of the lipid bodies in yeast are continuously depleted, but once plasma membrane proliferation ceases sterol esters can no longer be hydrolyzed and lipid bodies start to build up. It was suggested for yeast that the sterol esters are then available to support the rapid resumption of growth as soon as environmental conditions improve (see Murphy, 2001), a situation that seems comparable to that of higher plant meristems which are released from endo-dormancy.

It is of interest that the 'dormancy hormone' abscisic acid (ABA) influences the formation of spherosomes, and the accumulation of proteins in their half membranes that play a role in the stabilization and

FIGURE 8. Origin and fate of spherosomes (oil bodies). **(A)** Spherosomes bud off from specific microdomains of the endoplasmic reticulum (ER) by the deposition of triacylglycerols between the inner and outer leaflet of the ER-membrane. They may bud off from the outer leaflet of the endoplasmic reticulum ER or, alternatively, they may degrade the inner leaflet. The resulting half-unit membrane carries a particular set of proteins, that relates to the putative functions of the spherosomes. Hydrolases may be located within the half-unit membrane, or they may be attached to the other proteins. **(B)** Young developing spherosome of birch (*Betula pubescens* Ehrh), one to two weeks after the start of the short photoperiod. Note the shrinking inner membrane leaflet (i), and the position of the hardly visible half-unit membrane (o). **(C)** Young developing spherosomes from *Populus tremula* × *tremuloides* under short photoperiod. The invaginations at the membrane (arrows) might indicate fusion areas (Murphy 2001). **(D)** Spherosomes in chilled meristems of birch. Membraneous structures and an internal vacuole (v) often appear when the spherosome docks at a plasmodesmal site (arrow). cw, cell wall. **(E)** Spherosomes in chilled meristems of birch engulf structures that might belong to a sphincter-complex. Arrows point at plasmodesmata. Bars = 200 nm. v, vacuole. cw, cell wall.

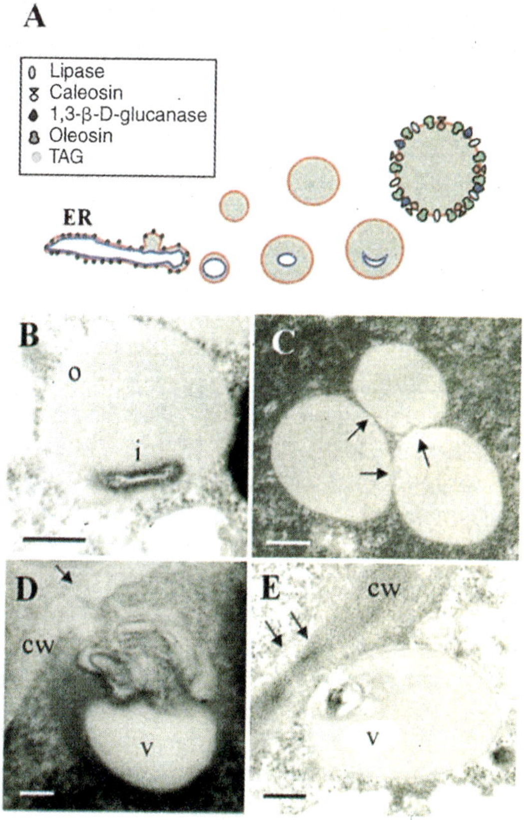

intracellular trafficking of spherosomes (Frandsen, Mundy, and Tzen, 2001; Murphy, 2001). ABA levels are increased by short photoperiod (and low temperature) during dormancy induction in *Betula* and other species (e.g., Lång et al., 1994; Rinne, Welling, and Kaikuranta, 1998), but this has commonly been related to the development of freezing tolerance. In general, ABA is involved in the cellular adaptation to stress, particularly to those stresses that cause dehydration of the cytoplasm, such as drought, salinity and freezing (Levitt, 1980; Ingram and Bartels, 1996). In view of the present data, it might be speculated that in trees the photoperiod-induced increase in ABA could prepare the meristem for survival through winter and, indirectly, for spherosome-mediated release from endo-dormancy.

The uniform size of spherosomes is thought to be due to the presence of negatively charged ABA-inducible proteins, oleosins, in the half-unit membrane. Oleosins prevent coalescence with other oleosin-carrying spherosomes, particularly in dehydrated tissues (Murphy, 2001). ABA-insensitive mutants of *Arabidopsis* have a delayed oleosin production, larger spherosomes, and an altered fatty acid composition (Frandsen, Mundy, and Tzen, 2001). Keeping the size of spherosomes restricted may provide a sufficient surface area for the attachment of lipases. This would ensure the rapid mobilization of triacylglycerides, for example, during the germination of seeds (Frandsen, Mundy, and Tzen, 2001). We propose that one of the additional roles of spherosomes might be to function as vehicles for the efficient delivery of hydrolytic enzymes like 1,3-β-D-glucanase to the sphincter-complexes during cold exposure. The restricted size of the spherosomes in the endo-dormant buds of birch and populus could make the delivery of 1,3-β-D-glucanase to the minute sphincter-complexes more effective. The diameter of the external sphincter ring is estimated to be about 0.10 μm, while the spherosomes are 5 to 20 times larger. Considering that at one cell-cell interface hundreds of plasmodesmata are present, numerous small-sized spherosomes would be required for the simultaneous restoration of a sufficient number of plasmodesmata. Spherosomes larger than 0.5-2.0 μm would significantly lower the number of tangential contacts between spherosomes and plasmodesmata.

Since dormancy can also be broken by various near-lethal stresses (Wisniewski et al., 1996b; Rinne et al., 1997), it seems possible that these stresses unspecifically promote the restoration of the symplasmic network of the meristem. The cytoskeleton is reported to be involved in the trafficking of spherosomes (Murphy, 2001), and it seems possible that cold as well as near-lethal stresses influence the movement of spherosomes via configurational changes in the cytoskeleton. In addition, near-lethal stresses might enhance the *de novo* production of free cytoplasmic 1,3-β-D-glucanases as is documented for, for example, pathogen-induced stresses (Stone and Clarke, 1992).

Despite the likelihood that the spherosomes deliver their 1,3-β-D glucanases to the sphincter complexes, it is presently unclear how the enzyme comes into contact with its substrate because it has to pass the plasma membrane. Recently it has been shown, however, that in many plant species the half-membrane of the spherosomes carries caleosins, ABA-inducible calcium-binding proteins that are thought to mediate fission and fusion processes of membranes (Frandsen, Mundy, and Tzen, 2001; Murphy, 2001). If the presence of caleosins in spherosomes is a general feature, the spherosomes in the birch meristem might have the capacity to fuse with the plasma membrane, where it covers the sphincter ring, in order to directly release the enzyme into the ring space. The docking of spherosomes at the plasmodesmata might be facilitated by yet another calcium-binding protein, calreticulin, which has recently been localized to plasmodesmata (Baluška et al., 1999, 2001). Interaction between spherosome-attached caleosin and plasmodesmal calreticulin seems unlikely as long as the spherosomes remain dispersed in the cells, as is the case in the endo-dormant meristem. The chilling-induced displacement of spherosomes into the vicinity of the plasma membrane (Rinne, Kaikuranta, and van der Schoot, 2001) might suffice, however, to bring the spherosomes to their docking sites at the plasmodesmata. Such a simple, biophysically induced change in the location of the spherosomes seems an appropriate way to exert an environmental effect on a relatively inert endo-dormant system.

Spherosomes in birch were occasionally seen to form membranous connections to the plasmodesmata, but it is presently unclear whether this allows the passage of enzymes to the sphincter ring. Alternatively, enzymes might pass from the spherosome to the sphincter ring during the recycling of membrane lipids or phagy of entire pieces of plasma membrane (Figure 8D,E). This would explain the presence of myelin-like bodies in plasmodesma-associated spherosomes (Figure 8D), characteristic for autophagous vacuoles, for example, in root meristems and flower organs (Matile and Moore, 1968; Matile and Winkenbach, 1971). In addition, more substantial membrane configurations could be found in some spherosomes, which might have represented cortical ER-sacculi that were removed from the plasmodesmal entrance to make closer contact possible between an spherosome and a plasmodesma. That spherosomes can digest ER-sacculi seems plausible, considering that spherosomes can digest the inner leaflet of the ER membrane from which they arise (Figure 8A,B). This mechanism could be analogous to that suggested for sieve plate maturation, where the local removal of membrane lipids precedes the digestion of the 1,3-β-D-glucan collar around the plasmodesmata by 1,3-β-D-glucanase (Lucas, Ding, and van der Schoot, 1993). Autophagic activities would be expected considering that spherosomes are classically included in the lytic compartment of plant cells (Matile, 1975), and that their half-unit membrane can contain lipases among other hydrolases (Murphy, 2001).

DIAGNOSING AND CONTROLLING ENDO-DORMANCY

Although much phenological knowledge is available with regard to growth cessation, bud set, and bud burst, this information is of restricted value for the management of dormancy-related phenomena in agriculture and horticulture. It is well known that seedlings are most resistant to lifting, transplanting, storage, and other handling procedures when they have entered the endo-dormant state prior to these treatments (Lavender, 1991). A precise knowledge of meristem states would greatly improve the successful timing of the various practises. Assessing the endo-dormant state from phenological knowledge is further complicated by the fact that many tree species have photoperiodic eco-types that are adapted to local climatic conditions and become endo-dormant at different times (Pauley and Perry, 1954; Vaartaja, 1959; Howe et al., 1995).

Endo-dormancy not only plays a role in woody perennials but also in tubers and seeds, where meristems can enter endo-dormancy as a part of a developmental continuum (van Es and Hartmans, 1987; Bewley, 1997; Kaplan and Cooke, 1997). Endo-dormancy in tubers is of great importance for their keepability and storage. For example, in potato storage conditions are optimised to maintain the natural dormancy period as well as to delay sprouting when dormancy is lost (van Es and Hartmans, 1987; Burton, 1989; Burton, van Es, and Hartmans, 1992). Sprouting is not desired as it is accompanied by enhanced respiration and evaporation, and increased infection by pests and diseases, resulting in a considerable loss of weight and overall quality. Similarly to trees, prediction of dormancy breaking in potato tubers is difficult as the length of the dormancy period depends on the cultivar, the maturity of the tuber, the soil type and even the weather conditions during growth (van Es and Hartmans, 1987; Burton, van Es, and Hartmans, 1992).

As illustrated by the above examples the pinpointing of the inception and termination of the various phases of dormancy has been problematic due to the lack of features that are specific for any of these phases (Saure, 1985). Some recent studies have investigated whether cellular features could be used as a marker for the dormant state. For example, expression of cell-division-related genes was found to be very low in dormant axillary meristems of *Populus* (Rohde et al., 1997). Similarly, in five other woody species, beta-tubulin, which is required for cell division activities (Hepler and Hush, 1996), was found to be present in high amounts in proliferating apices but it was virtually absent in the 'dormant' buds (Bergervoet et al., 1999). In contrast, in pea certain mRNAs (PsDRM2) are highly expressed in dormant axillary buds (and other non-growing tissues) but considerably less in proliferating apices (Stafstrom et al., 1998). In two of the above studies (Rohde et al., 1997; Stafstrom et al., 1998) the axillary meristems were para-dormant, according to the standard terminology. They might be vulnerable to stress as para-dormant buds commonly lack tolerance to, for example, freezing stress (Rinne, Welling, and Kaikuranta, 1998). In case of the five woody species studied (Bergervoet et al., 1999), it is unclear whether the apical and axillary buds were endo-dormant, as the plants were stored at 2-4°C for 6 weeks or more before the meristems were analyzed. Chilling periods equal in temperature and duration were reported to release birch meristems from endo-dormancy to eco-dormancy (Rinne, Kaikuranta, and van der Schoot, 2001). Although these eco-dormant buds can remain highly stress and freezing tolerant, they loose tolerance very easily when conditions improve (Proebsting, 1963; Lavender, 1991). In conclusion, cell cycle based markers as well as mRNAs characteristic of non-growing tissues might permit discrimination between non-growing and proliferative states but, unfortunately, they cannot pinpoint the endo-dormant state.

In our view, the lack of a comprehensive theory of dormancy might have prevented the discovery of the specific characters of endo-dormancy. We proposed a theory that explains dormancy in terms of suspended cell-cell communication, and identified observable characters for the endo-dormant state in the form of sphincter-complexes (Figure 9). Since the basic 1,3-β-D-glucan sealing mechanism of the symplasm is evolutionary highly conserved (Stone and Clarke, 1992; Lucas, Ding,

FIGURE 9. Network model of dormancy cycling. Environmental cues impinge on cell biological processes that alter the cell communication patterns in specific ways. During the 'online' state cells are symplasmically coupled in patterns that reflect the organization of the meristem. Arrival of a photoperiodically generated signal results in the activation of 1,3-β-D-glucan synthase at the plasmodesmata and the symplasmic uncoupling of all meristem cells, resulting in an 'offline' state. Simultaneously, 1,3-β-D-glucanase-containing spherosomes are produced, scattered in the cytoplasm of all cells. In birch, this situation is reversible by return to long photoperiod during the first two weeks. After two weeks the meristem becomes fixed in a state of endo-dormancy. An adequate chilling period displaces the 1,3-β-D-glucanase-containing spherosomes into the vicinity of the plasma membrane and the plasmodesmal sphincter-complexes. In addition, free 1,3-β-D-glucanase is present in the cortical cytoplasm. Subsequently, the sphincter-complexes disappear from the plasmodesmata, which return to their conductive state. A subsequent rise in temperature and water availability then amplifies signal exchange between the re-coupled cells to give rise to the networking activity of the 'online' state.

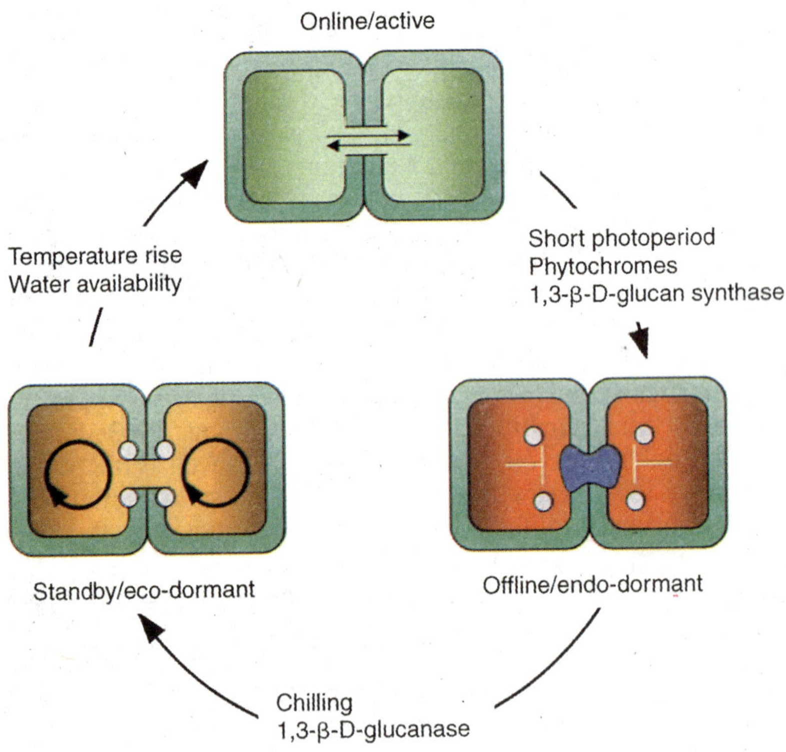

and van der Schoot, 1993), it might be expected that all higher plants that display endo-dormancy exploit plasmodesmal sphincter-complexes. If so, then the sphincter-complexes might represent the long awaited 'specific observable characters' (Saure, 1985). In principle, however, any trait uniquely associated with 'endo-dormancy sphincters' could function as such. For the purpose of diagnosing the endo-dormant state the challenge for the future is to develop an easily applicable test that could be the basis for deciding, for example, when to lift seedlings or trees, when to commence the industrial processing of ware potatoes, and when to plant seed potatoes. The potential universality of this lock mechanism–the sphincter-complex–also makes it an obvious target for the manipulation by molecular technologies. In principle, endo-dormancy could be prevented, shortened or lengthened by manipulating the production or targeting of the relevant enzymes that control the sphincter-complexes. Once the generation and the nature of the putative signals that trigger these mechanisms are known they could be subject to control as well. Against this background it is of interest to establish whether the development of freezing tolerance would also be influenced, as in the natural environment endo-dormancy and freezing-tolerance occur in tandem.

CONCLUSIONS

The above results show that plasmodesmata in the shoot apical meristem, like those in other parts of the plant (see McLean, Hempel, and Zambryski, 1997), are highly responsive to environmental conditions. Considering the vital function of the symplasmic network in meristem functioning, it is not surprising that environmental modulations of this network evoke developmental transitions such as found in dormancy cycling and flowering (Rinne and van der Schoot, 1998; Gisel et al., 1999; Rinne, Kaikuranta, and van der Schoot, 2001). Understanding how this is achieved in the case of dormancy cycling requires the identification of the cell biological processes that regulate the formation and the breakdown of sphincter-complexes. In addition, future investigations need to resolve the nature of the 'environmentally activated' input pathways, such as phytochrome–and temperature mediated signal cascades that evoke the installation of sphincter-complexes during endo-dormancy development. These signal cascades might be partly different in plant species that are adapted to different natural environments.

REFERENCES

Arora, R. and M.E. Wisniewski. (1994). Cold acclimation in genetically related (sibling) deciduous and evergreen peach (*Prunus persica* [L.] Batsch). II. A 60 kDa bark protein in cold acclimated tissues of peach is heat stable and related to the dehydrin family of proteins. *Plant Physiology* 105: 95-101.

Arora, R., M.E. Wisniewski and R. Scorza. (1992). Cold acclimation in genetically related (sibling) deciduous and evergreen peach (*Prunus persica* [L.] Batsch). I. Seasonal changes in cold-hardiness and polypeptides of bark and xylem tissues. *Plant Physiology* 99: 1562-1568.

Atkins, D., R. Hull, B. Wells, K. Roberts, P. Moore and R.N. Beachy. (1991). The tobacco mosaic virus 30K movement protein in transgenic tobacco plants is localized to plasmodesmata. *Journal of General Virology* 72: 209-248.

Badelt, K., R.G. White, R.L. Overall and M. Vesk. (1994). Ultrastructural specialisations of the cell wall sleeve around plasmodesmata. *American Journal of Botany* 81: 1422-1427.

Baluška, F., J. Šamai, R. Napier and D. Volkman. (1999). Maize reticulin localizes preferentially to plasmodesmata in root apex. *Plant Journal* 19: 481-488.

Baluška, F., F. Cvrèková, J. Kendrick-Jones and D. Volkman. (2001). Sink plasmodesmata as gateways for phloem unloading. Myosin VIII and calreticulin as molecular determinants of sink strength? *Plant Physiology* 126: 39-46.

Beebe, D.U. and W.A. Russin. (1999). Plasmodesmata in the phloem-loading pathway. Pp. 261-293. In: A.J.E. van Bel and W.J.P. van Kesteren (eds.), *Plasmodesmata. Structure, Function, Role in Cell Communication*. Springer-Verlag, Berlin-Heidelberg, New York.

Bergervoet, J.H.W., H-C. Jing, J.W.E. van den Hout, R. Delmondez de Castro, B.P.A.M. Kunneman, R.J. Bino and S.P.C. Groot. (1999). Expression of beta-tubulin during dormancy induction and release in apical and axillary buds of five woody species. *Physiologia Plantarum* 106: 238-245.

Bergmans, A.C.J., A.D. de Boer, J.W.M. Derksen and C. van der Schoot. (1997). The symplasmic coupling of L2-cells diminishes in early floral development of Iris. *Planta* 203: 245-252.

Berggren, B. (1984). Ultrastructure of the histological zones in growing vegetative buds of *Salix* spp. *Nordic Journal of Botany* 4: 771-789.

Bewley, J.D. (1997). Seed germination and dormancy. *Plant Cell* 9: 1055-1066.

Blackman, L.M., J.D.I. Harper and R.L. Overall. (1999). Localization of a centrin-like protein to higher plant plasmodesmata. *European Journal of Cell Biology* 78: 297-304.

Burton, W.G. (1989). *The Potato*. Longman, UK.

Burton, W.G., A. van Es and K.J. Hartmans. (1992). The physics and physiology of storage. In: P. Harris (ed.), *The Potato Crop*. Chapman & Hall, London.

Carpenter, R. and E.S. Coen. (1995). Transposon induced chimeras show that floricaula, a meristem identity gene, acts non-autonomously between cell layers. *Development* 121: 19-26.

Citovsky, V., M.L. Wong, A.L. Shaw, B.V. Venkataram Prasad and P. Zambryski. (1992).Visualization and characterization of tobacco mosaic virus movement protein binding to single-stranded nucleic acids. *The Plant Cell* 4: 397-411.

Clark, S.E., M.P. Running, and E.M. Meyerowitz. (1995). *CLAVATA3* is a specific regulator of shoot and floral meristem development affecting the same processes as *CLAVATA1*. *Development* 121: 2057-2067.

Clark, S.E., R.W. Williams and E.M. Meyerowitz. (1997). The *CLAVATA1* gene encodes a putative receptor kinase that controls shoot and floral meristem size in Arabidopsis. *Cell* 89: 575-585.

Cleland, R.E., T. Fujiwara and W.J. Lucas. (1994). Plasmodesmal-mediated cell-to-cell transport in wheat roots is modulated by anaerobic stress. *Protoplasma* 178: 81-85.

Cline, M.G. (2000). Execution of the auxin replacement apical dominance experiment in temperate woody species. *American Journal of Botany* 87: 182-190.

Cooke, T.J., M.S. Tilney and L.G. Tilney. (1996). Plasmodesmal networks in apical meristems and mature structures: Evidence for both primary and secondary formation of plasmodesmata. Pp. 471-488. In: Smallwood, M., Knox, J.P. and Bowles, D.J. (eds) *Membranes: Specialised Functions in Plants*. Bios Scientific Publishers, Oxford.

Coville, F.V. (1920). The influence of cold in stimulating the growth of plants. *Journal of Agricultural Research* 20: 151-160.

Crawford, K.M. and P.C. Zambryski. (1999). Plasmodesmata signaling: many roles, sophisticated statutes. *Current Opinion in Plant Biology* 2: 382-387.

Crawford, K.M. and P.C. Zambryski. (2001). Non-targeted and targeted protein movement through plasmodesmata in leaves in different developmental and physiological states. *Plant Physiology* 125: 1802-1812.

Delmer, D.P. (1987). Cellulose biosynthesis. *Annual Review of Plant Physiology* 38: 259-290.

Delmer, D.P., M. Volokita, M. Solomon, U. Fritz, W. Delphendahl and W. Herth. (1993). A monoclonal antibody recognises a 65-kDa higher plant membrane polypeptide which undergoes cation-dependent association with callose synthase in vitro and colocalizes with sites of high callose deposition in vivo. *Protoplasma* 176: 33-42.

Dennis, F.G. (1994). Dormancy. What we know (and don't know). *HortScience* 29: 1249-1255.

Ding, B., M.O. Kwon and L. Warnberg. (1996). Evidence that actin filaments are involved in controlling the permeability of plasmodesmata in tobacco mesophyll. *The Plant Journal* 10: 157-164.

Ding, B. (1997). Cell-to-cell transport of macromolecules through plasmodesmata: A novel signalling pathway in plants. *Trends in Cell Biology* 7: 5-9.

Ding, B. (1999). Tissue preparation and substructure of plasmodesmata. Pp. 37-49. In: A.J.E. van Bel and W.J.P. van Kesteren (eds.), *Plasmodesmata. Structure, Function, Role in Cell Communication*. Springer-Verlag, Berlin-Heidelberg, New York.

Emilsson, B. (1949). Studies on the rest period and dormant period in the potato tuber. *Acta Agricultarae Suecana* 3: 189-284.

Esau, K. (1977). *Anatomy of Seed Plants*. John Wiley, New York.

Esau, K. and J. Thorsch. (1985). Sieve plate pores and plasmodesmata, the communication channels of the symplast: Ultrastructural aspects and developmental relations. *American Journal of Botany* 72: 1641-1653.

Evans, M.M.S. and M.K. Barton. (1997). Genetics of angiosperm shoot apical meristem development. *Annual Review of Plant Physiology and Plant Molecular Biology* 48: 673-701.

Flemming, A.J., T. Mandel, S. Hofmann, P. Sterk, S.C. de Vries and C. Kuhlemeier. (1992). Expression patterns of a tobacco lipid transfer protein gene within the shoot apex. *The Plant Journal* 2: 855-862.

Flemming, A.J., T. Mandel, I. Roth and C. Kuhlemeier. (1993). The patterns of gene expression in the tomato shoot apical meristem. *The Plant Cell* 5: 297-309.

Fletcher, J.C. and E.M. Meyerowitz. (2000). Cell signalling within the shoot meristem. *Current Opinion in Plant Biology* 3: 23-30.

Frandsen, G.I., J. Mundy and J.T.C. Tzen. (2001). Oil bodies and their associated proteins, oleosin and caleosin. *Physiologia Plantarum* 112: 301-307.

Furner, I.J., J.F-X. Ainscough, J.A. Pumfrey and L.M. Petty. (1996). Clonal analysis of the late flowering fca mutant of *Arabidopsis thaliana*: Cell fate and cell autonomy. *Development* 122: 1041-1050.

Garrill, A. G.P. Findlay and S.D. Tyerman. (1996). Mechanosensitive ion channels. Pp. 247-260. In: Smallwood, M., Knox, J.P. and Bowles, D.J. (eds.), *Membranes: Specialised Functions in Plants*. Bios Scientific Publishers, Oxford.

Ghoshroy, S., R. Lartey, J. Sheng, and V. Citovsky. (1997). Transport of proteins and nucleic acids through plasmodesmata. *Annual Review of Plant Physiology and Plant Molecular Biology 48*: 27-50.

Gisel, A., S. Barella, F.D. Hempel and P.C. Zambryski. (1999). Temporal and spatial regulation of symplastic trafficking during development in *Arabidopsis thaliana* apices. *Development* 126: 1879-1889.

Grabski, S., A.W. de Feijter and M. Schindler. (1993). Endoplasmic reticulum forms a dynamic continuum for lipid diffusion between contiguous soybean root cells. *The Plant Cell* 5: 25-38.

Hallé, F., R.A.A. Oldeman and P.B. Tomlinson. (1978). Tropical trees and forests: An architectural analyzes. Springer, Berlin.

Hantke, S.S., R. Carpenter and E.S. Coen. (1995). Expression of floricaula in single layers of periclinal chimeras activates downstream homeotic genes in all layers of floral meristems. *Development* 121: 27-35.

Hayashi, T., S.M. Read, J. Bussel, M. Thelen, F-C. Lin, R.M. Brown Jr. and D.P. Delmer. (1987). UDP-glucose: 1,3-β-glucan synthases from mung bean and cotton. Differential effects of Ca^{2+} and Mg^{2+} on enzyme properties and on macromolecular structure of the glucan product. *Plant Physiology* 83: 1054-1062.

Heide, O.M. (1993). Dormancy release in beech buds (*Fagus sylvatica*) requires both chilling and long days. *Physiologia Plantarum* 89: 187-191.

Hepler, P.K. (1982). Endoplasmic reticulum in the formation of the cell plate and plasmodesmata. *Protoplasma* 111: 121-133.

Hepler, P.K. and J.M. Hush. (1996). Behavior of microtubules in living plant cells. *Plant Physiology* 112: 455-461.

Holdaway-Clarke, T.L., N.A. Walker, P.K. Hepler and R.L. Overall. (2000). Physiological elevations in cytoplasmic free calcium by cold or ion injection result in transient closure of higher plant plasmodesmata. *Planta* 210: 329-335.

Howe, G.T., W.P. Hacket, G.R. Furnier and R.E. Klevorn. (1995). Photoperiodic responses of a northern and a southern ecotype of black cottonwood. *Physiologia Plantarum* 30: 695-708.

Imlau, A., E. Truernit and N. Sauer. (1999). Cell-to-cell and long-distance trafficking of the green fluorescent protein in the phloem and symplastic unloading of the protein into sink tissues. *The Plant Cell* 11: 309-322.

Ingram, J. and D. Bartels. (1996). The molecular basis of dehydration tolerance in plants. *Annual Review of Plant Physiology and Plant Molecular Biology* 47: 377-403.

Jackson, D., B. Veit and S. Hake. (1994). Expression of maize KNOTTED1 related homeodomain genes in the shoot apical meristem predicts patterns of morphogenesis in the vegetative shoot. *Development* 120: 405-413.

Kakimoto, T. and H. Shibaoka. (1992). Synthesis of polysaccharides in phragmoplasts isolated from tobacco BY-2 cells. *Plant and Cell Physiology* 33: 353-361.

Kaplan, D.R. and T.J. Cooke. (1997). Fundamental concepts in the embryogenesis of dicotyledons: A morphological interpretation of embryo mutants. *The Plant Cell* 9: 1903-1919.

Kauss, H. (1985). Callose biosynthesis as a Ca^{2+} regulated process and possible relations to the induction of other metabolic changes. *Journal of Cell Science* (supplement) 2: 89-103.

Kauss, H. (1996). Callose synthesis. In: M. Smallwood, J.P. Knox and D.J. Bowles (eds.), *Membranes: Specialized Functions in Plants*. Bios scientific publishers, Oxford, UK.

Kawase, M. (1961). Growth substances related to dormancy in Betula. *Proceedings of the American Society of Horticultural Science* 78: 532-544.

Krabel, D., W. Eschrich, S. Wirth and G. Wolf. (1993). Callase-(1,3-β-D-glucanase) activity during spring reactivation in deciduous trees. *Plant Science* 93: 19-23.

Lang, G.A. (1987). Endo-, para-and eco-dormancy: Physiological terminology and classification for dormancy research. *HortScience* 22: 371-377.

Lang, G.A. (1994). Dormancy. The missing links: Molecular studies and integration of regulatory plant and environmental interactions. *HortScience* 29: 1255-1263.

Lång, V., E. Mäntylä, B. Welin, B. Sundberg and E.T. Palva. (1994). Alterations in water status, endogenous abscisic acid content, and expression of rab 18 gene during the development of freezing tolerance in *Arabidopsis thaliana*. *Plant Physiology* 104: 1341-1349.

Laufs, P., C. Jonak and J. Traas. (1998). Cells and domains: Two views of the shoot apical meristem in *Arabidopsis*. *Plant Physiology and Biochemistry* 36: 33-45.

Laufs, P., O. Grandjean, C. Jonak, K. Kiêu and J. Traas. (1998). Cellular parameters of the shoot apical meristem in *Arabidopsis*. *The Plant Cell* 10: 1375-1389.

Lavender, D.P. (1991). Measuring phenology and dormancy. Pp. 403-422. In: J.P. Lassoie, and T.M. Hinckley (eds.), *Techniques and Approaches in Forest Tree Ecophysiology*. CRC Press, Boca Raton, FL.

Levitt, J. (1980). Responses of Plants to Environmental stress. Vol. 1. Chilling, Freezing and high temperature stresses. 2nd edn. Academic Press, New York.
Lu, P., R. Porat, J.A. Nadeau and S.D. O'Neill. (1996). Identification of a meristem L1 layer-specific gene in *Arabidopsis* that is expressed during embryonic pattern formation and defines a new class of homeobox genes. *The Plant Cell* 8: 2155-2168.
Lucas, W.J., B. Ding and C. van der Schoot. (1993). Plasmodesmata and the supracellular nature of plants. *The New Phytologist* 125: 435-476.
Lucas, W.J., S. Bouche-Pillon, D.P. Jackson, L. Nguyen, L. Baker, B. Ding and S. Hake. (1995). Selective trafficking of KNOTTED1 homeodomain protein and its mRNA through plasmodesmata. *Science* 270: 1980-1983.
Matile, Ph. (1975). *The Lytic Compartment of Plant Cells.* Wien: Springer.
Matile, Ph. and H. Moore. (1968). Vacuolation: origin and development of the lysosomal apparatus in root-tip cells. *Planta* 80: 159-175.
Matile, Ph. and F. Winkenbach. (1971). Function of lysosomes and lysosomal enzymes in the senescing corolla of the morning glory (*Ipomea purpurea*). *Journal of Experimental Botany* 22: 759-771.
McLean, B.G., F.D. Hempel and P.C. Zambryski (1997). Plant intercellular communication via plasmodesmata. *The Plant Cell* 9: 1043-1054.
Medford, J.I. (1992). Vegetative apical meristems. *The Plant Cell* 4: 1029-1039.
Menzel, G., K. Apel and S. Melzer. (1996). Identification of two MADS box genes that are expressed in the apical meristem of the long-day plant *Sinapis alba* in transition to flowering. *The Plant Journal* 9: 399-408.
Metzger, J.D. (1988). Localisation of the site of perception of thermoinductive temperatures in *Thlaspi arvense* L. *Plant Physiology* 88: 424-428.
Metzger, J.D. (1996). A physiological comparison of vernalization and dormancy chilling requirement. Pp. 147-155. In: G.A. Lang (ed.), *Plant Dormancy. Physiology, Biochemistry and Molecular Biology.* CAB International, Wallingford, UK.
Meyerowitz, E.M. (1999). Plants, animals and the logic of development. *Trends in Cell Biology* 9: 65-68.
Mollenhauer, H. and J. Morré. (1987). Some unusual staining properties of tannic acid in plants. *Histochemistry* 88: 17-22.
Müller, W.A. (1997). *Developmental Biology.* Springer, New York.
Murphy, D. (2001). The biogenesis and functions of lipid bodies in animals, plants and microorganisms. *Progress in Lipid Research* 40: 325-438.
Muthalif, M.M. and L.J. Rowland. (1994). Identification of dehydrin-like proteins responsive to chilling in floral buds of blueberry (*Vaccinium*, section *Cyanococcus*). *Plant Physiology* 104: 1439-1447.
Napoli, C.A., C.A. Beveridge and K.C. Snowden. (1999). Reevaluating the concepts of apical dominance and the control of axillary bud outgrowth. *Current Topics in Developmental Biology* 44: 127-169.
Newman, I.V. (1965). Pattern in the meristems of vascular plants. III. Pursuing the patterns in the apical meristem where no cell is a permanent cell. *Journal of the Linnean Society* (Botany) 59: 185-214.
Nishimura, A., M. Tamaoki, Y. Sato and M. Matsuoka. (1999). The expression of tobacco knotted1-type class 1 homeobox genes corresponds to regions predicted by the cytohistological zonation model. *The Plant Journal* 18: 337-347.

Noodén, L.D. and J.A. Weber. (1978). Environmental and hormonal control of dormancy in terminal buds of plants. Pp. 221-268. In: M.E. Cutter (ed.), *Dormancy and Developmental Arrest*. Experimental Analysis in Plants and Animals. Academic Press, New York.

Northcote, D.H., R. Davey and J. Lay. (1989). Use of antisera to localise callose, xylan and arabinogalactan in the cell-plate, primary and secondary walls of plant cells. *Planta* 178: 353-366.

Ohana, P., M. Benziman and D.P. Delmer. (1993). Stimulation of callose synthesis in vivo correlates with changes in intracellular distribution of the callose synthase activator β-furfuryl-β-glucoside. *Plant Physiology* 101: 187-191.

Ohana, P., D.P. Delmer, J.C. Steffens, D.E. Matthews, R. Mayer and M. Benziman. (1991). β-furfuryl-β-glucoside: An endogenous activator of higher plant UDP-glucose: 1,3-β-glucan synthase. *Journal of Biological Chemistry* 266: 13742-13745.

Ohana, P., D.P. Delmer, G. Volman, J.C. Steffens, D.E. Matthews and M. Benziman. (1992). β-furfuryl-β-glucoside: an endogenous activator of higher plant UDP-glucose: 1,3-β-glucan synthase. Biological activity, distribution, and in vivo synthesis. *Plant Physiology* 98: 708-715.

Olesen, P. (1979). The neck constriction in plasmodesmata. Evidence for a peripheral sphincter-like structure revealed by fixation with tannic acid. *Planta* 144: 349-358.

Olesen, P. (1980). A model of a possible sphincter associated with the plasmodesmal neck region. *European Journal of Cell Biology* 22: 250.

Olesen, P. and A.W. Robards. (1990). The neck region of plasmodesmata: general architecture and some functional aspects. Pp. 145-170. In: A.W. Robards, W.J. Lucas, J.D. Pitts, H.J. Jongsma, and D.C. Spray (eds.), *Parallels in Cell-to-Cell Junctions in Plants and Animals*. NATO ASI Series, H46, Springer-Verlag, Berlin-Heidelberg, New York.

Oparka, K.J., A.G. Roberts, P. Boevink, S. Santa Cruz, I. Roberts, K.S. Pradel, A. Imlau, G. Kotlizky, N. Sauer and B. Epel. (1999). Simple, but not branched, plasmodesmata allow the nonspecific trafficking of proteins in developing tobacco leaves. *Cell* 97: 743-754.

Ormenese, S., A. Havelange, R. Deltour and G. Bernier. (2000). The frequency of plasmodesmata increases early in the whole shoot apical meristem of *Sinapis alba* L. during floral transition. *Planta* 211: 370-375.

Ormenese, S., A. Havelange, G. Bernier and C. van der Schoot. (2002). The shoot apical meristem of *Sinapis alba* L. expands its central symplasmic field during the floral transition. *Planta* 215: 67-68.

Overall, R.L. and L.M. Blackman. (1996). A model of the macromolecular structure of plasmodesmata. *Trends in Plant Science* 1: 307-311.

Overall, R.L. (1999). Substructure of plasmodesmata. Pp. 129-148. In: A.J.E. van Bel and W.J.P. van Kesteren (eds.), *Plasmodesmata. Structure, Function, Role in Cell Communication*. Springer-Verlag, Berlin-Heidelberg, New York.

Pauley, S.S. and T.O. Perry. (1954). Ecotypic variation of the photoperiodic response in *Populus*. *Journal Arnold Arboretum* 35: 167-188.

Pétel, G. and M. Gendraud. (1996). Processes at the plasma membrane and plasmalemma ATPase during dormancy. Pp. 233-243. In: G.A. Lang (ed.), *Plant Dormancy*.

Physiology, Biochemistry and Molecular Biology. CAB International, Wallingford, UK.

Pennazio, S., P. Roggero and M. Conti. (1999). Plasmodesmata and viruses. A centenary story. *Microbiologia* 22: 389-404.

Perbal, M-C., G. Haughn, H. Seadler and Z. Schwarz-Sommer. (1996). Non-cell autonomous function of the Antirrhinum floral homeotic proteins DEFICIENS and GLOBOSA is exerted by their polar cell-to-cell trafficking. *Development* 122: 3433-3441.

Powell, L.E. (1987). Hormonal aspects of bud and seed dormancy in temperate-zone woody plants. *HortScience* 22: 845-850.

Proebsting, E.L. (1963). The role of air temperatures and bud development in determining hardiness of dormant Elberta peach fruit buds. *Proceedings of the American Society for Horticultural Science* 83: 259-269.

Purvis, O.N. (1940). Vernalization of fragments of embryo tissues. *Nature* 145: 462.

Purvis, O.N. (1961). The physiological analysis of vernalization. *Encyclopaedia of Plant Physiology* 16: 76-122.

Radford, J.E., M. Vesk and R.L. Overall. (1998). Callose deposition at plasmodesmata. *Protoplasma* 201: 30-37.

Rappaport, L. and N. Wolf. (1969). The problem of dormancy in potato tubers and related structures. In: H.W. Woolhouse (ed.), *Dormancy and Survival. Symposium of the Society for Experimental Biology* (no. XXIII), Cambridge University Press, Cambridge, UK.

Rinne, P., H. Hänninen, P. Kaikuranta, J.E. Jalonen and T. Repo. (1997). Freezing exposure releases bud dormancy in *Betula pubescens* and *B. pendula*. *Plant, Cell and Environment* 20: 1199-1204.

Rinne, P., A. Welling and P. Kaikuranta. (1998). Onset of freezing tolerance in birch (*Betula pubescens* Ehrh.) involves LEA proteins and osmoregulation and is impaired in an ABA-deficient genotype. *Plant, Cell and Environment* 21: 601-611.

Rinne, P.L.H., P.L.M. Kaikuranta, L.H.W. van der Plas and C. van der Schoot. (1999). Dehydrins in cold acclimated apices of birch (*Betula pubescens* Ehrh.): Production, localization and potential role in rescuing enzyme function during dehydration. *Planta* 209: 377-388.

Rinne, P.L.H. and C. van der Schoot. (1998). Symplasmic fields in the tunica of the shoot apical meristem coordinate morphogenetic events. *Development* 125: 1477-1485.

Rinne, P.L.H., P.M. Kaikuranta and C. van der Schoot. (2001). The shoot apical meristem restores its symplasmic organization during chilling-induced release from dormancy. *The Plant Journal* 26: 249-264.

Robards, A.W. (1976). Plasmodesmata in higher plants. Pp. 15-57. In: B.E.S. Gunning and A.W. Robards (eds.), *Intercellular Communications in Plants: Studies on Plasmodesmata*. Springer, Heidelberg.

Roberts, K. (1994). The plant extracellular matrix: In a new expansive mood. *Current Opinion in Cell Biology* 6: 688-694.

Rohde, A., M. van Montagu, D. Inze, and W. Boerjan. (1997). Factors regulating the expression of cell cycle genes in individual buds of Populus. *Planta* 201: 43-52.

Romberger, J.A. (1963). Meristems, growth and development in woody plants. *USDA Technical Bulletin*. 1293.

Sachs, T. (1991). *Pattern Formation in Plant Tissues.* Cambridge University Press, New York, USA.

Sakai, A. and W. Larcher. (1987). *Frost Survival of Plants: Responses and Adaptations to Freezing Stress.* Springer, New York.

Samish, R.M. (1954). Dormancy in woody plants. *Annual Review of Plant Physiology* 5: 183-203.

Saure, M.C. (1985). Dormancy release in deciduous fruit trees. *Horticultural Reviews* 7: 239-300.

Sawhney, V.K., P.J. Rennie and T.A. Steeves. (1981). The ultrastructure of the central zone cells of the shoot apex of *Helianthus annuus. Canadian Journal of Botany* 59: 2009-2015.

Schulz, A. (1999). Physiological control of plasmodesmal gating. Pp. 173-204. In: A.J.E. van Bel and W.J.P. van Kesteren (eds.), *Plasmodesmata. Structure, Function, Role in Cell Communication.* Springer-Verlag, Berlin-Heidelberg, New York.

Seeley, S.D. (1994). Dormancy. The Black Box. *HortScience* 29: 1248.

Sessions, A. Yanofsky, M.F., and D. Weigel. (2000). Cell-cell signaling and movement by the floral transcription factors LEAFY and APETALA1. *Science* 289: 779-781.

Shih, C.Y. and L.R. Rappaport. (1971). Regulation of bud rest in tubers of potato, *Solanum tuberosum* L. VIII. Early effects of gibberellin A3 and abscisic acid on ultrastructure. *Plant Physiology* 48: 31-35.

Stafstrom, J.P., B.D. Ripley, M.L. Devitt and B. Drake. (1998). Dormancy-associated gene expression in pea axillary buds. *Planta* 205: 547-552.

Steeves, T.A. and I.A. Sussex. (1989). *Patterns in Plant Development.* 2nd ed. Cambridge University Press, New York, USA.

Stewart, W.W. (1981). Lucifer dyes: highly fluorescent dyes for biological tracing. *Nature* 292: 17-21

Stone, B.A. and A.E. Clarke. (1992). *Chemistry and Biology of 1,3-β-Glucans.* La Trobe University Press, Australia.

Storms, M.M.H., C. van der Schoot, M. Prins, R. Kormelink, J.W.M. van Lent and R.W. Goldbach. (1998). A comparison of two methods of microinjection for assessing altered plasmodesmal gating in tissues expressing viral movement proteins. *The Plant Journal* 13: 131-140.

Sussex, I.M. (1989). Developmental programming of the shoot apical meristem. *Cell* 56: 225-229.

Sussex, I.M. and N.M. Kerk. (2001). The evolution of plant architecture. *Current Opinion in Plant Biology* 4: 33-37.

Thompson, W.W. and K. Platt-Aloia. (1985). The ultrastucture of the plasmodesmata of the salt glands of Tamarix as revealed by transmission and freeze-fracture electron microscopy. *Protoplasma* 125: 13-23.

Tomenius, K., D. Clapham and T. Meshi. (1987). Localization by immunogold cytochemistry of the virus-coded 30K protein in plasmodesmata of leaves infected with tobacco mosaic virus. *Virology* 160: 363-371.

Tucker, E.B. and W.F. Boss. (1996). Mastorpan induced intracellular Ca^{2+} fluxes may regulate cell-to-cell communication in plants. *Plant Physiology* 111: 459-467.

Turner, A., B. Wells and K. Roberts. (1994). Plasmodesmata of maize root tips: structure and composition. *Journal of Cell Science* 107: 3351-3361.

Vaartaja, O. (1959). Evidence for photoperiodic ecotypes in trees. *Ecological Monographs* 29: 91-111.

van Bel, A.J.E., S. Günther and W.J.P. van Kesteren. (1999). Plasmodesmata, a maze of questions. Pp. 1-26. In: A.J.E. van Bel and W.J.P. van Kesteren (eds.), *Plasmodesmata. Structure, Function, Role in Cell Communication.* Springer-Verlag, Berlin-Heidelberg, New York.

van der Schoot, C. (1994). The role of symplasmic organization and cell-cell communication in dormancy. P. 30. In: G.A. Lang (ed.), *1st International Symposium on Plant Dormancy*, 7-10 August 1994, Oregon State University, Corvallis, Oregon.

van der Schoot, C. (1996). Dormancy and symplasmic networking at the shoot apical meristem. Pp. 59-81. In: G.A. Lang (ed.), *Plant Dormancy. Physiology, Biochemistry and Molecular Biology.* CAB International, Wallingford, UK.

van der Schoot, C. and P. Rinne. (1999a). Networks for shoot design. *Trends in Plant Science* 4: 31-37.

van der Schoot, C. and P. Rinne. (1999b). The symplasmic organization of the shoot apical meristem. Pp. 223-242. In: A.J.E. van Bel and W.J.P. van Kesteren (eds.), *Plasmodesmata. Structure, Function, Role in Cell Communication.* Springer-Verlag, Berlin-Heidelberg, New York.

van Es, A. and K.J. Hartmans. (1987). Dormancy, sprouting and sprout inhibition. In: A. Rastovski, A. van Es et al., (eds). *Storage of Potatoes. Post-Harvest Behaviour, Store Design, Storage Practise, Handling.* Pudoc, Wageningen.

Vegis, A. (1964). Dormancy in higher plants. *Annual Review of Plant Physiology* 15: 185-224.

Waigman, E., A. Turner, J. Peart, K. Roberts and P. Zambryski. (1997). Ultrastructural analysis of leaf trichome plasmodesmata reveals major differences from mesophyll plasmodesmata. *Planta* 203: 75-84.

Weiser, C.J. (1970). Cold resistance and injury in woody plants. *Science* 169: 1269-1278.

Welling, A., P. Kaikuranta and P. Rinne. (1997). Photoperiodic induction of dormancy and freezing tolerance in *Betula pubescence.* Involvement of ABA and dehydrins. *Physiologia Plantarum* 100: 119-125.

White, R.G., K. Badelt, R.L. Overall and M. Vesk. (1994). Actin associated with plasmodesmata. *Protoplasma* 180: 169-184.

Wisniewski, M., T.J. Close, T. Artlip and R. Arora. (1996a). Seasonal patterns of dehydrins and 70-kDa heat shock proteins in bark tissues of eight species of woody plants. *Physiologia Plantarum* 96: 495-505.

Wisniewski, M., L.H. Fuchigami, J.J. Sauter, A. Shirazi and L. Zhen. (1996b). Near-lethal stress and bud-dormancy in woody plants. Pp. 201-210. In: G.A. Lang (ed.), *Plant Dormancy. Physiology, Biochemistry and Molecular Biology.* CAB International, Wallingford, UK.

Hormones and Endodormancy Induction in Woody Plants

Karen K. Tanino

SUMMARY. The scope of this review is limited to hormonal involvement in the induction of woody bud endodormancy and does not address dormancy maintenance or release. Knowledge of hormonal regulation of processes has become increasingly more complex, particularly with recent findings of auxin- and ethylene-triggered abscisic acid induction revealing many more responses mediated by ABA than originally considered. The intent of this review is to provide an account of the current state of knowledge of hormones in bud dormancy research within an historical context with a focus on promising avenues of research. While hormones are undoubtedly involved at some point during dormancy induction, meaningful progress in unraveling the mechanism of endodormancy-induction is difficult through this hormone-centered approach. The path to dormancy acquisition is a continuum. In many plants, this path begins far earlier than the traditionally-examined autumn period. Aside from more useful hormonal localization studies within the bud and utilization of mutants, continued gross analysis of hormone presence/absence during dormancy induction may not be fruitful. Examin-

---

Karen K. Tanio is affiliated with the Department Plant Sciences, 51 Campus Drive, College of Agriculture, University of Saskatchewan, Saskatoon, SK, S7N 5A8, Canada (E-mail: tanino@sask.usask.ca).

The author gratefully acknowledges the valuable editorial comments provided by Graham Simpson, Larry Gusta, and Erl Svendsen. Use of the figures contained in this review article is also appreciated.

[Haworth co-indexing entry note]: "Hormones and Endodormancy Induction in Woody Plants." Tanino, Karen K. Co-published simultaneously in *Journal of Crop Improvement* (Food Products Press, an imprint of The Haworth Press, Inc.) Vol. 10, No. 1/2 (#19/20), 2004, pp. 157-199; and: *Adaptations and Responses of Woody Plants to Environmental Stresses* (ed: Rajeev Arora) Food Products Press, an imprint of The Haworth Press, Inc., 2004, pp. 157-199. Single or multiple copies of this article are available for a fee from The Haworth Document Delivery Service [1-800-HAWORTH, 9:00 a.m. - 5:00 p.m. (EST). E-mail address: docdelivery@haworthpress.com].

http://www.haworthpress.com/web/JCRIP

Digital Object Identifer: 10.1300/J411v10n01_08

ing the regulation of the early molecular switch within key cells of the apical meristem or subtending tissues may provide greater insight into dormancy induction. Turgor pressure-dependent cellular differentiation within the apical meristem and the extrinsic and intrinsic factors regulating turgor may be central to that molecular switch. Simpler model systems may need to be utilized to test these hypotheses. Only then can the potential role of hormones in endodormancy induction be more clearly articulated.

KEYWORDS. Hormones, endodormancy, induction, buds, woody plants

The essence of nature is a true harmony of its parts. It is a balance between action and reaction at all levels of organization. Perhaps nowhere is this balance more evident than in the interaction of a plant with its environment. The stationary form of plant existence demands effective adaptation for survival. It is not surprising, therefore, that highly developed environmental sensor and hormonal response systems have evolved, and that the natural distribution of plant species reflects the degree of inherent plasticity and success of these adaptive systems (Tanino, 1990).

Branching is the most efficient adaptation to stationary life (Crabbe and Barnola, 1996). The subsequent evolution of the bud (about 400 million years ago–100 million years before the seed) provided a mechanism to limit branching and conserve energy (Crabbe, 1994; Crabbe and Barnola, 1996; Viemont and Crabbe, 2000). The induction, maintenance and release of bud dormancy are largely environmentally-regulated. Environmental mechanisms of dormancy regulation have been the topic of several thousand research articles to date. Since Hemberg (1949), hormones have been inextricably linked to woody bud dormancy and are implicated as a means by which plants respond to environmental cues. The path to endodormancy induction is a continuum, which in some plants, begins as early as bud break in the spring. While it has been tempting to "explain" bud dormancy on the basis of hormonal regulation alone, dormancy is controlled by numerous integrated plant structures and functions (Simpson, 1990; Crabbe, 1994).

Although considerable progress has been achieved, large functional gaps of knowledge still exist in bud dormancy induction research. The fundamental problem involves separating dormancy processes from those functions important to freezing/dehydration tolerance as well as distinguishing cause versus effect. Localization of hormones at the cellular level and assessing sensitivity vs. quantity of hormone are also key points. Doorenbos (1953) defined dormancy as when " a tissue predisposed to elongate does not do so." While the currently used dormancy terminology (Lang et al., 1987) seems more definitive and appealing from a conceptual basis, it is not reflective of cellular processes. Active cellular alterations in the apical meristem, including more refined knowledge of the mitotic cycle should be assessed within a dormancy context (Macdonald, 2000). Ultrastructural and cellular physiological changes of the developing bud should be examined in parallel with defined points along the dormancy continuum.

The scope of this review is limited to hormonal involvement in the induction of woody bud endodormancy and does not address dormancy maintenance or release. Knowledge of hormonal regulation of processes has become increasingly more complex, particularly with recent findings of auxin- and ethylene-triggered abscisic acid (ABA) induction revealing many more responses mediated by ABA than originally considered (Sharp et al., 2000; Grossman and Hansen, 2001). The intent of this review is to provide an account of the current state of knowledge of hormones in bud dormancy research in an historical context with a focus on promising avenues of research.

EARLY HORMONAL RESEARCH

Torsten Hemberg (1949) is generally credited as the first to demonstrate growth inhibiting substances in resting buds: they declined from October to February and decreased within both the outer two scales as well as the de-scaled bud. Samish (1954) proposed that dormancy induction and release were due to the interaction between the fluctuating inhibitor levels and the stable auxin concentrations. Wareing (1956) conducted experiments in which leaves and buds were independently exposed to long- or short-day treatments and found an inhibitory substance was transmitted from leaves to buds under short-day conditions, but not under long-days. Thus, leaves under short-days prevented apical growth and not as an indirect result of the failure to produce some factor under long-day conditions. Nitsch (1957a), however, proposed that the arrest of growth under short-days could be caused either by the synthesis of an active inhibitor or by the lack of a growth-stimulating substance. Recent evidence involving GA suggests this latter case may be true (Olsen et al., 1995c).

Phillips and Wareing (1958) suggested bud scales had a dormancy-breaking mechanism since scales were alive and grew with elongating sycamore buds. The water content of the terminal bud decreased from as early as June 1, reaching a low by November 1. The growth inhibitor quantitatively changed through the course of the year with greatest inhibiting activity in early winter and least coinciding with active growth in May-June. Dennis and Edgerton (1961) reported no correlation between inhibitors and rest (endodormancy). However, inhibitor levels, largely confined to the bud scales, decreased with dormancy release. An inhibitor β was identified from leaf extracts of *Acer pseudoplatanus* under short-day conditions (Robinson et al., 1963). The term 'dormin' was later proposed for substances that appeared to function as endogenous dormany-inducers (Eagles and Wareing, 1963). Additional evidence in support of 'dormin' regulation of dormancy was provided from experiments where the formation of resting buds was induced by applying the short-day extract of *Betula* leaves to actively growing seedlings (Eagles and Wareing, 1964; Cornforth et al., 1965).

These initial studies were followed by a series of publications in the following three decades which monitored endogenous levels of hormones within whole buds, leaves, stems, cambium and root tissues under natural fall and dormancy-inducing controlled-environment conditions. The β-inhibitor dormin was later identified as abscisic acid (ABA). Although recent evidence indicates ABA may not be produced in the roots (Shinozaki, pers. comm.), the existing general view is ABA biosynthesized in both roots and leaves (Sauter et al., 2001). Long-distance signaling from root to leaf tissues in terms of ABA flux within the plant is regulated by pH changes and environmental stress. Freundl et al. (2000) showed root uptake of ABA is reduced by Casparian band formation in the hypodermis. Under hydroponic or tissue culture conditions, ABA may be lost to the medium if it is alkaline relative to the root cortex. Lack of response to exogenous ABA application may be a function of these effects. In addition, acidification of the rhizosphere and the root apoplast increased apoplastic transport of ABA, which was proposed to cross the endodermis through the transpiration stream (Freundl et al., 1998). Therefore, root-zone pH and ABA concentration may modify root-to-shoot signaling as they affect apoplastic transport of ABA. Sauter et al. (2001) also make important points including the significance of assessing ABA in specific cells. For example, older lite ture quantified ABA in fully expanded leaf laminae rather than in the growing cells of leaf bases. Thus, the correlation between ABA accumulation in fully expanded cells and growth may not be relevant. An interesting study by Strauβ et al. (2001) examined the subcellular distribution of ABA using a single-chain ABA-binding antibody in potato and determined leaves of young plants developed in the absence of ABA while leaves of older plants developed in the presence of ABA. Importantly, they demonstrated that exogenously applied ABA was distributed differently from compartmentalized endogenous ABA within the cell. ABA-binding proteins might exist in the cytosol and/or endoplasmic reticulum and prevent ABA distribution based on cellular pH gradient alone. Therefore, localization of hormones such as ABA and GA and examination of pH alterations during endodormancy induction in the apex are critical aspects to understanding the regulation of growth and dormancy. Discrepancies also occur in plants from regions with relatively mild winters are compared with cold winters (Dennis and Edgerton, 1961). Consistent with this concept, endogenous ABA content was higher in Scotland in both normal and barley mutants compared to Spain (Romagosa et al., 2001). Additional problems include the use of lateral versus terminal buds; unseparated determinate and indeterminate growth; differential photoperiodic response of young versus mature leaves; long-day photoperiod treatment just above the critical photoperiod (Eagles and Wareing, 1963; Hocking and Hillman, 1975); use of whole buds versus partitioned bud tissues; and undefined stage of dormancy at the time of sampling (Fuchigami et al., 1982; Fuchigami and Wisniewski, 1997). Results were also complicated by the finding that hormone levels varied from basal to apical portions and changed from the fall through winter (Saure, 1985; Rinne et al., 1994b) and the determination of whether the observed changes were regulating factors or were simply a result of dormancy induction. While many detailed studies have been performed, it is against rigorous criteria that the literature should be examined.

EVIDENCE SUPPORTING ABA INVOLVEMENT IN DORMANCY INDUCTION

The majority of evidence for ABA involvement in dormancy induction originates from monitoring endogenous hormone levels under either natural field or controlled-environment conditions. Until leaf abscission, dormant peach flower buds increased their ABA and total acid fractions suggesting an interaction between ABA and growth promoting substances (Corgan and Peyton, 1970). End of rest was correlated with gibberellin-like activity and a reduction in inhibitor level in apricot (Ramsay and Martin, 1970). The ratio of free vs. bound ABA varied on an annual cycle with the highest ratio in autumn and lowest coinciding with bud burst in blackcurrant and beech buds (Wright, 1975). Esterified ABA progressively increased while free ABA decreased between mid-September to mid-March in birch buds (Harrison and Saunders, 1975). Seeley and Powell (1981) measured ABA in whole apple buds and in shoot tips and found the highest concentrations in shoot tips. ABA increased in the growing shoot tips to a high of 900 µg/kg after which shoot elongation ceased and terminal bud formation was induced. Rodriguez and Sanchez-Tames (1986) found the greatest concentration of free ABA in hazel buds occurred in October (Spain) approximately at the onset of dormancy followed by a decline starting in December, reaching its lowest value at bud burst in March. The reverse was true for IAA. Using phenological observations in the field, and assaying *Corylus* shoot tissue, the IAA/ABA balance favored ABA during mid winter but shifted to IAA in the spring (Rodriguez et al., 1991). While absolute levels of ABA were not correlated with dormancy throughout the season, ABA increased in both lateral buds, leaves and xylem sap of *Salix viminalis* during the summer prior to endodormancy induction (Barros and Neill, 1989). Although the specific dormancy stage was not quantified, Pongsomboon et al. (1990) reported high levels of ABA and ABA-like inhibitors in flower buds of 3 peach cultivars at the onset of dormancy; they subsequently decreased after 2 weeks. Similarly, ABA concentration was high during the early stage of dormancy then declined in flower buds of peach (Ramina et al., 1995). Quamaruddin et al. (1993) assayed ABA in needles of a southern and northern population of *Picea abies* during dormancy induction under 8-hour photoperiods for 6 wk followed by lowered night temperatures for the final 6 wk. ABA increased after 3 d of inductive conditions in the southern population but ABA levels of the northern population were most responsive to the lowered night temperatures. Colauzzi et al. (1993) found ABA increased during the initial part of dormancy and levels were enhanced with decreasing temperature. Rinne et al. (1994b) differentiated between the response of basal versus terminal buds of *Betula pubescens*. Terminal buds expressed a cyclic seasonal rhythm of ABA content with maximum levels detected between August to November. By contrast, the ABA content of basal buds was relatively constant during this period. Their results also supported a relationship between tissue water status and ABA concentration. ABA levels in both leaves and lateral buds of *Betula pubescens* increased under short days and water stress (Rinne et al., 1994a). Aneja et al. (1996) compared ABA levels and examined its catabolism in mature cocoa leaves and young flush leaves. Mature leaves expressed higher ABA catabolism and higher endogenous ABA levels compared to young leaves during the dormant period.

Xylem sap and stem tissues have also been monitored. Davison and Young (1974) followed ABA levels in the xylem sap of peach throughout one season: ABA peaked with leaf fall and prior to bud swell after which ABA content markedly declined. ABA content of *Vitis* canes and sap was high in November and decreased through to bud burst (Kubota et al., 1986). By contrast, gibberellic acid (GA)-like substances and cytokinins were low in November and increased until bud burst. However, dormancy stages were not identified. Alvim et al. (1976) followed seasonal changes in ABA and cytokinins (zeatin riboside-like compounds) in willow xylem sap and found the onset of dormancy was associated with low levels of cytokinin and high levels of ABA. In various species of wild grape that varied in timing of dormancy acquisition, Mochioka et al. (1996) determined that low shoot water content and the highest ABA levels coincided with the time of deepest endodormancy in each species. Tafazoli and Beyl (1993) reported highest ABA concentration coincided with deepest dormancy in *Actinidia arguta* 'Ananasnaja' stem tissue. The GA biosynthetic inhibitor, paclobutrazol was more effective than uniconizole in increasing ABA concentration: ABA content was lowest in March at budbreak.

Exogenous application of ABA can induce dormant bud development. Foliar application of aqueous solutions of D,L abscisin fed continuously onto leaves of *Betula pubescens*, *Acer pseudoplatanus*, and *Ribes nigrum* induced formation of typical resting buds under long-day conditions (El-Antably et al., 1967). Treatment of hydroponically grown *Alnus glutinosa* plants with ABA induced formation of resting buds after 30 to 60 d under 16 h photoperiods (Watts et al., 1987). Free *cis*-ABA accumulated in shoot apices and leaves while inactive bound *trans*-ABA accumulated in roots and nodules. According to Blake et al. (1990), ABA hastened terminal bud formation in *Picea engelmannii* Parry and *Pinus contorta* Dougl. when exposed to short day (SD) and low temperature conditions. *In vitro* levels of ABA in microcuttings of *Cedrus libani* were higher in dormant buds compared to actively growing buds (Piola, 1998). ABA application also induced formation of dormant buds at temperatures normally conducive to growth. Le Bris et al.

(1999) found endodormant cells of *Rosa hybrida* were arrested in the G1 phase. By contrast, ABA arrested cells in the G2 phase and buds remained dormant.

The reduced form of glutathione has been implicated in breaking endodormancy (Fuchigami and Nee, 1987). Glutathione has also been associated with bud dormancy induction where ABA induced both GSSG and dormancy development in grape buds (Tohbe et al., 1998). Exogenous GSSG application inhibited budbreak. In addition, GST (glutathione S-transferase) has 9-*cis*-epoxycarotenoid dioxygenase activity (Iuchi et al., 2000) which is involved in ABA biosynthesis (Taylor et al., 2001). Both the accumulation of ABA and 9-*cis*-epoxycarotenoid dioxygenase were induced under drought stress in cowpea (Iuchi et al., 2000).

EVIDENCE OF ABA INVOLVEMENT IN OTHER PLANT SYSTEMS

Karssen et al. (1983) utilized crosses with mutant lines of *Arabidopsis thaliana* exhibiting an absence of seed dormancy and low levels of ABA. They demonstrated a dual origin of ABA in seeds and determined the onset of dormancy correlated well with the presence of the embryonic ABA fraction but not with the maternal tissue ABA. By contrast, Hole et al. (1989) reported ABA was required for dormancy induction in intact embryos in kernels of *Zea mays* but suggested maternal ABA was involved. Karssen et al. (1983) inferred exogenously-applied ABA may not be effective unless the hormone reaches the site of dormancy regulation within the embryo: once dormancy was induced, ABA was no longer required. Le Page-Degivry (1990) also demonstrated that dormancy induction was dependent upon ABA synthesis and that ABA was not necessarily correlated with the entire dormancy cycle in *Helianthus annuus* seeds. Pinfield et al. (1987) separated the effects of testa versus embryo-imposed dormancy and reported chilling temperatures relieved testa-imposed dormancy but not embryo dormancy. They further suggested ABA may be involved in the regulation of embryo dormancy which was differentiated from testa-imposed dormancy. Using fluridone, an ABA inhibitor, Kim et al. (1994) promoted leaf formation and prevented dormancy development in *Lilium speciosum* plantlets under tissue culture. ABA application induced scale formation and required a chilling period for subsequent re-growth. In aseptic cultures of temperate grass species, Jamsran et al. (1999) showed dormant bud formation was only induced with high exogenous ABA treatment when combined with elevated sucrose levels. Growth cessation and tillering was observed in plants with lower ABA, sucrose and mannitol levels.

EVIDENCE AGAINST ABA INVOLVEMENT IN DORMANCY INDUCTION

ABA declined under both warm and cold chilling requirement conditions although only the latter alleviated dormancy. Lenton et al. (1972) did not find any increase in ABA content of *Betula pubescens*, *Acer pseudoplatanus*, and *Acer rubrum* tissue upon transfer to dormancy-inducing conditions. Loveys et al. (1974) found short day inductive conditions did not increase ABA or its metabolites in leaves of mature *Betula lutea*. However, a decrease in leaf diffusive resistance was noted upon transfer of plants to short day conditions. During and Bachmann (1975) reported maximal ABA values coincident with maximum endodormancy (October 16). However, when bud break was significantly reduced (September 16), ABA levels were relatively low. Powell (1976) did not detect increased ABA levels in shoot tips of *Malus hupehensis* Rehd., *Malus domestica* Borkh. 'Northern Spy', and *Betula papyrifera* Marsh grown under short days.

In grape buds, extractable ABA content was not correlated with rest and changes in free versus bound ABA also did not correlate with bud response (Iwasaki and Weaver, 1977). ABA levels in buds and stems of *Acer saccharum* were highest in July and August and progressively declined through to October but increased following leaf fall and subsequently declined again after December (Dumbroff et al., 1979). They concluded that ABA was not the primary endogenous factor responsible for initiation of the dormant state. Alvim et al. (1979) measured ABA under short photoperiod in roots, sap, leaves and apical tissues of *Salix viminalis* L. but could not detect any correlation with dormancy induction. Similarly, neither ABA nor IAA cambial concentrations appeared related to dormancy (Savidge and Wareing, 1984). According to Wignall and Browning (1988), decreased concentrations of ABA after mid-August indicated the level of this hormone was not responsible for the entry of the cambium into winter dormancy in *Quercus robur* L. but may be related to stress-induced ABA epicormic bud dormancy. Dormancy induction and break were not correlated with ABA levels in whole buds of *Salix viminalis* L. (Barros and Neill, 1986, 1989). In these studies, peak ABA levels in lateral buds, leaves and xylem sap occurred in July, preceding bud dormancy induction. Therefore, the possibility exists that ABA is simply involved very early in the dormancy cycle–before the typical reduction of bud break can be observed. Thus, understanding subtle cellular changes in the developing buds well before endodormancy induction is critical to determining what role ABA plays in dormancy processes.

Moreover, changing tissue sensitivity may be as important as hormonal level. According to Nadel et al. (1991), bud growth response to tissue culture media containing no hormone, IAA, GA_3 or BA varied from November to February with the most enhanced effect in February. Borkowska and Powell (1982) found response of bud tissue to 10^{-4} M ABA tissue culture medium varied depending on stage of bud growth and may relate to differential sensitivity of the tissue to hormones. Rinne et al. (1998) also reported buds to be more sensitive to ABA application in fall and least sensitive in spring.

Perry and Hellmers (1973) and Hocking and Hillman (1975) were unable to induce dormant bud formation through application of ABA. However, apical growth was inhibited. Exogenously applied ABA under long-day conditions to leaves of *Betula pubescens* and *Alnus glutinosa* did not induce resting buds (Hocking and Hillman, 1975). According to Heide (1986), ABA application to apical buds of *Picea abies* L. Karst. under 24 h photoperiod induced growth cessation but not normal endodormant buds. In *Pinus contorta*, ABA levels peaked in the cambium in April and declined thereafter. IAA levels increased in midsummer and also subsequently declined. Welling et al. (1997) sprayed ABA onto *Betula pubescens* Ehrh. leaves and increased endogenous ABA content but did not induce growth cessation or dormancy. ABA appeared to be more related to increasing freezing tolerance than growth regulation.

In *Betula pubescens,* SD induced a transient ABA elevation that did not occur under 95% relative humidity or with fluridone application (Welling et al., 1997). However, after 21 d at 95% RH and SD conditions, dormancy was still induced. Perhaps the most compelling evidence against ABA involvement comes from studies that utilize ABA-deficient mutants. Rinne et al. (1998) studied *Betula pubescens* ABA-deficient mutants and reported the wild type expressed elevated ABA levels before acclimation under SD or natural field conditions. This was accompanied by tissue desiccation, osmotic adjustment, and accumulation of Group 2 LEA proteins. By contrast, the mutant type had reduced water loss, defective osmoregulation, absence of Group 2 LEA proteins and reduced tolerance to low temperature stress. However, SD-conditions were still able to induce bud dormancy in ABA-deficient mutants.

Rodriguez et al. (1994) utilized evergreen peach in which lateral but not terminal buds become dormant. This continued apical growth is considered to be controlled by a single homozygous recessive gene. It would be interesting to determine ABA/GA levels in these buds under inducing and non-inducing daylength conditions. According to Arora et al. (1992) and Rowland and Arora (1997), a 16 kDa protein was associated with the deciduous type but not the mutant evergreen peach. This 16 kDa protein appears to be a bark-storage protein and not directly related to dormancy.

EVIDENCE SUPPORTING GA INVOLVEMENT

'Dormin' appeared to antagonize the activity of GA in coleoptile tests (Thomas et al., 1965). While the coleoptiles did not respond to GA in the absence of the inhibitor, in the presence of the inhibitor the coleoptiles became GA-requiring. Maage (1986) sprayed raspberry cv. Veten plants with paclobutrazol (GA biosynthetic inhibitor) and induced growth cessation and early dormancy acquisition. In 'Red Haven' peach, GA, IAA, and cytokinins exhibited different concentrations over the season with lowest levels between August and October (Psota et al., 1989).

GA_1-like and GA_8-like compounds were associated with intermittent flushing while GA_9-like and GA_{32}-like compounds increased during bud dormancy (cessation of extension) and floral initiation (Chen, 1994). Plasmamembrane-bound G proteins have now been identified in plants (see Fujisawa et al., 2001 for a review). These proteins interact with receptor molecules; in mammals they regulate vesicle trafficking and are associated with the nuclear membrane and the cytoskeleton. Literature is still sparse in plants, but G proteins appear to participate in gibberellin signaling at low concentrations and are involved in normal organ development in rice.

EVIDENCE AGAINST GA INVOLVEMENT

Root pruning and $GA_{4/7}$ + root pruning treatments retarded shoot elongation and terminal bud development (Owens et al., 1985). $GA_{4/7}$ alone did not affect shoot or apical development but did enhance flowering and the authors speculated on the role of nutrient diversion to the apex in flower induction. Root pruning did not affect cell division but reduced cell expansion, slowed mitotic activity and apical growth. Growth versus dormancy: results from their study did not support the hypothesis that the rate of apical meristem growth has a strong effect on bud development. In Douglas fir, apical meristems from both control and $GA_{4/7}$-treated trees were similar and followed the normal sequence of bud-scale initiation, differentiation, and leaf initiation (Owens et al., 1986).

HORMONES AND A TRANSLOCATABLE DORMANCY FACTOR

From the time of Hemberg (1949) and Wareing (1956), several lines of evidence have supported a translocatable factor from leaves during dormancy induction: Fuchigami et al. (1971) demonstrated a translocatable cold hardiness promoter in grafted ecotypes of *Cornus sericea* L. exposed to short or long days. By contrast, Perry and Hellmers (1973), found no translocatable dormancy factor was detected when Massachusetts and Florida red maple ecotypes were grafted. This response may have been species-specific. Defoliation of apple after harvest contributes to both dormancy avoidance (Janick, 1974) and reduction of chilling requirement of blueberry vegetative buds (Darnell et al., 1992). Mechanical defoliation of sour cherry trees prior to the onset of dormancy prevented free and bound ABA accumulation in buds but had no subsequent effect on rest (Mielke and Dennis, 1978). Barros and Neill (1989) defoliated willow from August to mid-September and reduced both ABA concentration and dormancy levels in buds. A detailed analysis of free and bound ABA (*cis*-ABA and *cis*-ABA-GE) during the dormancy cycle in *Vitis vinifera* leaves, buds and internodes found *cis*-ABA was initially induced in leaves during dormancy induction and subsequently remobilized to buds and internodes (Koussa et al., 1998). However, the collective literature does not clearly indicate that ABA is the translocatable dormancy factor.

Dormancy can be induced in plants by photoperiod, temperature, and water stress. Furthermore, molecular connections between dehydration and low temperature signaling pathways involve both ABA-dependent and ABA-independent pathways. Therefore, the various environmental triggers of dormancy induction will be addressed within a hormonal context in the following sections.

DORMANCY INDUCTION: PHOTOPERIODIC ECOTYPES AND PHYTOCHROME SIGNAL TRANSDUCTION

The importance of photoperiod in growth responses of woody plants has long been documented (Klebs, 1914 cited by Phillips, 1941; Garner and Allard, 1924; Sylven, 1942 cited by Wareing, 1956; Nitsch, 1957a,b; Sarvas, 1967 cited by Koski and Sievanen, 1984; Dormling et al., 1968; Weiser, 1970; Perry, 1971; Juntilla, 1980). Extensive studies of photoperiodic ecotypes include *Betula*, *Salix*, *Populus*, and *Cornus* species. Vegetative growth and the onset of dormancy are both controlled by photoperiod. The tissues responsive to photoperiod are in young leaves, just reaching full expansion, rather than old mature leaves (Nitsch, 1957b). This response is not universal, however. While dormancy was primarily controlled by the shoot apex in *Betula pubescens,* the situation was reversed in *Acer pseudoplatanus* and *Rhus pseudacacia* where induction of dormancy was determined by conditions imposed on mature leaves (Wareing, 1956).

Irving (1969) reported ABA-like levels (R_f values) were highest when *Acer negundo* was treated with either short days or long days plus 5°C nights. Conversely, GA-like activity was highest in leaves under long-day treatment. ABA in needles increased on exposure to combined SD and 20°/10°C day/night temperatures compared to constant 20°C conditions for northern populations but not southern populations of *Picea abies* (Qamaruddin et al., 1993). Although a one day exposure to SD was sufficient to induce 100% budset in a northern population of *Picea abies*, a transient peak of free ABA in needles was only observed after 4 d of SD exposure (Qamaruddin et al., 1995). By contrast, free ABA levels did not peak until 8 d of SD exposure in southern populations. A less abrupt photoperiod change but one still below the critical photoperiod induced a less defined peak for the northern populations. Wake and Fennell (1996) reported protein formation during dormancy induction in *Vitis riparia*. ABA did not appear to be involved in the photoperiodic control of dormancy in *Betula pubescens* but was more directly involved with freezing tolerance (Welling et al., 1997). However, support for photoperiodically-induced dehydrins (34 and 36 kDa) was found. These proteins were less abundant in buds of plants under 95% RH or fluridone treatment.

According to Johansen et al. (1986), exogenous ABA application did not prevent apical growth and no elevation of endogenous ABA was observed in whole plant extracts (leaves, stems and buds) from either SD or LD-grown *Salix pentandra* plants from two latitudes. Perry and Hellmers (1973) found exogenous ABA application induced growth cessation but did not produce normal bud scales or terminal bud formation in ecotypes of red maple.

Reciprocal grafts among different photoperiodic ecotypes of *Salix pentandra* indicate the principle regulation of growth response resides in the shoots (Juntilla, 1988). However, this response was also modified by roots. A follow-up study by Olsen et al. (1994) found similar concentrations of GA_8 and GA_{19} in both roots and shoots while GA_1 and its immediate precursor GA_{20} were lower in roots than in shoots. The levels of GA_{19}, GA_1, and GA_{20} were quantitatively correlated with SD-induced cessation of apical growth of *Salix pentandra* (Olsen et al., 1995a). GA_1 and GA_{20} levels were reduced within two days of SD treatment in both leaf and stem tissue. Further, GA_1 was higher in young leaves and GA_{19} was 5-10 times higher in stem tissue than leaves and roots under both long and short days. GA_{19} levels in the upper stem were more sensitive to SD treatment than the lower stem and levels decreased by 50% after 14 d exposure. Photoperiod did not affect the metabolism of GA_{19} and GA_1 (Olsen et al., 1995b). Localization studies were performed and maximum levels of GA_1 were found 5-20 mm below the apex which represented twice that of other segments (Olsen et al., 1995c). This response occurred within 5 d exposure to a short photoperiod, before growth cessation could be observed. Throughout 15 d of SD, IAA levels gradually declined in segments from the apex to 20 mm below this tissue and was proposed to be another factor in dormancy induction (Olsen et al., 1997a). Overexpression of an oat phytochrome A gene (PHYA) transformed into *Populus tremula x tremuloides* caused continued growth under SD conditions (Olsen et al., 1997b). IAA, GA_{53}, GA_{19}, GA_{20}, GA_1, and GA_8 levels were similar under 24 h and 10 h photoperiods in the transformed plants compared with reduced levels of these hormones in the wild type. Photoperiod and GA_1 influenced the cell divisions of the subapical tissues below the eumeristem (Olsen et al., 1999).

The phytochrome family of photoreceptors control photoperiodic induction of vegetative bud set, growth cessation, dormancy induction, and cold hardiness in tree species (Howe et al., 1998). Neff et al. (2000) outline 4 major approaches used to identify phytochrome signaling components. These include: (a) the direct and rapid responses of ion fluxes at the plasmamembrane on exposure to red/far red light; (b) pharmacological approaches identifying downstream elements in the phytochrome signal transduction pathway of heterotrimeric G proteins, cyclic

GMP, calcium nucleotide diphosphate kinase 2, calcium and transcriptional regulators (Briggs and Olney, 2001); (c) yeast hybrid screens identifying PKS1 and NDPK2. PKS1 appears to be a negative regulator of phytochrome B and NDPK1 to be a positive regulator of phytochromes A and B; and (d) mutants with altered responses to phytochrome signaling. In addition to the cytoplasm, phytochromes have now been localized to the nucleus (Sakamoto and Nagatani, 1996) as well as chloroplasts, and mitochondria (Moysset et al., 2001). Kircher et al. (1999) showed that red light treatment induced nuclear translocation of both phyA-GFP and phyB-GFP from the cytosol in transgenic tobacco and that far red inhibited phyB-GFP nuclear import but not phyA-GFP. Evidence is increasing for a cytosolic-based physiologically inactive Pr form in the dark and light-dependent nuclear import of Pfr upon light conversion from Pr. Thus, light-regulated differential nucleocytoplasmic partitioning of phytochrome A and B may represent another control point for growth regulation. Phytochromes A and B migrate into the nucleus under various stimuli (Neff et al., 2000). Ni et al. (1999) demonstrated that once in the nucleus, the protein PIF3 physically interacts only with the Pfr form of phytochrome and this complex dissociates if the Pfr form is converted back to Pr by red light. Furthermore, Pfr activates transcription factors in the cytoplasm (Merkle, 2001). Light-dependent nuclear translocation of transcription factors was shown to be dependent upon phosphorylation status (Harter et al., 1994 cited by Merkle, 2001). These transcription factors were termed Common Plant Regulatory Factors (CPRFs) which bind to the G-box promoter elements and play regulatory roles in expression of light-dependent genes. A phytochrome-binding protein, Phytochrome Kinase Substrate 1 (PKS1) was identified as a substrate for light-regulated phytochrome kinase activity *in vitro* (Fankhauser et al., 1999). *In vivo* tests indicate PKS1 is phosphorylated (light-dependent process) by phytochrome and appears to be a negative regulator of phytochrome B signaling. Understanding of the signal transduction from phytochrome A (sensitive to FR light) to bud dormancy induction might be achieved through the use of new *Arabidopsis* mutants (Ballesteros et al., 2001; Bagnall and King, 2001). Gene products of phytochrome A signaling are categorized according their cellular localization in the cytoplasm, plastids or nucleus (Ballesteros et al., 2001). The wide range of protein responses and locations indicate complex signaling cascades by FR induction of phytochrome A.

Merkle (2001) reviewed nuclear import/export of plant proteins and the regulation of signaling by regulation of nuclear transport. In addition to activity of transcription factors, changes in transcription factor localization (and their interacting proteins) between the nucleus and cytoplasm are also proposed as molecular switches in response to intra- and extra-cellular signals that affect growth, development or differentiation. Protein kinases are evolutionarily conserved, ubiquitous signaling molecules (Carroll et al., 2001). This group reported that cyclin-dependent kinase Pho85 controls the expression of genes also involved in reserve carbohydrate metabolism and stress such as antioxidants, HSP chaperones, redoxins, factors involved in degradation and vacuolar function. They suggest inhibition of Pho85 could be an important component of the Environmental Stress Response (ESR) where whole genome expression has been profiled under stress (Gasch et al., 2000).

Cyclin D2 gene expression itself is controlled by the transcriptional factor, Myc, and transcriptional activation by Myc was dependent on histone acetylation at the cyclin D2 promoter (Bouchard et al., 2001). Cell-cycle checkpoints serve to detect damaged or abnormal DNA and coordinate cell cycle progression to allow for DNA repair (Abraham, 2001). Cell-cycle signaling is thought to occur through ATM and ATR kinases in response to various environmental stress agents and thus indirectly are signal transduction pathways controlling plant growth and development (Figure 1). Check points regulate cell cycle progression at the G1, S_2 and G2 stages and have been a focus of cancer research. A key pathway controlling cell division and differentiation in animal cells is mediated by the retinoblastoma (RB) family of tumor suppressors which regulate the passage from G1 to S and through S phase (Umen and Goodenough, 2001). For the first time, they report the presence of RB in *Chlamydomonas reinhardtii*, a unicellular alga considered to be a precursor to land plants. RB in *Chlamydomonas* affects size-dependent cell cycle points: the cell cycle is initiated at a subnormal size with extra rounds of S phase/mitosis. Utilizing human colon cancer cells (Tanino et al., unpublished) found low physiological doses of ABA (which is not considered to play a functional role in humans) suppressed tumor cell growth by 72% and expressed a synergistic response in combination with calcium-channel blockers to completely halt growth under cell culture conditions.

Rohde et al. (1997) transformed poplar with cell-cycle genes and determined SD treatment triggered the most marked change in the cell cycle. Both photoperiod and temperature had opposite effects on the expression of the cell cycle promoters and suggested differential regulation of the cell cycle where photoperiod may accumulate more cells in G1 and temperature-accumulating cells in the G2 stage.

FIGURE 1. A generic cell-cycle checkpoint signaling pathway. (ROI) Reactive oxygen intermediate. Reprinted with permission from Cold Spring Harbor Laboratory Press. Abraham, R.T. (2001). Cell cycle checkpoint signaling through the ATM and ATR kinases. Genes and Development 15:2177-2196.

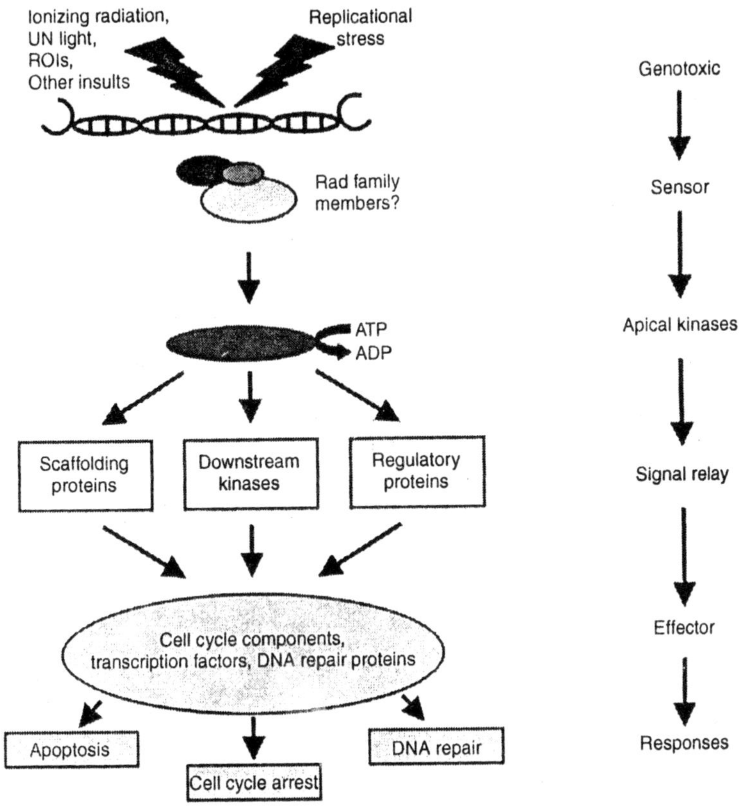

DORMANCY INDUCTION: TEMPERATURE

Recent quantitative genetic studies indicate that genetic differences in photoperiodic responses play a less significant role in explaining genetic differences in the timing of bud set under natural conditions and that factors such as temperature may play a larger part than first proposed (Howe et al., 2000). Earlier models describing the timing of growth cessation have taken both heat sum and night length into account (Koski and Sievanen, 1984). Faust et al. (1997) and Erez (2000) propose IAA and ABA influence correlative inhibition during dormancy induction. By contrast, endodormancy was suggested to be induced primarily by low temperature responses resulting in polyunsaturated FA membrane compositional and communication changes between the bud and the surrounding tissue.

Either short photoperiod or temperature (day/night temperature shifts) under long days could induce growth reduction in northern ecotypes but not southern ecotypes of *Salix pentandra* L. (Juntilla, 1980). Other results (Stevenson, 1994) also support these observations where northern dogwood ecotypes (*Cornus sericea* L.), can be induced into endodormancy under long days with high day/night temperature differences. The greater the day/night temperature shift, the earlier the dormancy induction. Southern dogwood ecotypes required short photoperiod for dormancy induction but day/night temperature shifts enhanced this response. The model system of *Cornus sericea* L. latitudinal ecotypes is based on the groundwork of many others (see Weiser, 1970 for a review; Hummel, 1982).

The differential dormancy response reported between field versus greenhouse conditions (Chen et al., 2000) may be attributed to the additional large diurnal temperature changes commonly occurring under natural conditions. Although ecotypic differences in timing of dormancy induction are similar between growth chamber and field conditions (Stevenson, 1994), the field acted to cluster northern from southern ecotype responses whereas the growth chamber evenly distributed ecotype dormancy induction. The dormancy cycle of each of the 192 individuals of this segregating F2 population has been now profiled and parental types are being characterized in terms of differential phytochrome expression and dormancy markers under both photoperiodic and temperature induction.

DORMANCY INDUCTION: MOISTURE STRESS

In order to understand the mechanism of dormancy, it may be important to understand the historical climatological context within which this phenomenon may have evolved. Baskin and Baskin (1998) provide an extensive account of the evolution, ecology and biogeography of seed dormancy. Since the Devonian period, temperatures have fluctuated and with these temperature shifts, environmental stress such as drought was evident. They consider that seed dormancy has developed in some species as a consequence of this dry season.

While angiosperms originated in humid tropical regions where temperature, daylength, and precipitation were stable through the year, one of the major evolutionary forces differentiating plant species was environmental change (Okubo, 2000). Some deciduous temperate-zone tree species probably expanded from these more tropical regions due to the cyclic wet/dry season and the subsequent adaptation of leaf abscission during the dry period. Frankie et al. (1974) found that 75% of the 113 tropical deciduous forest trees of Costa Rica lose their leaves during the dry season.

Water status of meristems of droughted plants is probably well buffered against drought stress (Sauter et al., 2001). They cite Thomas et al. (1988) where maintenance of turgor in intercalary meristems of grass under drought stress was demonstrated. Nevertheless, plant tissues are extremely sensitive to minor alterations in water potential (Levitt, 1980). Much evidence supports a relationship between ABA concentration and bud water content under either SD or water stress-induced dormancy in *Betula pubescens* (Rinne et al., 1994 a,b; Welling et al., 1997) and *Vitis vinifera* 'Merlot Noir' (Koussa et al., 1994). They concluded ABA to be involved in dormancy regulation through internal water stress. Depth of dormancy has been proposed to be related to endogenous levels of ABA (Tamura et al., 1993) and may be related to maintaining this water status. In *Betula pubescens,* SD induced a transient ABA elevation that did not occur under 95% relative humidity or with fluridone application (Welling et al., 1997). However after 21 d at 95% RH and SD conditions, dormancy was still induced. Water content of buds decreased after 4 d of SD exposure but remained stable throughout the 21 d sampling period in LD plants. No difference in percent water content was observed in leaves of either SD or LD-treated leaves. The initial decline in the growth rate of field-grown *Salix viminalis* trees was not induced by photoperiod but was coincident with declining leaf-water potentials. This was accompanied by a transient increase in ABA. Controlled environment SD treatments induced growth cessation but did not induce an increase in ABA. In seeds of *Arabidopsis thaliana* (L.) Heynh, limited access to water appeared to be the primary trigger for developmental arrest (Karssen et al., 1983).

Using magnetic resonance imaging, Faust et al. (1991) found endodormant buds had less free water than ecodormant buds, suggesting chilling requirement satisfaction was associated with the conversion of water from the bound to free state. Bound to free water changes occurred during chilling requirement satisfaction in the absence of any change in total water content (Parmentier et al., 1998). They reported blueberry cultivars with the deepest dormancy and longest chilling requirement also had the most bound water. Bound water was subsequently shown to increase with endodormancy and freezing tolerance in peach buds (Erez et al., 1998) and either short photoperiods or low temperatures could induce this response. They concluded that bound water status was associated with low temperature stress tolerance rather than directly to dormancy itself. Gardea et al. (1994) developed an improved NMR method to distinguish different stages of dormancy in *Vitis vinifera* 'Pinot Noir' in terms of bud water status. Using ^{1}H-NMR, Fennell et al. (1996) showed changes in bud water state occurred after 2 weeks of short day photoperiod exposure in *Vitis riparia* Michx. Fennell and Line (2001) determined increasing bound water with endodormancy in both the grape buds and cortex/gap tissue adjacent to the bud.

The dehydrin family of proteins is a diverse group of hydrophilic proteins ranging in size from 15 to 150 kDa found in a number of plant species and organs (Close et al., 1993). Various abiotic environmental stresses and dehydration stress in particular, as well as ABA can induce these proteins. A 60 kDa cold regulated dehydrin protein was monitored in both deciduous (dormant) and mutant evergreen (non-dormant) peach (Arora and Wisniewski, 1992). This protein was only present with cold acclimation under controlled-environment and field conditions. Welling et al. (1997) demonstrated short day treatment induced unique 34 kDa and 36 kDa dehydrin proteins which were less abundant under SD in the presence of fluridone or 95% relative humidity. Under LD treatment, these 2 dehydrins were not induced with ABA or water stress. Khanizadeh et al. (1994) determined that an elevation of hydrophilic amino acids and a corresponding reduction in hydrophobic amino acids occurred during dormancy induction in apple floral buds.

The observed reduction in plant water potential in bud tissue during dormancy induction may signal gene expression of both functional and regulatory proteins (Figure 2). While common genes are induced by both dehydration and low temperature, only certain pathways under dehydration induction are ABA-dependent (Figure 3). Further, the transient activation of the Arabidopsis MAP kinases ATMPK4 and ATMPK6 are both induced variously by low temperature, low humidity, hyper-osmolarity, touch and wounding but not by ABA (Ichimura et al., 2000). Seki et al. (2001) prepared a cDNA microarray using 1300 full length Arabidopsis cDNAs to target genes controlled by the DREB1A transcription factor (ABA-independent) under cold or drought stress. Roughly 40 proteins including aquaporin homologs, calcium binding EF hand proteins, Group II LEA proteins, heat shock dnaJ homolog,

FIGURE 2. Drought stress-inducible genes and their possible functions in stress tolerance and response. Gene products are classified into two groups. The first group includes proteins that probably function in stress tolerance (function proteins; open boxes), and the second group contains protein factors involved in further regulation of signal transduction and gene expression that probably function in stress response (regulatory proteins; shadowed boxes. (Reprinted with permission from Elsevier Science, Shinozaki K, Yamaguchi-Shinozaki K. *Molecular Responses to Drought Stress.* In Shinozaki K, Yamaguchi-Shinozaki, eds. Molecular Responses to Cold, Drought, Heat and Salt Stress in Higher Plants. Austin: R.G. Landes Co., 1999: Figure 2.2, p. 28).

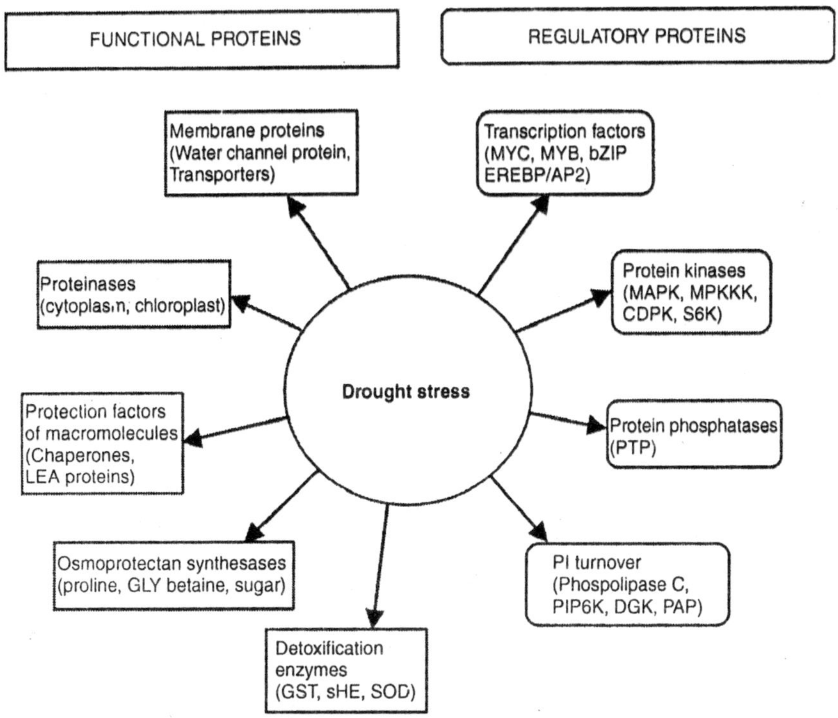

catalase 3, were induced by DREB1A and some proteins were induced by both, or neither, cold or drought stress.

Plant responses at the whole plant level to stress such as desiccation may occur faster than first thought. Recently, Pickard (2001) reported plant response of 200 μs, or less, to electrical events on tomato leaves. ABA application lowered the event rate while desiccation stress increased it. The electrical events were thought to be associated with water column fracture and retraction of xylem vessels.

FIGURE 3. Cellular signal transduction pathways between the initial drought-stress or cold-stress signal and gene expression in *Arabidopsis.* There are at least six signal transduction pathways: two (b, c) are ABA-dependent and four (a, d, e, f) are ABA-independent. Thin and thick arrows represent the minor and major signaling pathways that are involved in dehydration-responsive gene expression, respectively. Broken arrows represent the signaling pathways that are involved in low temperature stress responses (Reprinted from Current Opinions in Plant Biology, vol. 3, Shinozaki and Yamaguchi-Shinozaki, Molecular responses to dehydration and low temperature: difference and cross-talk between two stress signaling pathways, pp. 217-223, 2000, with permission from Elsevier Science).

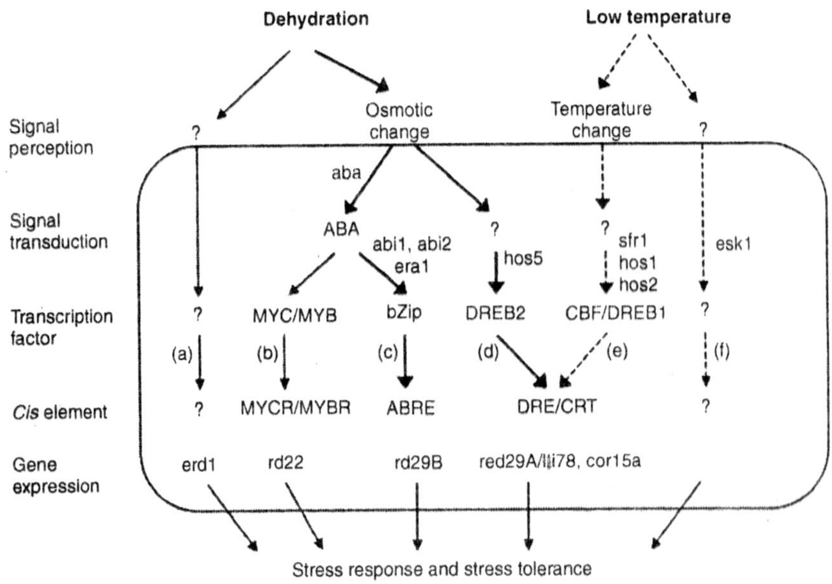

Accumulation of ABA in wilted leaves is generally a result of synthesis rather than release from a conjugated form or inhibition of ABA metabolism (Pierce and Raschke, 1981). Recent evidence indicates ABA treatment induces disruption of guard cell actin cytoskeleton through inactivation of the GTPase protein, AtRac1 (Lemichez et al., 2001). Root hydraulic conductivity and transpiration are two key elements determining water relations in plants. ABA had no effect on root hydraulic conductivity but demonstrated a pronounced effect on reducing stomatal conductance under non-stress conditions (Wan and Zwiazek, 2001). Aquaporins are a relatively new class of membrane-bound water channel proteins which can increase water transport across membranes 20 fold over diffusion alone (see Maurel, 1997 for a review). ABA, GA3, and desiccation induce these plasmamembrane intrinsic proteins (PIP) in *Arabidopsis* while salt stress down regulated PIP in *Mesembryanthemum crystallinum*. It is not known if dormancy induction may be in part regulated through dehydration stress via alterations in the quantity or activity of aquaporins. Localization of aquaporins in apical, bud scale and sub-apical meristematic regions will also be an important component of future research.

BUD SCALES

Cessation of growth is associated with changes occurring in the terminal growing point (Nitsch, 1957a). These changes vary widely and may include cases where: (a) terminal growth is switched into inflorescence (e.g., *Hibiscus syriacus*); (b) the terminal growing point dies and abscises (e.g., *Rhus typhina*, *Syringa vulgaris*, *Vitis* sp.); and (c) the terminal growing point remains alive but becomes enclosed in scales (e.g., *Populus*, *Platanus*, and *Cornus* sp.).

Many reports have shown dormancy to be an intrinsic property of the bud itself (Denney and Stanton, 1928; Johannsen, 1906 cited by Pollock, 1953; Schneider, 1968) and associated vascular connections to host tissues in both burst and unburst buds (Halim et al., 1988). The annual transition between foliage leaves and cataphyll or bud scales in black hickory was detailed by Foster (1932). The basic morphologic divergence between foliage leaf and cataphyll formation was determined during black hickory's morphogenetic phase of development: acropetal formation on their ventral margins of lateral leaflet primordia (leaf) versus ventral marginal growth with resultant wing formation (cataphyll). Foster suggested this differentiation between cataphyll and foliage leaf development was an inherent quality within the bud. This contrasts with earlier work by Klebs and Goebel in the 1880s in which initially identical primordia diverge in morphology as they mature (Goffinet and Larson, 1981). The findings of Goffinet and Larson (1981, 1982) and most others cited in this review support the latter view of morphological differentiation controlled by external factors.

A detailed study by Owens and Molder (1973) in Douglas fir indicated that mitotic activity and dormancy was related to carbohydrate levels within the bud and subtending shoot. Lack of sufficient carbohydrates may lead to the accumulation of cells in the G1 stage. In Douglas fir, the reduction of nuclei at the 4C level is characteristic of dormancy onset. The most important developmental period in determining the pathway of axillary bud development was late bud-scale initiation. Cell divisions first appear in the peripheral zone during bud scale initiation in Douglas fir (Owens and Molder, 1973) and white spruce (Owens et al., 1977). The peripheral zone of the apical meristem possesses the highest frequency distribution of DNA over all developmental stages throughout the year. The process of early to late bud scale initiation followed by early to late leaf initiation culminates in bud dormancy induction in these coniferous species. Dormancy itself is the end product of a long sequence of events, which in the case of conifers originates with bud break in the spring. Macdonald and Owens (1993) demonstrated a strong genetic control of bud scale initiation in response to short day treatment in Douglas fir. Moisture stress also induced the transformation from neo-formed leaves to bud scales even under long day conditions. This was considered a response to desiccation protection and short days were required for further bud development.

While it is clear that the transition from leaf to bud scale or cataphyll development is one of the first morphologic shifts in buds of woody plants, it is less certain whether this transition represents a key element in dormancy induction or whether hormones may be involved. Depending on plant species, the presence/absence of vascular connections or the presence of physical barriers from the bud to either the bud scales and/or subtending tissues are important to facilitate supercooling (Sakai, 1979; Ashworth, 1984; Chalker-Scott, 1992; Quamme et al., 1995; Kang et al., 1997; Jones et al., 2000). Whether these features are also important in dormancy acquisition or maintenance merits further investigation. Views of Champagnat (1989) and others (for review see Crabbe and Barnola, 1996; Faust et al., 1997) propose a mechanism of dormancy induction based on a communication block or permeability barriers between the bud and adjacent tissues. Champagnat (1989) determined a correlation between cell pH and the competitive ability of the bud to act as a sink tissue.

Cottonwood terminal shoots only form one bud-scale leaf under SD, controlled-environment conditions while under field conditions, 2-5 bud scales are produced (Goffinet and Larson, 1981, 1982). They suggest the prolonged dormancy under photoperiod and day/night temperature differences in the field permit more bud scale leaves to form. Thus, could the divergent ecotype responses reflect diversity in rates of bud scale maturation? Uemoto (1992) indicated the northern migration of *Camellia japonica* wild species was possible through the acquisition of cold resistance manifested by intensified bud dormancy. This intensified bud dormancy was reflected in an increase in bract number (4-5 in tropical populations to 7-12 in temperate types) as well as a reduction in auxin content. Baldwin et al. (2000a, Figure 4) induced bud scale maturation on *in vitro*-grown *Amelanchier alnifolia* Nutt. with 50 μM ABA and 5.5 μM BA. However, ABA alone only halted growth but did not produce mature bud scales. The absence of the auxin, NAA was a requirement for bud scale maturation in this system.

Goffinet and Larson (1982) indicate the leaf primordium first originating in the shoot apex after transfer to SD conditions ultimately differentiates into the first bud-scale leaf. The morphologic plasticity of this leaf to resume growth upon return to long day conditions is gradually lost with dormancy induction. The coordinated anatomical and cellular responses between bud scale maturation and apical meristematic alterations during dormancy induction may be a fruitful area of research.

According to Schneider (1968), the onset and duration of rest was dependent on bud scales in *Rhododendron* flower buds. When scales were removed before the onset of rest, flowers continued to elongate resulting in anthesis whereas intact buds stopped growing and remained in rest for 4 months. Removal of scales after the onset of rest induced bud break in *Potamogeton nodosus* Poir (Frank, 1966), floral elongation in

FIGURE 4. Induction of bud scale maturation in *Amelanchier alnifolia* Nutt. in tissue culture, (a) terminal bud scale maturation under 50 μM ABA, 5.5 μM BA; (b) lateral bud maturation under 50 μM ABA, 5.5 μM BA; (c) control lateral bud under 5.5 μM BA (adapted from Baldwin et al., 2000a).

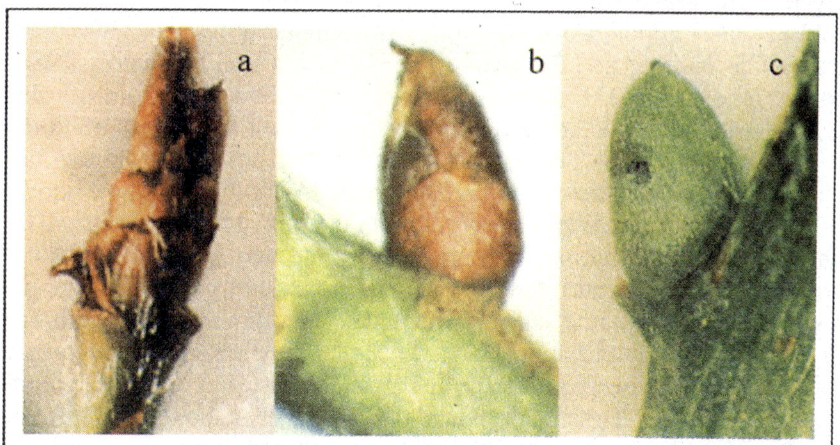

Rhododendron (Schneider, 1968), bud break in *Actinidia* (Lionakis and Schwabe, 1984), *Vitis* (Iwasaki and Weaver, 1977), and *Malus* (Swartz et al., 1984). Pre-mature scale removal also prevented further floral development.

Whether it is the removal of the bud scale itself or a result of the increased oxygen (Pollock, 1953) or light which promotes bud break is unclear. Wareing (1953) showed light-dependent bud break when de-scaled buds were exposed to varying levels of light. Wareing also reported the light penetration of the bud scales to be about 7 lux which varied from the base of the scales (35% light transmission) to the upper portions (7% light transmission).

HORMONAL INVOLVEMENT IN BUD SCALE DEVELOPMENT

Acid growth inhibitors were found in both the bud scales and the inner portions of *Fraxinus* buds and disappeared on release from rest (Hemberg, 1958). Growth inhibitors were confined to the bud scales and did not appear in floral organs until bud break in peach (Dennis and Edgerton, 1961). Inhibitors were found in both bud scales and apricot floral organs while gibberellin-like activity was confined to floral portions (Ramsay and Martin, 1970). Demaggio and Freeberg (1969) suggested under natural conditions, bud scales may inhibit apical activity since de-scaled buds grew in culture under both long and short days. ABA inhibited both GA-stimulated and photoperiodically-stimulated apical growth in culture.

Bud scales appear to house ABA (Tinklin and Schwabe, 1970; Spiers, 1973; Mielke and Dennis, 1975; Lionakis and Schwabe, 1984; Swartz et al., 1984). However, ABA levels in bud scales did not decrease with treatments inducing termination of rest in peach floral buds (Freeman and Martin, 1981). Mist treatments did advance rest breaking and also significantly decreased ABA content in floral primordia whereas ABA levels in bud scales were equivalent to controls. Less bound ABA was present in the bud scales compared to the bud meristems of *Vitis vinifera* L. (Iwasaki and Weaver, 1977). The reduction of ABA associated with chilling in bud scales, petal and gynoecium tissues was not related to dormancy release in azalea since plants which did not receive a cold treatment flowered equally rapidly (Pemberton et al., 1985). In a detailed study, Tinklin and Schwabe (1970) suggested bud inhibition in blackcurrant was induced by an inhibitor produced in the leaves which accumulated in the bud scales. Bud scales appeared to be the main source of inhibition of lateral bud break even before leaf fall. Chilling treatments depressed inhibitor content of the entire bud, including bud scales. Partial removal of bud scales indicated lack of oxygen and wounding were not the main cause of bud growth inhibition. Baldwin et al. (2000b) reported bud scales did not appear to regulate endodormancy in itself but may be more involved with inhibiting rapid spring bud break. Nevertheless, the initial morphologic shift from foliage leaf to bud scale induction probably represents the first committed step along the dormancy path.

ABA AND SIGNAL TRANSDUCTION

A review of ABA signal transduction by Leung and Giraudat (1998) reveals significant progress in downstream signaling elements. Specifically, reversible phosphorylation, regulation of cytosolic calcium levels and pH appear to be intermediates in ABA signal transduction. Recent findings by Xiong et al. (2001) identified an *Arabidopsis* mutant (fiery 1) which provided evidence that phosphoinositols (PI) also mediate ABA and stress signaling (cold, drought and salt stress) transduction. Turnover of PI was critical for attenuating ABA and stress signaling.

Endodormancy-associated proteins in woody plants appear to fall into two categories: (a) bark storage proteins or (b) dehydrins (see Rowland and Arora, 1997 for a review). cDNA from pea may be useful to assess dormancy in woody plants where PsDRM1 appears to have potential as a dormancy marker (Stafstrom et al., 1998). While PsDRM2 expression is not associated with dormancy, it is found in cold- and ABA-stimulated clones of alfalfa. Rohde et al. (1998) reported the isolation expression of ABI3 (ABscisic acid-Insensitive 3) homologous gene from *Populus trichocarpa*. In *Arabidopsis*, ABI3 is expressed in stipules, abscission zones of the siliques and the shoot apex during induction of ecodormancy. Their further studies will examine ABI3 involvement in poplar bud dormancy.

APICAL MERISTEM AND THE REGULATION OF GROWTH

Developmental signals within the shoot apical meristem are influenced by signals from one cell layer to another layer (Szymkowiak and Sussex, 1992 as reported by Blackman and Overall, 2001). Molecular trafficking between these layers may be regulated by plasmodesmata.

Such studies involving cell to cell signaling and regulation of flow within the apical meristem are important to develop a cellular understanding of dormancy induction. Recently, Blackman and Overall (2001) reviewed plasmodesmatal (PD) structure and function and concluded that the 1kDa selective transport of molecules is inaccurate. Movement of proteins up to 50 kDa including transcription factors is possible. Pressure differentials between cells can close PD and the cytoskeleton may provide active cell-cell transport, regulating permeability of the PD.

Jian et al. (2000) reported differential plasmodesmatal connections between various cells in the apical meristem of poplar buds during dormancy development. In addition, active alterations in the localization of calcium in the meristem during dormancy induction were observed. Van der Schoot (1996) also examined symplasmic networking in the shoot apical meristem of potato and reported differences in concentric ER between endodormant and non-dormant tissue.

The giant unicellular organism, *Acetabularia acetabulum* has potential for studies of the control of growth and differentiation (Mandoli, 1998; Von Dassow et al., 2001). This organism is 10 centimeters at maximum height and progresses through discrete vegetative and reproductive stages. The nodes, internodes and needle-like hairs resemble the nodes, internodes and leaves of plants. It is theoretically possible to regulate patterning of the elaborate body plan simply by locally controlling differentiation of the apices (Mandoli, 1998). Turgor-pressure wall stress in turn, is postulated to modulate apical cell wall growth and thus regulation of morphology. If the evolutionary mechanism of apical meristem differentiation has been conserved, this ancient organism with its 570 million year history may offer unique insights into environmental regulation of woody bud differentiation. *Acetabularia* and other simpler organisms may provide useful indices of fundamental control mechanisms altering growth and differentiation in woody plants.

Microtubules and microfilaments are indispensable for the determination of cell shape, nuclear division, and cytokinesis (Kumagai and Hasezawa, 2001). Orientation of cortical microtubules is considered to regulate orientation of cell wall microfibrils and therefore direction of cell expansion and growth. Iwata et al. (2001) demonstrated turgor pressure regulation of cortical microtubule orientation in *Spirogyra* sp. under various mannitol solutions. Geoffrey Findlay (2001) provides a comprehensive review on membranes and electrophysiology of turgor regulation using marine algae as a model system. Alterations of components such as potassium flux, calcium influx, chloride channels, sucrose transporters, osmotica, aquaporins and proton pumps all serve directly or indirectly in regulating turgor. ABA mediates K^+ transport across plant roots by altering K^+ channel activity and membrane potential in maize roots and *Arabidopsis* (Roberts and Snowman, 2000). Root cell response was clearly differentiated from guard cell responses where ABA favored activation of K_{in} channels and inhibited K_{out} channel activity in root cells.

Utilizing *Chara corallina* as a model system, Proseus et al. (2000) have also shown rapid changes in growth were independent of cell wall alterations but dependent upon turgor pressure and temperature. When temperature was lowered, growth was immediately inhibited due to a decrease in turgor pressure (via osmotic potential). However, returning the cell back up to the original turgor while maintaining the lower temperature did not produce cell growth. Thus, they concluded growth is both a turgor pressure and temperature-sensitive process. They also suggest the rapidity of this response excludes the participation of changes in gene expression. In *Arabidopsis*, BON1 and BAP1 genes are regulated by temperature and may directly regulate cell expansion and division at lower temperatures (Hua et al., 2001). BON1 contains a Ca^{2+}-dependent phospholipid-binding domain associated with the plasmamembrane.

Cyclin-CDK complexes represent key regulators of the cell cycle in plants and their distribution is directly linked with microtubule structure (Mews et al., 1997 cited by Kumagai and Hasezawa, 2001). Le Bris et al. (1999) reported endodormant cells of *Rosa hybrida* were arrested in the G1 phase. Fluridone inhibited ABA biosynthesis and induced new leaf primordia development. ABA arrested cells in the G2 phase and buds remained dormant. Kupila-Ahvenniemi et al. (1984) found exposure of *Pinus sylvestris* L. branches to 5 ppm ABA solution for 10 d slightly restricted the mitotic frequency of the bud cells. During endodormancy, nucleoli structure was altered and ER cisternae were more regularly aligned, circular in certain cases and increased in number.

Water deficit arrests pea root meristems in the G2 phase of the cell cycle (Bracale et al., 1997). A 40 kDa protein was associated with this dehydration stress and also induced by ABA, low temperature and mannitol. However, the cell cycle was not clearly arrested in a specific stage under these treatments but a lowering of the proportion of cells entering the S phase was detected.

CONCLUSION

Hormones are undoubtedly involved at some point during dormancy induction. However, meaningful progress in unraveling the mechanism of endodormancy-induction will probably not be made through this hormone-centered approach. The path to dormancy acquisition is a continuum. In many plants, this path begins far earlier than the traditionally-examined autumn period. Aside from more useful hormonal localization studies within the bud and utilization of mutants, continued gross analysis of hormone presence/absence during dormancy induction will not be fruitful. Instead, first examining the regulation of the molecular switch from active foliage leaf formation to bud scale induction within key cells of the apical meristem may provide greater insight into dormancy induction. Specifically, turgor pressure-dependent cellular differentiation within the apical meristem and the extrinsic and intrinsic factors regulating turgor may be central to that molecular switch. Simpler model systems may need to be utilized to test these hypotheses. Only then can the potential role of hormones in endodormancy induction be more clearly articulated.

REFERENCES

Abraham, R.T. (2001). Cell cycle checkpoint signaling through the ATM and ATR kinases. *Genes and Development* 15: 2177-2196.

Alvim, R., E.W. Hewett and P.F. Saunders. (1976). Seasonal variation in the hormone content of willow. I. Changes in abscisic acid content and cytokinin activity in the xylem sap. *Plant Physiology* 57: 474-476.

Alvim, R., P.F. Saunders and R.S. Barros. (1979). Abscisic acid and the photoperiodic induction of dormancy in *Salix viminalis* L. *Plant Physiology* 63: 774-777.

Aneja, M., T. Gianfagna and E. Ng. (1996). Abscisic acid metabolism and episodic growth in cocoa. *Plant Growth Regulation* 20: 209-214.

Arora, R., M.E. Wisniewski and R. Scorza. (1992). Cold acclimation in genetically related (sibling) deciduous and evergreen peach (*Prunus persica* [L.] Batsch). *Plant Physiology* 99: 1562-1568.

Arora, R. and M.E. Wisniewski. (1994). Cold acclimation in genetically related (sibling) deciduous and evergreen peach (*Prunus persica* L. Batsch). *Plant Physiology* 105: 95-101.

Ashworth, E.N. (1984). Xylem development in Prunus flower buds and relationships to deep supercooling. *Plant Physiology* 74: 862-865.

Bagnall, D.J. and R.W. King. (2001). Phytochrome, photosynthesis and flowering of *Arabidopsis thaliana*: photophysiological studies using mutants and transgenic lines. *Australian Journal of Plant Physiology* 28: 401-408.

Baldwin, B.D., M.S. Bandara and K.K. Tanino. (2000a). Bud scale maturation in saskatoon berry (*Amelanchier alnifolia* Nutt.) plantlets following in vitro hormonal treatments. In: eds., L.H.W. Plas and G.J. de Klerk. *Acta Horticulturae* 520: 203-208.

Baldwin, B.D., M.S. Bandara and K.K. Tanino. (2000b). Role of bud scales and phytohormones in the maintenance and release of endo-dormancy in field-grown Saskatoon berry floral buds. *Acta Horticulturae* 520: 209-218.

Ballesteros, M.L., B. Cordelia, L.M. Lois, J.M. Moore, J.-P. Vielle-Calzada, U. Grossniklaus and N.-M. Chua. (2001). LAF1, a MYB transcription activator for phytochrome A signaling. *Genes and Development* 15: 2613-2625.

Barros, R.S. and S.J. Neill. (1986). Periodicity of response to abscisic acid in lateral buds of willow (*Salix viminalis* L.). *Planta* 168: 530-535.

Barros, R.S. and S.J. Neill. (1987). Shoot growth in willow (*Salix viminalis*) in relation to abscisic acid, plant water status and photoperiod. *Physiologia Plantarum* 70: 708-712.

Barros, R.S. and S.J. Neill. (1989). The status of abscisic acid in willow as related to the induction of bud dormancy. *Acta Physiologiae Plantarum* 11(2): 117-123.

Baskin, C.C. and J.M. Baskin. (1998). *Seeds. Ecology, biogeography, and evolution of dormancy and germination.* New York: Academic Press.

Bewley, J.D. (1997). Seed germination and dormancy. *The Plant Cell* 9: 1055-1066.

Blackman, L.M. and R.L. Overall. (2001). Structure and function of plasmodesmata. *Australian Journal of Plant Physiology* 28: 709-727.

Blake, T.J., E. Bevilacqua, G.A. Hunt and S.R. Abrams. (1990). The effects of abscisic acid and its acetylenic alcohol on dormancy, root development and transpiration in three conifer species. *Physiologia Plantarum* 80(3): 371-378.

Borkowsda, B. and L.E. Powell. (1982/83). Abscisic acid relationships in dormancy of apple buds. *Scientia Horticulturae* 18: 111-117.

Bouchard, C., O. Dittrich, A. Kiermaier, K. Dohmann, A. Menkel, M. Eilers and B. Luscher. (2001). Regulation of *cyclin D2* gene expression by the Myc/Max/Mad network: Myc-dependent TRRAP recruitment and histone acetylation at the *cyclin D2* promoter. *Genes and Development* 15: 2042-2047.

Bracale, M., M. Levi, C. Savini, W. Dicorato and M.G. Galli. (1997). Water deficit in pea root tips: Effects on the cell cycle and on the production of dehydrin-like proteins. *Annals of Botany* 79: 593-600.

Brand, U., J.C. Fletcher, M. Hobe, E.M. Meyerowitz and R. Simon. (2000). Dependence of stem cell fate in *Arabidopsis* on a feedback loop regulated by *CLV3* activity. *Science* 289:617-619.

Briggs, W.R. and M.A. Olney. (2001). Photoreceptors in plant photomorphogenesis to Date. Five phytochromes, two cryptochromes, one phototropin, and one superchrome. *Plant Physiology* 125: 85-88.

Carroll, A.S., A.C. Bishop, J.L. DeRisi, K.M. Shokat and E.K. O'Shea. (2001). Chemical inhibition of the Pho85 cyclin-dependent kinase reveals a role in the environmental stress response. *Proceedings of the National Academy of Sciences USA* 98: 12578-12583.

Chalker-Scott, L. 1992. Disruption of an ice-nucleation barrier in cold-hardy Azalea buds by sublethal heat stress. *Annals of Botany* 70: 409-418.

Champagnat, P. (1989). Rest and activity in vegetative buds of trees. *Annales des Sciences Forestieres* 46 (suppl.): 9-26.

Chen, T.H.H., J. Davis, B.E. Frewen, G.T. Howe and H.D. Bradshaw Jr. (2000). Molecular genetic analysis of bud dormancy-related traits in *Populus*. In: eds., J.-D. Viemont and J. Crabbe. *Dormancy in plants.* New York, USA: CABI Publishing.

Chen, W.-S. (1994). Gibberellins in buds of *Euphoria longana* Lam. During different stages of growth. *Botanical Bulletin of Academia Sinica* 35: 39-43.

Close, T.J., R.D. Fenton and F. Moonan. (1993). A view of plant dehydrins using antibodies specific to the carboxy terminal peptide. *Plant Molecular Biology* 23: 279-286.

Corbineau, F. and D. Come. (2000). Dormancy of cereal seeds as related to embryo sensitivity to ABA and water potential. In: eds., J.-D. Viemont and J. Crabbe. *Dormancy in plants*. New York, USA: CABI Publishing.

Corgan, J.N. and C. Peyton. (1970). Abscisic acid levels in dormant peach flower buds. *Journal of the American Society for Horticulture Science* 95(6): 770-774.

Cornforth, J.W., B.V. Milborrow, G. Ryback and P.F. Wareing. (1965). Chemistry and physiology of 'dormins' in sycamore. *Nature* 205: 1269-1270.

Crabbe, J. (1994). Dormancy. In: Arntzen, C.J. and Ritter, E.M. (eds). *Encyclopedia of agricultural sciences*, Academic Press, New York. vol. 1: 597-611.

Crabbe, J. and P. Barnola. (1996). A new conceptual approach to bud dormancy in woody plants. In: ed. G.A. Lang. *Plant dormancy, physiology, biochemistry and molecular biology*. Wallingford, UK: CAB International.

Darnell, R.L., G.W. Stutte, G.C. Martin, G.A. Lang and J.D. Early. (1992). Developmental physiology of rabbiteye blueberry. *Horticultural Reviews* 13: 339-405.

Davison, R.M. and H. Young. (1974). Seasonal changes in the level of abscisic acid in xylem sap of peach. *Plant Science Letters* 2: 79-82.

Demaggio, A.E. and J.A. Freeberg. (1969). Dormancy regulation: hormonal interaction in maple (*Acer platanoides*). *Canadian Journal of Botany* 47: 1165-1169.

Dennis, F.G. and L.J. Edgerton. (1961). The relationship between an inhibitor and rest in peach flower buds. *American Society for Horticultural Science* 77: 107-116.

Denney, F.E. and E.N. Stanton. 1928. Localization of response of woody tissues to chemical treatments that break the rest period. *American Journal of Botany* 15: 337-344.

Dormling, I., A. Gustafsson and D. von Wettstein. (1966). The experimental control of the life cycle in *Picea abies* (L.) Karst. *Silvae Genetica* 17: 44-64.

Doorenbos, J. (1953). *Review of the literature on dormancy in buds of woody plants*. Meded. Landbouwhogeschool Wageningen 53: 1-24.

Dumbroff, E.B., D.B. Cohen and D.P. Webb. (1979). Seasonal levels of abscisic acid in buds and stems of *Acer saccharum*. *Physiologia Plantarum* 45: 211-214.

During, H. and O. Bachmann. (1975). Abscisic acid analysis in *Vitis vinifera* in the period of endogenous bud dormancy by high pressure liquid chromatography. *Physiologia Plantarum* 34: 201-203.

Eagles, C.G. and P.F. Wareing. (1963). Dormancy regulators in woody plants. *Nature* 199: 874-875.

Eagles, C.F., and P.F. Wareing. (1964). The role of growth substances in the regulation of bud dormancy. *Physiologia Plantarum* 17: 697-709.

Ekberg, I., M. Qamaruddin, D. Clapham, I. Dormling, G. Eriksson and E. Tillberg. (1993). Abscisic acid content and gene activity at defined levels of bud dormancy and frost tolerance in two contrasting populations of *Picea abies* grown in a phytotron. *Agricoltura Ricerca* 15:146.

El-Antably, H.M., P.F. Wareing and J. Hillman. (1967). Some physiological responses to D,L abscisin (dormin). *Planta* 73: 74-90.

Erez, A., M. Faust and M.J. Line. (1998). Changes in water status in peach buds on induction, development and release from dormancy. *Scientia Horticulturae* 73: 111-123.

Erez, A. (2000). Bud dormancy: A suggestion for the control mechanism and its evolution. In: eds., J.-D. Viemont and J. Crabbe. *Dormancy in plants*. New York, USA: CABI Publishing.

Fankhauser, C., K.-C. Yeh, J. Clark Lagarias, H. Zhang, T.D. Elich and J. Chory. (1999). PKS1, a substrate phosphorylated by phytochrome that modulates light signalling in *Arabidopsis*. *Science* 284: 1539-1541.

Faust, M., D. Liu, M.M. Millard and G.W. Stutte. (1991). Bound versus free water in dormant apple buds–a theory for endodormancy. *HortScience* 26(7): 887-890.

Faust, M., A. Erez, L.J. Rowland, S.Y. Wang and H.A. Norman. (1997). Bud dormancy in perennial fruit trees: Physiological basis for dormancy induction, maintenance and release. *HortScience* 32(4): 623-629.

Fennell, A., C. Wake and P. Molitor. (1996). Use of ^{1}H-NMR to determine grape bud water state during the photoperiodic induction of dormancy. *Journal of the American Society for Horticulture Science* 121: 1112-1116.

Fennell, A. and M.J. Line. (2001). Identifying differential tissue response in grape (*Vitis riparia*) during induction of endodormancy using nuclear magnetic resonance imaging. *Journal of the American Society for Horticulture Science* 126(6): 681-688.

Findlay, G.P. (2001). Membranes and the electrophysiology of turgor regulation. *Australian Journal of Plant Physiology* 28: 617-634.

Foster, A.S. (1932). Investigations on the morphology and comparative history of development of foliar organs III. Cataphyll and foliage-leaf ontogeny in the black hickory (*Carya buckleyi* var. Arkansana). *American Journal of Botany* 19: 75-99.

Frank, P.A. (1966). Dormancy in winter buds of American pondweed, *Potamogeton nodosus* Poir. *Journal of Experimental Botany* 17: 546-555.

Frankie, G.W., H.G. Baker and P.A. Opler. (1974). Comparative phenological studies of trees in tropical wet and dry forests in the lowlands of Costa Rica. *Journal of Ecology* 62:881-919.

Freeman, M.W. and G.C. Martin. (1981). Peach floral bud break and abscisic acid content as affected by mist, light and temperature treatments during rest. *Journal of the American Society for Horticultural Science* 106: 333-336.

Freundl, E., E. Steudle and W. Hartung. (1998). Water uptake by roots of maize and sunflower affects the radial transport of abscisic acid and its concentration in the xylem. *Planta* 207: 8-19.

Freundl, E., E. Steudle and W. Hartung. (2000). Apoplastic transport of abscisic acid through roots of maize: effect of the exodermis. *Planta* 210: 222-231.

Fuchigami, L.H., D.R. Evert and C.J. Weiser. (1971). A translocatable cold hardiness promoter. *Plant Physiology* 47: 164-167.

Fuchigami, L.H., C.J. Weiser, K. Kobayashi, R. Timmis and L.V. Gusta. (1982). A degree growth stage (oGS) model and cold acclimation in temperate woody plants. In: P.H. Li and A. Sakai, eds. *Plant cold hardiness and freezing stress*. New York: Academic Press.

Fuchigami, L.H. and C.C. Nee. (1987). Degree growth stage model and rest-breaking mechanisms in temperate woody perennials. *HortScience* 22: 836-845.

Fuchigami, L.H. and M. Wisniewski. (1997). Quantifying bud dormancy: physiological approaches. *HortScience* 32(4): 618-623.

Fujisawa, T., H. Kato and Y. Iwasaki. (2001). Structure and function of heterotrimeric G proteins in plants. *Plant Cell Physiology* 42(8): 789-794.

Gardea, A.A., L.S. Daley, R.L. Kohnert, A.H. Soeldner, L. Ning, P.B. Lombard, A.N. Azarenko. (1994). Proton NMR signals associated with eco-and endodormancy in winegrape buds. *Scientia Horticulturae* 56: 339-358.

Garner, W.W. and H.A. Allard. (1923). Further studies in photoperiodism, the response of the plant to relative length of day and night. *Journal of Agricultural Research* 23: 871-920.

Gasch, A.P., P.T. Spellman, K.M. Kao, O. Carmel-Harel, M.B. Eisen, G. Storz, D. Botstein and P.O. Brown. (2000). Genomic expression programs in the response of yeast cells to environmental changes. *Molecular Biology of the Cell*. 11: 4241-4257.

Goffinet, M.C. and P.R. Larson. (1981). Structural changes in *Populus deltoides* terminal buds and in the vascular transition zone of the stems during dormancy induction. *American Journal of Botany* 68(1): 118-129.

Goffinet, M.C. and P.R. Larson. (1982). Lamina abortion in terminal bud-scale leaves of *Populus deltoides* during dormancy induction. *Botanical Gazette* 143(3): 331-340.

Grossmann, K. and H. Hansen. (2001). Ethylene-triggered abscisic acid: A principle in plant growth regulation? *Physiologia Plantarum* 113: 9-14.

Halim, H., G.R. Edwards, B.G. Coombe and D. Aspinall. (1988). The dormancy of buds of *Citrus sinensis* (L.) Osbeck inserted into rootstock stems: factors intrinsic to the inserted bud. *Annals of Botany* 61: 525-529.

Hansen, E., J.E. Olsen and O. Junttila. (1999). Gibberellins and subapical cell divisions in relation to bud set and bud break in *Salix pentandra*. *Journal Plant Growth Regulation* 18(4): 167-170.

Harrison, M.A. and P.F. Saunders. (1975). The abscisic acid content of dormancy birch buds. *Planta* 123: 291-298.

Heide, O.M. (1986). Effects of ABA application on cessation of shoot elongation in long-day grown Norway spruce seedlings. *Tree Physiology* 1: 79-83.

Hemberg, T. (1949). Growth-inhibiting substances in terminal buds of *Fraxinus*. *Physiologia Plantarum* 2: 37-44.

Hemberg, T. (1958). The occurrence of acid inhibitors in resting terminal buds of Fraxinus. *Physiologia Plantarum* 11: 610-614.

Hocking, T.J. and J.R. Hillman. (1975). Studies on the role of abscisic acid in the initiation of bud dormancy in *Alnus glutinosa* and *Betula pubescens*. *Planta* 125: 235-242.

Hole, D.J., J.D. Smith and B.G. Cobb. (1989). Regulation of embryo dormancy by manipulation of abscisic acid in kernels and associated cob tissue of *Zea mays* L. cultured *in vitro*. *Plant Physiology* 91: 101-105.

Howe, G.T., P.A. Bucciaglia, W.P. Hackett, G.R. Furnier, M.-M. Cordonnier-Pratt and G. Gardiner. (1998). Evidence that the phytochrome gene family in black cottonwood has one *PHYA* locus and two *PHYB* loci but lacks members of the *PHYC/F* and *PHYE* subfamilies. *Molecular Biology Evolution* 15(2): 160-175.

Howe, G.T., P. Saruul, J. Davis and T.H.H. Chen. (2000). Quantitative genetics of bud phenology, frost damage, and winter survival in an F2 family of hybrid poplars. *Theoretical and Applied Genetics* 101: 632-642.

Hua, J., P. Grisafi, S.-H. Cheng and G.R. Fink. (2001). Plant growth homeostasis is controlled by the *Arabidopsis BON1* and *BAP1* genes. *Genes and Development* 15: 2263-2272.

Hummel, R.C., P.D. Ascher and H.M. Pellett. (1982). Inheritance of the photoperiodically induced cold acclimation response in *Cornus sericea* L. Red Osier Dogwood. *Theoretical and Applied Genetics* 62(4): 385-394.

Ichimura, K., T. Mizoguchi, R. Yoshida, T. Yuasa and K. Shinozaki. (2000). Various abiotic stresses rapidly activate *Arabidopsis* MAP kinases ATMPK4 and ATMPK6. *The Plant Journal* 24(5): 655-665.

Irving, R.M. (1969). Characterization and role of an endogenous inhibitor in the induction of cold hardiness in *Acer negundo. Plant Physiology* 44: 801-805.

Iuchi, S., M. Kobayashi, K. Yamaguchi-Shinozaki and K. Shinozaki. (2000). A stress-inducible gene for 9-cis-epoxycarotenoid dioxygenase involved in abscisic acid biosynthesis under water stress in drought-tolerant cowpea. *Plant Physiology* 123: 553-562.

Iwasaki, K. and R.J. Weaver. (1977). Effects of chilling, calcium cynamide, and bud scale removal on bud break, rooting, and inhibitor content of buds of 'Zinfandel' grape (*Vitis vinifera* L.). *Journal of the American Society for Horticultural Science* 102(5): 584-587.

Iwata, K., M. Tazawa and T. Itoh. (2001). Turgor pressure regulation and the orientation of cortical microtubules in *Spirogyra* cells. *Plant Cell Physiology* 42(6): 594-598.

Jamsran, U., K. Fujino, Y. Kikuta and H. Nakashima. (1999). Dormant bud formation in temperate grass species (Poaceae) cultured *in vitro* system. *Journal of the Faculty of Agriculture Hokkaido University* 69(3): 151-169.

Janick, J. (1974). The apple in Java. *HortScience* 9: 13-15.

Jensen, E., L. Rivier, O. Juntilla and A. Crozier. (1986). Identification of abscisic acid from shoots of *Salix pentandra. Physiologia Plantarum* 66: 406-408.

Jian, L.-C., J.H. Li, P.H. Li and T.H.H. Chen. (2000). Intercellular communication channels and intracellular calcium levels involved in the dormancy development of poplar (*Populus deltoides*) plants. In: eds., J.-D. Viemont and J. Crabbe. *Dormancy in plants.* New York, USA: CABI Publishing.

Johannsen, W. (1906). Das aether-verfahren beim fruhtrieben. Jena.

Johansen, L.G., P.-C. Oden and O. Juntilla. (1986). Abscisic acid and cessation of apical growth in *Salix pentandra. Physiologia Plantarum* 66: 409-412.

Jones, K.S., B.D. McKersie and J. Paroschy. (2000). Prevention of ice propagation by permeability barriers in bud axes of *Vitis vinifera. Canadian Journal of Botany* 78: 3-9.

Junttila, O. (1980). Effect of photoperiod and temperature on apical growth cessation in two ecotypes of *Salix* and *Betula. Physiologia Plantarum* 48: 347-352.

Juntilla, O. (1988). Effect of rootstock on photoperiodic control of elongation growth in grafted ecotypes of *Salix. Physiologia Plantarum* 74: 39-44.

Kang, S.K., H. Motosugi, K. Yonemori and A. Sugiura. (1997). Exothermic characteristics of dormant buds of persimmon (*Diospyros kaki* Thunb.) in relation to cold hardiness. *HortScience* 32(5): 840-843.

Karssen, C.M., D.L.C. Brinkhorst-van der Swan, A.E. Breekland and M. Koornneef. (1983). Induction of dormancy during seed development by endogenous abscisic acid: studies on abscisic acid deficient genotypes of *Arabidopsis thaliana* (L.) Heynh. *Planta* 157: 158-165.

Kelner, J.-J., S. Lachaud and J.-L. Bonnemain. (1993). Seasonal variations of the tissue distribution of [^{3}H]ABA and [^{3}H]nutrients in apical buds of beech. *Plant Physiology and Biochemistry* 31(4): 531-539.

Khanizadeh, S., D. Buszard and C.G. Zarkadas. (1994). Seasonal variation of hydrophilic, hydrophobic and charged amino acids in developing apple flower buds. *Journal of Plant Nutrition* 17: 2025-2030.

Kim, K.S., E. Davelaar and G.J. de Klerk. (1994). Abscisic acid controls dormancy development and bulb formation in lily plantlets regnerated *in vitro*. *Physiologia Plantarum* 90(1): 59-64.

Kircher, S., L. Kozma-Bogner, L. Kim, E. Adam, K. Harter, E. Schafer and F. Nagy. (1999). Light quality-dependent nuclear import of the plant photoreceptors phytochrome A and B. *The Plant Cell* 11: 1445-1456.

Koski, V. and R. Sievanen. (1984). Timing of growth cessation in relation to the variations in the growing season. In: eds., P.M.A. Tigerstedt, P. Puttonen and V. Koski. *Crop Physiology of Forest Trees*. University of Helsinki, Finland.

Koussa, T., M. Broquedis and J. Bouard. (1994). Relations entre la teneur en acide abscissique des bourgeons latents de vigne (*Vitis vinifera* L. var. Merlot Noir) et leur teneur en eau. *Journal International des Sciences de la Vigne et du Vin* 28(3): 203-211.

Koussa, T., D. Zaoui and M. Broquedis. (1998). Relationship between the levels of abscisic acid in latent buds, in leaves and in internodes of *Vitis vinifera* L. cv. Merlot during the dormancy phase. *Journal International des Sciences de la Vigne et du Vin* 32(4): 203-210.

Kubota, N., N. Nitta, T. Egawa and K. Shimamura. (1986). Effects of root temperatures under forced conditions on growth of shoot and flower clusters, and endogenous regulators in Muscat of Alexandria vines. *Scientific Reports of the Faculty of Agriculture, Okayama University* 67: 1-9.

Kumagai, F. and S. Hasezawa. (2001). Dynamic organization of microtubules and microfilaments during cell cycle progression in higher plant cells. *Plant Biology* 3: 4-16.

Kupila-Ahvenniemi, S., A. Taanila-Kukka and L. Rautiainen. (1984). Effect of hormone treatments and unusual temperature conditions on the structure of cells in pine buds during the fall and early winter. *Aquilo Series of Botany* 20: 24-34.

Label, P., R. Maldiney, L. Sossountzov, D. Cornue and E. Miginiac. (1988). Endogenous levels of abscisic acid, indole-3-acetic acid and benzyladenine during *in vitro* bud growth induction of Wild Cherry (*Prunus avium* L.). *Plant Growth Regulation* 7: 171-180.

Lang, G.A., J.D. Early, G.C. Martin and R.L. Darnell. (1987). Endo-, para- and ecodormancy: physiological terminology and classification for dormancy research. *HortScience* 22: 371-377.

Le Bris, M., N. Michaux-Ferriere, Y. Jacob, A. Poupet, P. Barthe, J.-M. Guigonis and M.-T. Le Page-Degivry. (1999). Regulation of bud dormancy by manipulation of ABA in isolated buds of *Rosa hybrida* cultured *in vitro*. *Australian Journal of Plant Physiology* 26: 272-281.

Lemichez, E., Y. Wu, J.-P. Sanchez, A. Mettouchi, J. Mathur and N.-H. Chua. (2001). Inactivation of AtRac1 by abscisic acid is essential for stomatal closure. *Genes and Development* 15: 1808-1816.

Lenton, J.R., V.M. Perry and P.F. Saunders. (1972). Endogenous abscisic acid in relation to photoperiodically induced bud dormancy. *Planta* 106: 13-22.

Levitt, J. (1980). *Plant responses to environmental stresses*. Vol I. New York: Academic Press.

Leopold, C. (1996). Natural history of seed dormancy. In: ed., G.A. Lang. *Plant dormancy, physiology, biochemistry and molecular biology*. Wallingford, UK: CAB International.

Le Page-Degivry, M-T., P. Barthe and G. Garello. (1990). Involvement of endogenous abscisic acid in onset and release of *Helianthus annuus* embryo dormancy. *Plant Physiology* 92: 1164-1168.

Le Page-Degivry, M-T., J. Bianco, P. Barthe and G. Garello. (1996). Changes in hormone sensitivity in relation to onset and breaking of sunflower embryo dormancy. In: ed. G.A. Lang. *Plant dormancy, physiology, biochemistry and molecular biology*. Wallingford, UK: CAB International.

Leung, J. and J. Giraudat. (1998). Abscisic acid signal transduction. *Annual Review of Plant Physiology and Plant Molecular Biology* 49:199-222.

Lionakis, S.M. and W.W. Schwabe. (1984). Bud dormancy in the kiwi fruit, *Actinidia chinensis* Planch. *Annals of Botany* 54: 467-484.

Loveys, B.R., A.C. Leopold and P.E. Kriedemann. (1974). Abscisic acid metabolism and stomatal physiology in *Betula lutea* following alteration in photoperiod. *Annals of Botany* 38: 85-92.

MacDonald, J.E. and J.N. Owens. (1993). Bud development in coastal Douglas-fir seedlings in response to different dormancy-induction treatments. *Canadian Journal of Botany* 71: 1280-1290.

MacDonald, J.E. (2000). The developmental basis of bud dormancy in 1-year-old *Picea* and *Pseudostuga* seedlings. In: eds., J.-D. Viemont and J. Crabbe. *Dormancy in plants*. New York, USA: CABI Publishing.

Mandoli, D.F. (1998). Elaboration of body plan and phase change during development of *Acetabularia*: how is the complex architecture of a giant unicell built? *Annual Review of Plant Physiology and Plant Molecular Biology* 49: 173-198.

Maurel, C. (1997). Aquaporins and water permeability of plant membranes. *Annual Review of Plant Physiology and Plant Molecular Biology* 48: 399-429.

Merkle, T. (2001). Nuclear import and export of proteins in plants: a tool for the regulation of signalling. *Planta* 213: 499-517.

Mielke, E.A. and F.G. Dennis Jr. (1975). Hormonal control of flower bud dormancy in sour cherry (*Prunus cerasus* L.). II. Levels of abscisic acid and its water soluble complex. *Journal of the American Society for Horticultural Science* 100: 287-290.

Mielke, E.A. and F.G. Dennis Jr. (1978). Hormonal control of flower bud dormancy in sour cherry (*Prunus cerasus* L.). III. Effects of leaves, defoliation and temperature on levels of abscisic acid in flower primordia. *Journal of the American Society for Horticultural Science* 103: 446-449.

Mochioka, R., M. Tohbe, S. Horiuchi, T. Ogata, S. Shiozaki, K. Kawase, H. Kurooka and H. Matsui. (1996). The relationship between bud dormancy and the endogenous ABA and water contents of several wild grape species native to Japan. *Journal of the Japanese Society for Horticulture Science* 65(1): 49-54.

Moysset, L., E. Fernandez, N. Cortadellas and E. Simon. (2001). Intracellular localization of phyotchrome in *Robinia pseudoacacia* pulvini. *Planta* 213: 565-574.

Nadel, B.L., A. Altman, S. Pleban, R. Kocks, and A. Huettermann. (1991). *In vitro* development of mature *Fagus sylvatica* L. buds: II. Seasonal changes in the response to plant growth regulators. *Journal of Plant Physiology* 138(2): 136-141.

Neff, M.M., C. Fankhauser and J. Chory. (2000). Light: an indicator of time and place. *Genes & Development* 14: 257-271.

Ni, M., J.M. Tepperman and P.H. Quail. (1999). Binding of phytochrome B to its nuclear signalling partner PIF3 is reversibly induced by light. *Nature* 400: 781-784.

Nitsch, J.P. (1957a). Growth responses of woody plants to photoperiodic stimuli. *American Society for Horticultural Science Proceedings* 70: 512-525.

Nitsch, J.P. (1957b). Photoperiodism in woody plants. *American Society for Horticultural Science Proceedings* 70: 526-544.

Okubo, H. (2000). Growth cycle and dormancy in plants. In: eds., J.-D. Viemont and J. Crabbe. *Dormancy in plants*. New York, USA: CABI Publishing.

Olsen, J.E., T. Moritz, E. Jensen and O. Juntilla. (1994). Comparison of endogenous gibberellins in roots and shoots of elongating *Salix pentandra* seedlings. *Physiologia Plantarum* 90: 378-381.

Olsen, J.E., E. Jensen, O. Juntilla and T. Moritz. (1995a). Photoperiodic control of endogenous gibberellins in seedlings of *Salix pentandra*. *Physiologia Plantarum* 93: 639-644.

Olsen, J.E., T. Moritz, E. Jensen and O. Juntilla. (1995b). Lack of effect of photoperiod on metabolism of exogenous GA_{19} and GA_1 in *Salix pentandra* seedlings. *Physiologia Plantarum* 94: 522-528.

Olsen, J.E., O. Juntilla and T. Moritz. (1995c). A localised decrease of GA_1 in shoot tips of *Salix pentandra* seedlings precedes cessation of shoot elongation under short photoperiod. *Physiologia Plantarum* 95: 627-632.

Olsen, J.E., O. Juntilla and T. Moritz. (1997a). Long-day induced bud break in *Salix pentandra* is associated with transiently elevated levels of GA_1 and gradual increase in indole-3-acetic acid. *Plant Cell Physiology* 38(5): 536-540.

Olsen, J.E., O. Junttila, J. Nilsen, M.E. Eriksson, I. Martinussen, O. Olsson, G. Sandberg and T. Moritz. (1997b). Ectopic expression of oat phytochrome A in hybrid aspen changes critical daylength for growth and prevents cold acclimatization. *The Plant Journal* 12(6): 1339-1350.

Owens, J.N. and M. Molder. (1973). A study of DNA and mitotic activity in the vegetative apex of Douglas fir during the annual growth cycle. *Canadian Journal of Botany* 51: 1395-1409.

Owens, J.N., M. Molder and H. Langer. (1977). Bud development in *Picea glauca*. I. Annual growth cycle of vegetative buds and shoot elongation as they relate to date and temperature sums. *Canadian Journal of Botany* 55: 2728-2745.

Owens, J.N., J.E. Webber, S.D. Ross and R.P. Pharis. (1985). Interaction between gibberellin $A_{4/7}$ and root-pruning on the reproductive and vegetative processes in Douglas-fir. III. Effects on anatomy of shoot elongation and terminal bud development. *Canadian Journal of Forest Research* 15: 354-364.

Owens, J.N., J.E. Webber, S.D. Ross and R.P. Pharis. (1986). Interaction between gibberellin $A_{4/7}$ and root-pruning on the reproductive and vegetative processes in Douglas-fir. IV. Effects on lateral bud development. *Canadian Journal of Forest Research* 16: 211-221.

Parmentier, C.M., L.J. Rowland and M.J. Line. (1998). Water status in relation to maintenance and release from dormancy in blueberry flower buds. *Journal of the American Society for Horticulture Science* 123(5): 762-769.

Pemberton, H.B., M.L. Brenner and H.F. Wilkins. (1985). Endogenous ABA levels in floral bud parts of the evergreen azalea during floral development. *Journal of the American Society for Horticulture Science* 110: 737-742.

Perry, T.O. (1971). Dormancy of trees in winter. *Science* 171: 29-36.

Perry, T.O. and H. Hellmers. (1973). Effects of abscisic acid on growth and dormancy of two races of red maple. *Botanical Gazette* 134(4): 283-289.

Phillips, J.E. (1941). Effect of day length on dormancy in tree seedlings. *Journal of Forestry* 39: 55-59.

Phillips, I.D., and P.F. Wareing. (1958). Studies in dormancy of sycamore. I. Seasonal changes in the growth-substance content of the shoot. *Journal of Experimental Botany* 9(27): 350-364.

Pickard, W.F. (2001). A novel class of fast electrical events recorded by electrodes implanted in tomato shoots. *Australian Journal of Plant Physiology* 28: 121-129.

Pierce, M. and K. Raschke. (1981). Synthesis and metabolism of abscisic acid in detached leaves of *Phaseolus vulgaris* L. after loss and recovery of turgor. *Planta* 153: 156-165.

Pinfield, N.J., P.A. Stutchbury and S.M. Bazaid. (1987). Seed dormancy in *Acer*: Is there a common mechanism for all *Acer* species and what part is played in it by abscisic acid? *Physiologia Plantarum* 71: 365-371.

Piola, F., P. Label, P. Vergne, P. Von-Arderkas and R. Rohr. (1998). Effects of endogenous ABA levels and temperature on cedar (*Cedrus libani* Loudon) bud dormancy *in vitro*. *Plant Cell Reports* 18(3.4): 279-283.

Pollock, B.M. (1953). The respiration of *Acer* buds in relation to the inception and termination of the winter rest. *Physiologia Plantarum* 6: 47-64.

Pongsomboon, W., S. Subhadrabandhu and K.R. Chapman. (1990). Levels of inhibitors in flower buds during bud dormancy of three peach cultivars. *Acta Horticulturae* 279: 333-345.

Powell, L.E. (1976). Effect of photoperiod on endogenous abscisic acid in *Malus* and *Betula*. *HortScience* 11: 498-499.

Powell, L.E. (1987). The hormonal control of bud and seed dormancy in woody plants. In: ed. P.J. Davies. *Plant hormones and their role in plant growth and development*. Boston: Martinus Nijhoff Publishers.

Proseus, T.E., G.-L. Zhu and J.S. Boyer. (2000). Turgor, temperature and the growth of plant cells: using *Chara corallina* as a model system. *Journal of Experimental Botany* 51: 1481-1494.

Psota, V., S. Klicova, J. Sebanek, H. Vitkova and M. Jurasova. (1989). Changes in the levels of endogenous gibberellins, cytokinins and auxins in the apical parts of annuål shoots of two varieties of *Prunus persica*. *Acta Universitatis Agriculturae* XXXVII no. 1-2: 69-77.

Qamaruddin, M., I. Dormling, I. Ekberg, G. Eriksson and E. Tillberg. (1993). Abscisic acid content at defined levels of bud dormancy and frost tolerance in two contrasting populations of *Picea abies* grown in a phytotron. *Physiologia Plantarum* 87: 203-210.

Qamaruddin, M., I. Ekberg, I. Dormling, L. Norell, D. Clapham and G. Eriksson. (1995). Early effects of long nights on budset, bud dormancy and abscisic acid content in two populations of *Picea abies*. *Forest Genetics* 2(4): 207-216.

Quamme, H.A., W.A. Su and L.J. Veto. (1995). Anatomical features facilitating supercooling of the flower within the dormant peach flower bud. *Journal of the American Society for Horticulture Science* 120(5): 814-822.

Ramina, A., M. Colauzzi, A. Masia, A. Pitacco, T. Caruso, R. Messina, G. Calabrelli and G.A. Couvillon. (1995). Hormonal and climatological aspects of dormancy in peach buds. *Acta Horticulturae* 395: 35-46.

Ramsay, J. and G.C. Martin. (1970). Seasonal changes in growth promoters and inhibitors in buds of apricot. *Journal of the American Society for Horticulture Science* 95(5): 569-574.

Rey, M., C. Diaz-Sala and R. Rodriguez. (1994). Comparison of endogenous polyamine content in hazel leaves and buds between the annual dormancy and flowering phases of growth. *Physiologia Plantarum* 91: 45-50.

Rohde, A., M. Van Montagu, D. Inze and W. Boerjan. (1997). Factors regulating the expression of cell cycle genes in individual buds of *Populus*. *Planta* 201: 43-52.

Rhode, A., W. Ardiles-Diaz, M. Van Montagu and W. Boerjan. (1998). Isolation and expression analysis of an ABSCISIC ACID-INSENSITIVE 3 (ABI3) homologue from *Populus trichocarpa*. *Journal of Experimental Botany* 49: 1059-1060.

Rinne, P., A. Saarelainen and O. Junttila. (1994a). Growth cessation and bud dormancy in relation to ABA level in seedlings and coppice shoots of *Betula pubescens* as affected by a short photoperiod, water stress and chilling. *Physiologia Plantarum* 90(3): 451-458.

Rinne, P., H. Tuominen and O. Junttila. (1994b). Seasonal changes in bud dormancy in relation to bud morphology, water and starch content, and abscisic acid concentration in adult trees of *Betula pubescens*. *Tree Physiology* 14(6): 549-561.

Rinne, P., A. Welling and P. Kaikuranta. (1998). Onset of freezing tolerance in birch (*Betula pubescens* Ehrh.) involves LEA proteins and osmoregulation and is impaired in an ABA-deficient genotype. *Plant, Cell and Environment* 21: 601-611.

Robinson, P.M., P.F. Wareing and T.H. Thomas. (1963). Isolation of the inhibitor varying with photoperiod in *Acer pseudoplatanus*. *Nature* 199: 875-876.

Rodriguez, A. and R. Sanchez-Tames. (1986). Dormancy and seasonal changes of plant growth regulators in hazel buds. *Physiologia Plantarum* 66: 288-292.

Rodriguez, A., M.J. Canal and R. Sanchez-Tames. (1991). Seasonal changes of plant growth regulators in *Corylus*. *Journal of Plant Physiology* 138: 29-32.

Rodriguez, A.J., W.B. Sherman, R. Scorza, M. Wisniewski and W.R. Okie. (1994). 'Evergreen' peach, its inheritance and dormant behaviour. *Journal of the American Society for Horticulture Science* 119(4):789-792.

Romagosa, I., D. Prada, M.A. Moralejo, A. Sopena, P. Munoz, A.M. Casas, J.S. Swanston and J.L. Molina-Cano. (2001). Dormancy, ABA content and sensitivity of barley mutant to ABA application during seed development and after ripening. *Journal of Experimental Botany* 52: 1499-1506.

Rowland, L.J. and R. Arora. (1997). Proteins related to endodormancy (rest) in woody perennials. *Plant Science* 126: 119-144.

Sakai, A. (1979). Freeezing avoidance mechanism in primordial shoots of conifer buds. *Plant and Cell Physiology* 20: 1381-1390.

Sakamoto, K. and A. Nagatani. (1996). Nuclear localization activity of phytochrome B. *Plant Journal* 10: 859-868.

Samish, R.M. (1954). Dormancy in woody plants. *Annual Review of Plant Physiology* 5: 183-204.

Saure, M.C. (1985). Dormancy release in deciduous fruit trees. In: ed. J. Janick. *Horticultural Reviews*, volume 7. Westport, Conneticut: AVI Publishing Company.

Sauter, A., W.J. Davies and W. Hartung. (2001). The long-distance abscisic acid signal in the droughted plant: The fate of the hormone on its way from root to shoot. *Journal of Experimental Botany* 52: 1991-1997.

Savidge, R.A. and P.F. Wareing. (1984). Seasonal cambial activity and xylem development of *Pinus contorta* in relation to endogenous indole-3-yl-acetic and (S)-abscisic acid levels. *Canadian Journal of Forest Research* 14: 676-682.

Schneider, E.F. (1968). The rest period of *Rhododendron* flower buds. I. Effect of the bud scales on the onset and duration of rest. *Journal of Experimental Botany* 19: 817-824.

Seki, M., M. Narusaka, H. Abe, M. Kasuga, K. Yamaguchi-Shinozaki, P. Carninci, Y. Hayashizaki and K. Shinozaki. (2001). Monitoring the expression pattern of 1300 *Arabidopsis* genes under drought and cold stresses by using a full-length cDNA microarray. *The Plant Cell* 13: 61-72.

Seeley, S.D. and L.E. Powell. (1981). Seasonal changes of free and hydrolyzable abscisic acid in vegetative apple buds. *Journal of the American Society for Horticulture Science* 106(4): 405-409.

Sharp, R.E., M.E. LeNoble, M.A. Else, E.T. Thorne and F. Gherardi. (2000). Endogenous ABA maintains shoot growth in tomato independently of effects on plant water balance: Evidence for an interaction with ethylene. *Journal of Experimental Botany* 51: 1575-1584.

Shinozaki, K. and K. Yamaguchi-Shinozaki. (1999). Molecular responses to drought stress. In: eds. K. Shinozaki and K. Yamaguchi-Shinozaki. *Molecular responses to cold, drought, heat and salt stress in higher plants.* Austin, Texas, USA: R.G. Landes Company.

Shinozaki, K. and K. Yamaguchi-Shinozaki. (2000). Molecular responses to dehydration and low temperature: Differences and cross-talk between two stress signalling pathways. *Current Opinions in Plant Biology* 3: 217-223.

Simpson, G.M. (1990). *Seed dormancy in grasses.* Cambridge University Press: Cambridge.

Spiers, J.M. (1973). Isolation and identification of a growth inhibitor in tung bud scales. *Journal of the American Society for Horticulture Science* 98(3): 237-238.

Stafstrom, J.P., D.B. Ripley, M.L. Devitt and B. Drake. (1998). Dormancy-associated gene expression in pea axillary buds. Cloning and expression of PsDRM1 and PsDRM2. *Planta* 205(4): 547-552.

Stevenson, R.K. (1994). *Dormancy and acclimation in woody plants.* MSc dissertation, University of Saskatchewan, Saskatoon, Canada.

Strauβ, M., F. Kauder, M. Peisker, U. Sonnewald, U. Conrad and D. Heineke. (2001). Expression of an abscisic acid-binding single-chain antibody influences the subcellular distribution of abscisic acid and leads to developmental changes in transgenic potato plants. *Planta* 213: 361-369.

Swartz, H.J., A.S. Geyer, L.E. Powell and S.-H.C. Lin. (1984). The role of bud scales in the dormancy of apples. *Journal of the American Society for Horticultural Science* 109(5): 745-749.

Szymkowiak. E.J. and I.M. Sussex. (1992). The internal meristem layer (L3) determines floral meristem size and carpel number in tomato periclinal chimeras. *Plant Cell* 4: 1089-1100.

Tafazoli, E. and C. Beyl. (1993). Changes in endogenous abscisic acid and cold hardiness in *Actinidia* treated with triazole growth retardants. *Journal of Plant Growth Regulation* 12: 79-83.

Tamura, F., K. Tanabe and T. Ikeda. (1993). Relationship between intensity of bud dormancy and level of ABA in Japanese pear 'Nijisseiki'. *Journal of the Japanese Society for Horticultural Science* 62(1): 75-81.

Tanino, K.K. (1990). Abscisic acid induction of environmental stress tolerance in plant cells. PhD Thesis, Oregon State University, Corvallis, OR, USA.

Taylor, B., A. Burbidge and A.J. Thompson. (2000). Control of abscisic acid biosynthesis. *Journal of Experimental Botany*. 51: 1563-1574.

Thomas, T.H., P.F. Wareing and P.M. Robinson. (1965). Action of sycamore 'dormin' as a gibberellin antagonist. *Nature* 205: 1270-1272.

Thomas, A., Tomos, A.D., J.L. Stoddart, H. Thomas and C.J. Pollock. (1988). Cell expansion, temperature and turgor pressure in growing leaves of *Lolium temulentum*. *New Phytologist* 112: 1-5.

Thorneycroft, D., S.M. Sherson and S.M. Smith. (2001). Using gene knockouts to investigate plant metabolism. *Journal of Experimental Botany* 52: 1593-1601.

Tinklin, I.G. and W.W. Schwabe. (1970). Lateral bud dormancy in the blackcurrant *Ribes nigrum* (L.). *Annals of Botany* 34: 691-706.

Tohbe, M., R. Mochioka, S. Horiuchi, T. Ogata, S. Shiozaki and H. Kurooka. (1998). The role of glutathione on the onset of endodormancy of grape buds. *Journal of the Japanese Society for Horticultural Science* 67(6): 912-916.

Tyurina, M.M. (1979). Development of ideas about the state of dormancy in woody plants. Translated from *Fiziologiya Rastenii* 26(5): 899-907.

Uemoto, S. (1992). Ecological behavior of wild *Camellia japonica* population in northernmost boundary in the world. *International Camellia Journal* 24: 75-76.

Umen, J.G. and U.W. Goodenough. (2001). Control of cell division by a retinoblastoma protein homolog in *Chlamydomonas*. *Genes and Development* 15: 1652-1661.

Van der Schoot, C. (1996). Dormancy and symplasmic networking at the shoot apical meristem. In: ed., G.A. Lang. *Plant dormancy, physiology, biochemistry and molecular biology*. Wallingford, UK: CAB International.

Viemont, J.-D. and J. Crabbe. (2000). Preface. In: eds., J.-D. Viemont and J. Crabbe. *Dormancy in plants*. New York, USA: CABI Publishing.

Von Dassow, M., G.M. Odell and D.F. Mandoli. (2001). Relationships between growth, morphology and wall stress in the stalk of *Acetabularia acetabulum*. *Planta* 213: 659-666.

Wake, C.M.F. and A. Fennell. (1996). Protein expression in *Vitis* species during photoperiod induction of dormancy. *Plant Physiology* 111: 78 (abstracts).

Walton, D.C. (1980). Biochemistry and physiology of abscisic acid. *Annual Review of Plant Physiology* 31: 453-489.

Wareing, P.F. (1953). Growth studies in woody species V. Photoperiodism in dormant buds of *Fagus sylvatica* L. *Physiologia Plantarum* 6: 692-706.

Wan, X. and J.J. Zwiazek. (2001). Root water flow and leaf stomatal conductance in aspen (*Populus tremuloides*) seedlings treated with abscisic acid. *Planta* 213: 741-747.

Wareing, P.F. (1956). Photoperiodism in woody plants. *Annual Review of Plant Physiology* 7: 191-214.

Watts, S.H., C.T. Wheeler and J.R. Hillman. (1987). Abscisic acid and dormancy in root-nodulated *Alnus glutinosa* (L.) Gaertn. *The New Phytologist* 105: 459-468.

Weiser, C.J. (1970). Cold resistance and injury in woody plants. *Science* 169: 1269-1278.

Welling, A., P. Kaikuranta and P. Rinne. (1997). Photoperiodic induction of dormancy and freezing tolerance in *Betula pubescens*. Involvement of ABA and dehydrins. *Physiologia Plantarum* 100(1): 119-125.

Wignall, T.A. and G. Browning. (1988). Epicormic bud development in *Quercus robur* L. Studies of endogenous IAA, ABA, IAA polar transport and water potential in cambial tissues and effects of exogenous hormones on bud outgrowth from stem explants. *Journal of Experimental Botany* 39: 1667-1678.

Wright, S.T.C. (1975). Seasonal changes in the levels of free and bound abscisic acid in blackcurrent (*Ribes nigrum*) and beech (*Fagus sylvatica*) buds. *Journal of Experimental Botany* 26: 161-174.

Xiong, L., B. Lee, M. Ishitani, H. Lee, C. Zhang and J.-K. Zhu. (2001). FIERY1 encoding an inositol polyphosphate 1-phosphatase is a negative regulator of abscisic acid and stress signalling in *Arabidopsis*. *Genes and Development* 15: 1971-1984.

Freezing Tolerance and Injury in Grapevines

Anne Fennell

SUMMARY. Grapes, due to their wide distribution, are one of the temperate fruit crops most frequently damaged by freezing temperatures. Freezing injury can result in decreases in yield and substantial economic losses to grape growers, subsequently impacting fruit wholesalers, wineries, distributors, and related industries. Freeze damage is not limited to the northern or southern limits of the production range. Freezing injury can occur in spring, fall, or winter in many of the grape growing regions. An understanding of the mechanisms involved in freezing tolerance, acclimation, and deacclimation in grapevines is needed to match cultivars appropriately with growing sites, improve cultural practices that minimize freezing injury, and aid in breeding and selecting cultivars with improved freezing tolerance. The ability to avoid or tolerate freezing temperatures includes a complex set of traits that is influenced by the inherent genetic characteristics of the grapevine and its interaction with the environment. In the present review, the mechanisms of freezing tolerance in grapevines are summarized and discussed in relation to the influence of genotype, phenological development, and environmental factors. *[Article copies available for a fee from The Haworth Document Delivery Service: 1-800-HAWORTH. E-mail address: <docdelivery@haworthpress.com> Website: <http://www.HaworthPress.com> © 2004 by The Haworth Press, Inc. All rights reserved.]*

Anne Fennell is affiliated with the Horticulture, Forestry, Landscape, and Parks Department, 201 Northern Plains Biostress Laboratory, Box 2140A, South Dakota State University, Brookings, SD 57006.

[Haworth co-indexing entry note]: "Freezing Tolerance and Injury in Grapevines." Fennell, Anne. Co-published simultaneously in *Journal of Crop Improvement* (Food Products Press, an imprint of The Haworth Press, Inc.) Vol. 10, No. 1/2 (#19/20), 2004, pp. 201-235; and: *Adaptations and Responses of Woody Plants to Environmental Stresses* (ed: Rajeev Arora) Food Products Press, an imprint of The Haworth Press, Inc., 2004, pp. 201-235. Single or multiple copies of this article are available for a fee from The Haworth Document Delivery Service [1-800-HAWORTH, 9:00 a.m. - 5:00 p.m. (EST). E-mail address: docdelivery@haworthpress.com].

http://www.haworthpress.com/web/JCRIP

Digital Object Identifer: 10.1300/J411v10n01_09

KEYWORDS. Freezing tolerance, freezing injury, grapevine, *Vitis*, dormancy, ice nucleation, supercool, thermography, thermal analysis, low temperature exotherm (LTE), lethal temperature

INTRODUCTION

Annually, 7.5 million hectares of grapes are in production in 80 countries, yielding 60.7 million tons of fruit. Grape production exceeds that of other temperate fruit crops and ranks second to oranges in world fruit production (FAO, 1999). Although grape species are native to the northern hemisphere they have an extensive production range in the temperate climatic zones, extending to about 52° latitude in the northern and southern hemisphere (Alleweldt et al., 1990; Fuller and Telli, 1999; Hamed et al., 2000; Reisch and Pratt, 1996; Seyedbagheri and Fallahi, 1994). Freezing injury occurs in many of these production regions during the spring, fall, or winter and is a major factor reducing yield or limiting production of this versatile crop.

Distribution of traditional *Vitis vinifera* L. wine and table grapes is generally limited to regions with winter temperature minima above −25°C. However, even traditional *V. vinifera* production areas can have losses due to freezing injury (Barka and Audran, 1997; Boone and Durham, 1996; Lipe et al., 1992). In regions that are marginal for *V. vinifera* production, growers continue to optimize cultural practices, seek adapted cultivars, use hybrid cultivars, and breed and select new cultivars using North American or Asian species that can withstand −30°C and lower (*V. riparia* Michx. and *V. amurensis* Rupr.) in efforts to minimize freezing injury (Alleweldt and Possingham, 1988; Alleweldt et al., 1990; Hemstead and Luby, 2000; Reisch and Pratt, 1996).

Selecting cultivars for cold hardiness and adapting cultural practices requires an understanding of the inherent freezing tolerance of a grape cultivar at all stages of development and the interaction of environment with the acclimation and deacclimation characteristics of the grapevine. To broaden the understanding of freezing injury and tolerance of subzero temperatures in grapevines, this review summarizes studies that use controlled freezing techniques, identify differential tissue freezing characteristics, and characterize the influence of maturation, dormancy, and bud break timing. These genotypic characteristics will be discussed in relation to environmental factors and management practices that influence the cold hardiness of grapevines.

FREEZING TOLERANCE AND INJURY IN GRAPEVINES

Freezing injury is a serious problem that limits the productivity of grapevines even in cultivars that are considered cold hardy. Often differences between cultivars may only be 1 to 2°C; but this small difference can correspond to differences in field survival (Wolf and Cook, 1991). Injury can reduce crop yields, damage vine parts, promote secondary problems such as crown gall, or kill the whole vine. The extent of injury is influenced by the minimum temperature of a freeze event, length of time at a low temperature, and differences in response of the vine tissues to freezing temperatures. In the dormant season, the canes and trunk tolerate a certain level of extracellular freezing and intracellular desiccation, whereas the buds avoid freezing injury by supercooling. Canes and trunks may experience cytoplasmic dessication and be injured at subzero temperatures. In contrast, intracellular freezing can occur when the supercooled bud freezes, killing the primordia but often leaving adjacent cane tissues uninjured (Figure 1, Adrews et al., 1984; Pierquet and Stushnoff, 1980; Quamme, 1986). Controlled freeze studies provide information on the critical temperature ranges for injury in different tissues, cultivar differences and impact of environmental conditions on freezing tolerance. Observations of field grown vines following freezing events indicate that there is considerable variation in the freezing tolerance of different tissues within the vine, within the same cultivar between sites and years, as well as between different cultivars. In addition, the timing of the freezing event relative to the developmental stage of the vine and the temperature conditions the vine was exposed to prior to the freeze influence the extent of injury that results from a freezing episode (Tables 1 and 2). Damage to the primary bud may reduce yields but permit the vine to recover and grow; in contrast, multiple sites of injury or repeated injury may reduce vine vigor or kill the vine outright.

Root Freezing Tolerance

Roots are susceptible to freezing damage at temperatures of - 4°C and below (Okamoto et al., 2000). In spite of this, freezing injury in the roots is the least common form of vine injury. It does occur in the more northern and southern limits of production where extreme low winter temperatures are more frequent (Ahmedullah, 1985; Seyedbagheri and Fallahi, 1994). Young vines with shallow root systems are more susceptible than older established vines. Conditions such as the lack of snow

FIGURE 1. Grape bud freezing tolerance. a. Concord grape bud subjected to a sublethal freezing temperature (−4°C). b. Presence of freezing injury is indicated by tissue browning in dormant primary and secondary Concord grape buds subjected to lethal freezing temperatures (−8°C) during the early acclimation stage.

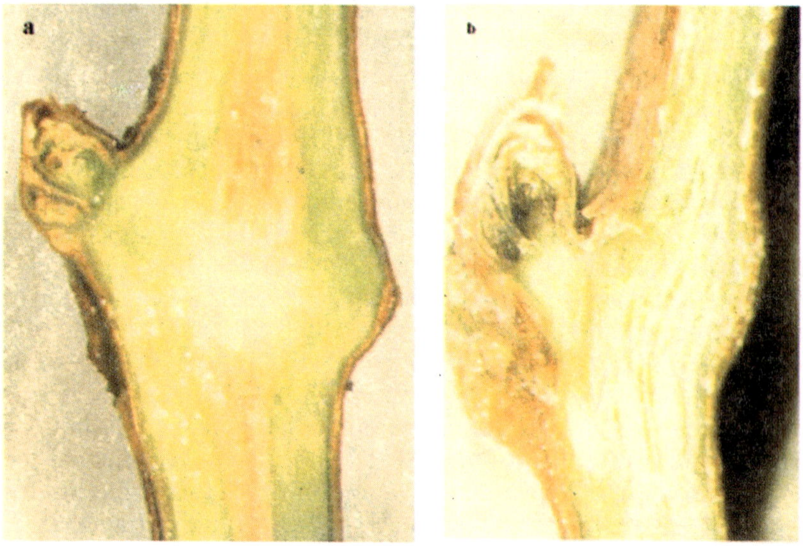

cover and very dry soil with sustained low temperatures can result in deep soil freezing and contribute to root damage. Root damage may be coupled with trunk and cane damage and be revealed by failure to grow or vine collapse during the growing season.

Trunk Freezing Tolerance

The trunk is generally the most freeze-tolerant portion of the vine, tolerating repeated freeze thaw episodes that can result in significant cane and bud damage without apparent trunk injury. However, trunk freezing damage and death does occur in many production sites, most commonly in regions with extreme or fluctuating winter temperatures, or areas with unseasonably low temperatures in the early fall (Ahmedullah, 1985; Hamman, 1993; Meiering et al., 1980; Mullins, 1986; Paroschy et al. 1980; Wolf and Warren, 2000). Late maturing vines that maintain high water contents are frequently more susceptible to this type of injury; however, with very cold temperatures even more hardy cultivars

TABLE 1. Range of grape bud freezing tolerance (°C) from dormancy induction to bud break in Northern Hemisphere. Bud freezing tolerance ranges were taken from studies that reported mean LTE or LT_{50} for more than one time point for field grown vines. Growing sites ranged from 34°N to 50°N.

Cultivar	August and September	October	November	December	January	February	March	Citation
***V. vinifera* L.**								
Cabernet Sauvignon	−5 to −8	−5 to −12	−9 to −22	−15 to −25	−17 to −26*	−18 to −25	−16 to −22	Andrews et al., 1984; Hamman and Dami, 2000; Proebsting et al., 1980; Wample, 1994; Wample and Bary, 1992; Wisniewski et al., 1996; Wolf and Cook, 1991, 1992, 1994.
Chardonnay		−10 to −15	−17 to −23	−20 to −29	−20 to −30	−17 to −24	−12 to −22	Andrews et al., 1984; Dami et al., 1996; Fennell unpubl. data; Jones et al., 2000; Jones et al., 1999; Wample, 1994; Wample and Wolf, 1996; Wolf and Pool, 1987, 1988; Wolf and Warren, 2000.
Riesling		−5 to −15	−10 to −25	−20 to −23	−22 to −26*	−20 to −27	−10 to −23	Andrews et al., 1984; Hubácková, 1996; Proebsting et al., 1980; Wample et al.,1993, Wisniewski et al., 1996; Wolf and Cook, 1994.
Viognier					−23 to −24	−22 to −25		Wolf and Warren, 2000;
Muscat Ottonel #1					−20 to −24	−20 to −24		Andrews et al., 1984; Wolf and Warren, 2000.
***V. labruscana* L. H. Bailey**								
Concord	−3 to −15	−12 to −22	−15 to −26	−20 to −29	−23 to −27*	−21 to −26	−10 to −25	Ahmedullah, 1985; Andrews et al., 1984; Fennell and Mathiason, 2002; Himelrick et al., 1991; Howell et al., 1978; Proebsting et al., 1980; Salzman et al., 1996; Stergios and Howell, 1977a; Wisniewski et al., 1996; Wolf and Cook, 1992,1994; Wolpert and Howell,1984, 1985a,b, 1986a,b.

TABLE 1 (continued)

Cultivar	August and September	October	November	December	January	February	March	Citation
***V. riparia* Michx.**	−8 to −12	−15 to −20	−15 to −22	−25 to −32	−30 to −42	−25 to −32		Fennell and Mathiason, 2002; Jones et al., 1980; Pierquet and Stushnoff, 1980; Pierquet, Stushnoff and Burke, 1977.
***V. aestivalis* Michx.**								
Norton (A)				−28 to −29	−24 to −25			Gu, 1999; Wolf and Cook, 1994.
Hybrids *Vitis* spp.**								
Carlos (M)			−11 to −13	−18 to −20	−21 to −22.5	−22 to −23		Clark and Watson, 1998; Clark et al., 1996.
Marechal Foch (V)			−20 to −22	−21 to −27	−28	−27		Andrews et al., 1984; Quamme, 1986.
Mars (L,V)		−10 to −15	−10 to −20	−19 to −25	−20 to −26	−15 to −25		Bourne and Moore, 1991; Bourne et al., 1991; Clark and Watson, 1998; Clark et al., 1996.
Saturn (L,V)		−5 to −10	−18	−20	−20	−20		Bourne and Moore, 1991; Bourne et al., 1991;
Seyval Blanc (V,Ru,A)		−21	−22	−23 to −26	−21 to −22	−17 to −18		Gu, 1999; Slater et al., 1991; Wisniewski et al., 1996; Wolf and Cook, 1994.
Summit (M,V)			−11 to −14	−18 to −20	−22 to −23	−22 to −23		Clark and Watson, 1998; Clark, Wolf and Warren, 1996.

* one report of very low temperature survival (−31, −37, and −40°C for Cabernet Sauvignon, Riesling, and Concord, respectively)
** Letters indicate predominate species in hybrid:
A = *V. aestivalis* Michx.
L = *V. labruscana* L. H. Bailey
M = *V. rotundifolia* Michx. (Muscadine)
V = *V. vinifera* L.
Ru = *V. rupestris* Scheele

TABLE 2. Relationship between seasonal freezing events, vine phenology, and resultant damage.

Timing of freeze event	Vine phenology	Primary injury	Factors that contribute to limiting the extent of damage
Early fall–prior to low temperature acclimation	Active growth/buds paradormant	Defoliation: (limit carbohydrate accumulation and acclimation), decreased fruit quality, shoot damage, bud kill	Early fruit and vine maturation, cultivars responsive to decreasing photoperiod, mild water stress, rootstocks, avoiding frost prone sites
Late fall–unseasonable low temperatures	Buds paradormant to endodormant	Kill primary buds, damage canes and trunks	Early fruit and vine maturation, cultivars responsive to decreasing photoperiod, rootstocks, site selection
Winter–low temperature extremes	Buds endodormant	Kills buds, canes, and trunks	Cultivar selection relative to growing site, use of hybrids containing *V. riparia* or *V. amurensis*, site selection, dual trunk training
Late winter–temperature fluctuations	Endodormant/ecodormant	Kills buds	Cultivar selection and site selection, late pruning after damage is determined
Early Spring–temperature fluctuations	Bud break	Kills primary buds	Late budding cultivars, fruitful secondary buds, dormant oils, evaporative cooling, site selection
Late Spring–unseasonable low temperatures	Shoot elongation	Shoot death	Slow developing cultivars, fruitful secondary buds, long growing season, avoiding frost prone sites

can be injured (Meiering et al., 1980; Paroschy et al., 1980). Damage ranges from death of young phloem cells in the trunk to death of the entire phloem tissue, cambium, and xylem ray cells (Pool, a). In the growing season following the injury, bud break may occur normally, but portions of the vine or an entire vine may collapse because of insufficient water transport during periods of high transpirational demand. Varying degrees of damage in the phloem tissue can result in stunted growth due to inability to mobilize carbohydrate reserves into the newly developing shoots, poor bud break, or complete failure to break bud. Freezing injury that results in bud death will further limit the ability of the vine to recover, since bud break and active bud growth stimulate vascular cambial division or development of new vascular cambium from the surviving phloem or xylem ray tissue (Pool, b). The recovery of a damaged vine is further influenced by the length of the growing season and whether there are repeated injuries in the following winters. A short-growing season will often not permit tissue maturation prior to the onset of low temperatures and repeated injury eventually kills the vine (Mullins, 1986).

Cane Freezing Tolerance

Various cane tissues either tolerate extracellular freezing or avoid freezing by supercooling (Hamman et al., 1990; Wolf and Pool, 1987; Pierquet et al., 1977; Pierquet and Stushnoff, 1980; Quamme, 1986, 1995). The lethal temperature varies with tissue type and stage of development of the cane. The phloem is the most susceptible and xylem the least susceptible to freeze injury (Pierquet and Stushnoff, 1980; Slater et al., 1991; Warmund et al., 1986). During the early fall, cane freezing tolerance increased from about −3°C to about −15°C in Concord (*V. Labruscana* L. H. Bailey) as the shoot matured into an overwintering cane (Howell and Shaulis, 1980; Stergios and Howell, 1977a,b; Wolpert and Howell, 1985a,b,1986a,b). With the onset of subzero temperatures the cane acclimated further and reached a maximum freezing tolerance during the winter dormant season. Freezing tolerance of the phloem during the winter ranged from −15 to −32°C and was influenced by the cultivar and prior exposure to subzero temperatures (Slater et al., 1991; Stergios and Howell, 1977a,b; Wolpert and Howell, 1984, 1985b, 1986a). The xylem had greater freezing tolerance than the phloem, with midwinter xylem freezing tolerance ranging from −20 to −50°C (Hamman et al., 1990; Pierquet et al., 1977; Rajashekar and Burke, 1996; Slater et al., 1991). After the fulfillment of the chilling requirement, cane freezing tolerance decreased with the onset of warm temperatures and bud break. The decrease in freezing tolerance was very dependent on temperature conditions and the progress of bud activity prior to a freeze event. Cane freezing tolerance in *V. vinifera*, *V. labruscana*, and hybrid cultivars during this period extended from −25 to −8°C (Hamman et al., 1990; Jones et al., 1999; Miller et al., 1988; Slater et al., 1991; Stergios and Howell, 1977a; Wolpert and Howell, 1984). As overwintering buds resumed growth, cane freezing tolerance reached a minimum around −4°C (Hamed et al., 2000).

Freezing injury in canes frequently occurs during unseasonably low temperatures in the fall when the canes are maturing and acclimating, particularly those canes in the interior of the canopy which mature more slowly (Howell and Shaulis, 1980; Miller et al., 1988; Striegler and Howell, 1991; Wolpert and Howell, 1985b, 1986a). The cane matures acropetally and the extent of maturation can be observed by the transition of the cane from green to yellow and brown with the development of periderm (Fennell and Hoover, 1991;Wolpert and Howell, 1986a). In normal seasonal weather patterns, the apical portion of the cane has the highest water content and is generally more sensitive to freezing injury than the basal portion of the cane (Wolpert and Howell, 1986a). Death of the apical portion of the cane is not a crop threatening injury as it is generally removed during annual pruning. An early fall freeze event, however, can result in death of all cane tissues in vegetatively immature green to yellow canes. Likewise, extreme low temperatures in the winter or early spring can injure the phloem or kill all cane tissue, thereby decreasing production potential in the following growing season.

Changes in freezing tolerance of the cane are directly correlated with decreased water content and starch and water soluble carbohydrate accumulation in the cane. Water content was most strongly correlated with increased freezing tolerance during the cane maturation phase in Concord (Wolpert and Howell, 1985a,b, 1986a). There was little correlation between water content and freezing tolerance during midwinter in *V. vinifera* or *V. labruscana*, as there was little change in water content during this period, but there was an increase in freezing tolerance (Hamman and Dami, 2000; Wolpert and Howell, 1984, 1985b). During the spring, there was little correlation between water content and freezing tolerance as the cane hydrated rapidly but lost freezing tolerance more slowly (Hamman et al., 1990).

Glucose, fructose, sucrose, raffinose, and stachyose increased in the canes of several *V. vinifera* cultivars during maturation and dormancy induction and with exposure to low temperatures (Hamman and Dami,

2000; Hamman et al., 1996; Jones et al., 1999; Wample and Bary, 1992; Wample et al.,1993; Warmund et al., 1986). Starch increased with cane maturation but decreased upon exposure to low temperatures (Jones et al., 1999; Wample and Bary, 1992), resulting in a negative correlation between starch accumulation and freezing tolerance. The decreased starch content was correlated with increased glucose in the canes upon exposure to low temperatures. In contrast, there was a strong correlation between increases in soluble sugars and increased cane freezing tolerance. A recent study used path analysis to describe interrelationship between soluble carbohydrates, starch, and cane freezing tolerance and distinguished direct effects of specific carbohydrates from the indirect effects of total soluble carbohydrates on freezing tolerance (Jones et al., 1999). Path analysis of carbohydrates indicated that soluble carbohydrates explained 93% of the variation in cane freezing tolerance, with fructose having the single largest effect and the other sugars having a smaller effect.

Bud Freezing Tolerance

Tolerance of freezing temperatures changes in the bud during maturation, dormancy, and bud break. Maturation and induction of dormancy in the overwintering buds proceed from the base of the cane to the apex. The apical buds, like the canes, are more susceptible to freeze injury than are the basal buds during this phase of development (Fennell and Hoover, 1991; Miller et al., 1988; Wolpert and Howell, 1985a,b, 1986a,b). During the dormant period, overwintering buds are more susceptible to freezing injury than are the canes or trunk (Ahmedullah, 1985; Jones et al., 1999; Miller et al., 1988; Quamme, 1986). The overwintering bud contains a primary, secondary, and tertiary bud. The primary and secondary buds are mixed buds containing leaf and flower primordia, and the tertiary bud is more commonly vegetative (Mullins et al., 1992). The primary bud is the most susceptible to freezing injury, followed by the secondary and tertiary buds, respectively (Hemstead and Luby, 2000; Howell et al., 1978; Pierquet and Stushnoff, 1980; Seyedbagheri and Fallahi, 1994; Stergios and Howell, 1977a,b; Wample and Wolf, 1996, Wolf and Cook, 1994). Thus, a damaging freeze event may result in injury of only the primary bud or some of the buds on a vine, permitting the vine to recover, grow, and fruit. A healthy productive grapevine typically produces more canes and buds than is necessary for optimal yields; therefore, annual balanced pruning removes excess buds and balances vegetative growth and fruit load. Many cultivars lost up to 40% of the primary buds without impacting yields when the freeze event occured prior to annual pruning (Wolf and Warren, 2000). In addition, secondary buds in some species and cultivars are productive and can partially offset losses of the primary bud crop.

Seasonal bud freezing tolerance ranges from −2 to −42°C, with the greatest freezing tolerance developing in the dormant bud upon exposure to subzero temperatures (Table 1; Bordelon et al., 1997; Fennell and Hoover, 1991; Jones et al., 1999; Miller et al., 1988). Bud freezing tolerance varied within the same cultivar over the course of a season, between cultivars growing on the same site, or with the same cultivars on different growing sites (Bordelon et al. 1997; Pool et al., 1992). There was also considerable variation in bud freezing tolerance from base to tip of the cane and among canes on the same vine (Wolpert and Howell, 1985a,b, 1986a,b). A difference in freezing tolerance of 5 to 6°C has been found in buds collected on the same day from similar vine positions throughout a cultivar planting (Bourne and Moore, 1991). The range of freezing tolerances shown in Table 1 reflect differences in temperature patterns in various growing regions and the interaction of cultivar with climatic conditions. Although there are similarities among some cultivars, differences in cultivar response are particularly evident in the early maturation and acclimation period. Cultivars from *V. vinifera*, *V. rotundifolia* Michx., and hybrids with a high proportion of *V. vinifera* develop freezing tolerance more slowly than the *V. riparia* and *V. labruscana* genotypes.

The developing bud has limited freezing tolerance during the growing season and is typically injured at −2 to −4°C. Bud dormancy is initiated at the base of the cane and develops acropetally (Fennell and Hoover, 1991; Wolpert and Howell, 1986b). Gradually decreasing daylengths and photoperiods less than 13 hours promoted initiation of bud dormancy and increased bud freezing tolerance 2 to 5°C in Concord and *V. riparia*, but did not appear to promote dormancy or acclimation in *V. vinifera* (Boone and Durham, 1996; Fennell and Hoover, 1991; Schnabel and Wample, 1987; Wake and Fennell, 2000; Wolpert and Howell, 1986a,b). The differences found in photoperiod induction of dormancy response may contribute to differences in freezing tolerance between cultivars during the early maturation and acclimation phase and ultimately influence winter survival (Table 1; Mullins, 1986). Low temperature exposure further increases bud freezing tolerance in the photoperiod-responsive genotypes and induces dormancy and acclimation in the nonphotoperiod-responsive cultivars (Meiering et al., 1980; Paroschy et al., 1980; Schnabel and Wample, 1987).

Changes in bud water content, soluble carbohydrates, and proteins are correlated with the increases in freezing tolerance during the transition from nondormant to dormant overwintering buds. Bud soluble proteins increased under short photoperiod during dormancy induction and a 27 kD LEA-like protein increased in buds with cold acclimation (Salzman et al., 1996; Wake and Fennell, 2000). While changes in proteins may contribute to freezing tolerance, a direct effect has not been established in grapevines.

Bud starch levels decreased and soluble carbohydrates increased with exposure to low temperatures (Jones et al., 1999; Wample and Bary, 1992; Wample et al., 1993). Increases in soluble carbohydrates were strongly correlated with increases in bud freezing tolerance (Hamman and Dami, 2000; Hamman et al., 1996; Jones et al., 1999; Seyedbagheri and Fallahi, 1994; Wample and Bary, 1992; Warmund et al., 1986). In field studies, cultivars with the lowest level of raffinose and other carbohydrates were the least freezing tolerant at leaf fall (Hamman et al., 1996). Levels of fructose, glucose, and sucrose were associated with increased freezing tolerance in Cabernet Sauvignon (*V. vinifera*); however, the relationship was not direct as there were large changes in carbohydrates and small changes in freezing tolerance during the midwinter period (Wample and Bary, 1992). Analysis of carbohydrates in Chardonnay and Riesling (*V. vinifera*) internodes indicated that high levels of glucose, fructose, raffinose, and stachyose were correlated with freezing tolerance, and that a high ratio of glucose and fructose to sucrose coincided with maximum freezing tolerance (Hamman et al., 1996). In contrast, raffinose and stachyose were the only sugars correlated with freezing tolerance in the hybrid cultivar Valiant (Stushnoff et al., 1993). Path analysis of the effect of individual soluble sugars indicated that fructose, sucrose, and raffinose had a direct effect on freezing tolerance. The effect of glucose was indirect and related to its role within other sugars. Path analysis of seasonal changes in carbohydrates indicated that the soluble carbohydrates explained 65% of the variation in bud freezing tolerance (Jones et al., 1999).

Bud water content decreases with induction of dormancy and acclimation (Fennell et al., 1996; Meiering et al., 1980; Salzman et al., 1996; Wolpert and Howell, 1984,1985a,b). Bud dormancy progresses from the basal to the apical buds in response to short photoperiods and/or low temperature (Fennell and Hoover, 1991; Wolpert and Howell, 1986a). Water content is lowest in the basal buds and highest in the apical buds in early fall. Water content continues to decrease in all buds with low temperature acclimation until early winter and remains fairly constant until deacclimation begins with the onset of warm temperatures in the spring (Salzman et al., 1996; Wolpert and Howell, 1984, 1985b). The changes in water content are strongly correlated with increases in freezing tolerance and are thought to contribute to an increased ability to supercool (Bourne and Moore, 1991; Pierquet et al., 1977; Quamme, 1995). In contrast to the mature cane, which can tolerate the presence of ice, the overwintering bud tolerates subzero temperatures by avoiding freezing or supercooling. Water in the bud primordium remains unfrozen at subzero temperatures as long as it is isolated from ice nucleation events; thus, the bud avoids dehydration stress as the water does not migrate to the extracellular ice in bud scales or cane (extraorgan freezing). When the supercooled bud does freeze, intracellular ice forms, resulting in bud death.

During the growing season the developing bud has limited ability to supercool and, like the actively growing shoot, it is susceptible to freeze injury at fairly high freezing temperatures. The decrease in water and increase in carbohydrates and proteins may contribute to the bud's ability to supercool; however, the bud water also needs to be isolated from ice nucleation, particularly from extracellular freezing events that occur in subjacent tissues of the cane. Ice nucleation is propagated readily in the vascular system of the cane. The dormant bud does not have a functional vascular connection to the cane; therefore it is isolated it from one potential source of ice nucleation (Barka et al., 1995; Kang et al., 1998; Quamme, 1995).

Although the bud is connected to cane tissues that freeze extracellularly, the tissues directly adjacent to the bud appear to contribute to the bud's isolation, inhibiting ice propagation from the cane. Thermal analysis studies indicated that overwintering buds did not achieve maximum supercooling (lack a low temperature exotherm) when they were excised without node interface tissue and frozen on ice nucleated moist filter paper (Wolf and Pool, 1987). If a small amount of cane tissue was excised with the bud, it retained its ability to supercool, even when the cane tissue had frozen, suggesting that there was a structural barrier to ice propagation into the bud (Andrews et al., 1984; Quamme, 1986; Wolf and Pool, 1987). During maturation and acclimation, evidence has been found for physical and biochemical changes in the bud and adjacent cane tissues that could be inferred to contribute to a barrier to ice propagation. In *V. riparia*, nuclear magnetic resonance imaging (MRI) indicated a change in the state of water (as measured by proton T_2 relaxation times) was associated with acclimation and dormancy development (Figure 2, Fennell and Line, 2001). Proton T_2 relaxation times decrease with increasing association of the water molecule with other molecules, such as carbohydrates or proteins, that restrict its motional freedom. In *V. riparia*, decreased proton T_2 relaxation times occurred in the tissues subjacent to the bud during short photoperiod induction of dormancy and acclimation. The decreased T_2 relaxation times in response to short days suggested biochemical changes occurred in these tissues during early acclimation (Figure 2; Fennell and Line, 2001). In addition, comparison of T_2 relaxation times over a series of decreasing temperatures allowed a composite image to be constructed that mapped

FIGURE 2. T_2 nuclear magnetic resonance images of cross section of *V. riparia* node at 2, 4, or 6 weeks of long photoperiod (a, c, or e, respectively; 15 h) or short photoperiod (b, d, or f, respectively; 8 h). T_2 relaxation times are color coded based on the range of T_2 relaxation times determined for the node sections, as follows: $T_2 < 11$ ms = blue; 11 ms $\leq T_2 < 16$ ms = yellow; $T_2 \geq 16$ ms = red. Tissues subjacent to the bud (indicated by arrow) show a decreased relaxation time after two weeks of short photoperiod.

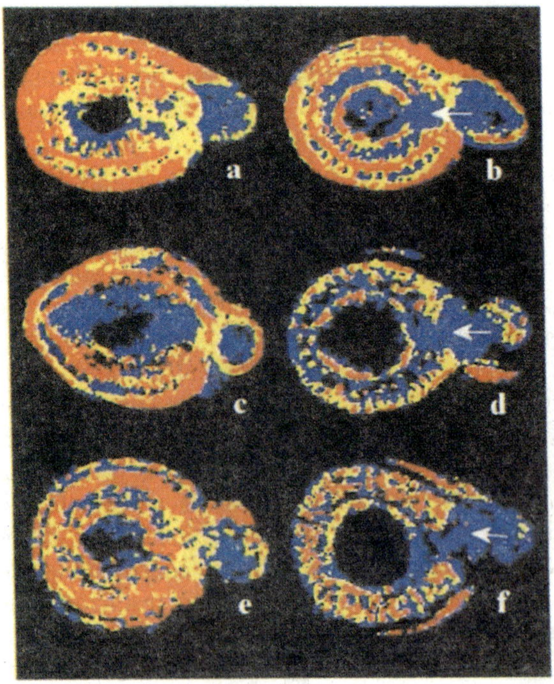

the freezing temperature of water in each tissue (Figure 3, Millard et al., 1995). When freezing composites of *V. riparia* were developed, differences in unacclimated (long day) or acclimating/dormant (short days) buds were very apparent (Figure 3). The majority of the water in short-day treated canes froze at lower temperatures than in long-day canes. In addition, differences in freezing temperature existed between cane tissues. The tissue subjacent to the bud froze at a lower temperature than other cane tissues. This suggested that biochemical changes occurred during maturation and acclimation and that there was a difference between the tissues in the bud axes and the rest of the cane. Biochemical changes are further indicated by the fact that cell wall strength and pore size increased during acclimation in grapevine cells (Rajashekar and Burke, 1996; Rajashekar and Lafta, 1996). Differentially sized dye exclusion tests indicated that the pore size and apoplastic permeability decreased in the bud axes tissues (Jones et al., 2000). Increased freezing tolerance and the ability to supercool were associated with this decreased permeability in the bud axes. Likewise, a decreased ability to supercool was associated with increased aploplastic permeability of the axes tissues. Treating the bud axes tissue with pectinase and phosphate removed pectin, increased apoplastic permeability and prevented supercooling (Jones et al., 2000). The MRI, thermal analysis, and dye exclusion studies indicated that biochemical changes occurred in the tissue resulting in a "barrier" with differential permeability that could limit ice

FIGURE 3. Representative nuclear magnetic resonance image of *V. riparia* grape buds frozen at 2°C/hour after exposure to 6 weeks of long photoperiod (a, 15 h) or 4 weeks of short photoperiod (b, 8h). Temperatures for freezing in the various tissue are color coded as red = −3 to −8°C, green = −8 to −12°C, gold = −13 to 17°C, and blue −17 to −22°C. Arrow indicates tissue subjacent to bud that has a very low freezing temperature.

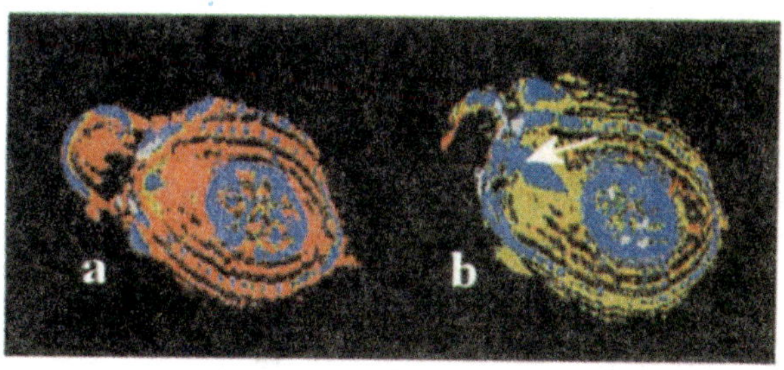

propagation from the cane into the bud, thereby promoting bud supercooling.

Although buds supercool and are most commonly injured by an intracellular freeze upon freezing of the supercooled water, there is some evidence that the bud can be injured by dehydration. The low temperature exotherm disappeared in dormant acclimated Chardonnay buds held for prolonged periods at −20°C without freezing (Wolf and Pool, 1987). The supercooled water was thought to migrate from the bud to tissues that had ice present such as the bud scales, thereby dehydrating the buds. Fifty percent of the Chardonnay buds were killed by these conditions whereas the acclimated bud could normally withstand transient freezing between −20 to −24°C without death. In regions that have sustained non-lethal freezing temperature dehydration injury could contribute to bud mortality.

In late winter, warm temperatures can promote deacclimation, and buds are injured when temperatures return rapidly to normal subzero conditions. Similarly, cultivars that commence growth early in the spring are susceptible to freeze damage when temperatures drop below zero. The bud deacclimates with the seasonal increase in temperatures at the end of winter, and there is little difference in bud freezing tolerance from basal and more distal positions during this period (Andrews et al., 1984; Barka et al., 1995; Hamman et al., 1990; Lipe et al., 1992; Slater et al., 1991; Wisniewski et al., 1996; Wolf and Cook, 1992). Freezing tolerance during the early spring is directly related to the air temperature, and buds that have not begun to break can reacclimate in response to low temperatures following brief warming periods (Hamman et al., 1990). Bud break is promoted by base temperatures of 4°C and as bud break proceeds, the bud is injured at progressively higher temperatures (Moncur et al., 1989). In Concord, the temperature at which 90% of the primary buds were killed in controlled freeze tests increased with increasing development from first swell, full swell, budburst, first leaf, and second to third leaf: −17°C, −12°C, −9°C, −6°C, and −5 to −3°C, respectively (Proebsting et al., 1980). Response of cultivars varied, with some cultivars commencing bud break early and rapidly but subsequently growing more slowly, while slow breaking cultivars develop more rapidly upon break (Anderson et al., 1980; Moncur et al., 1989; Wolf and Cook, 1992). Thus, the rate of bud break and the rate of development in the early spring determine vine susceptibility to spring freezes.

In field conditions, surface moisture on the buds may influence lethal temperatures. During radiation freezes when vines are resuming growth and buds are at first leaf stage, canes froze before buds, buds froze randomly, and many buds remained unfrozen (Fuller and Telli, 1999; Hamed et al., 2000). Thus a barrier to ice nucleation may still exist between the cane and bud at bud break when a functional vascular system is initially developing. Dry buds tolerated temperatures 3 to 4°C lower than wet buds as the surface moisture may have eliminated barriers to ice nucleation (Johnson and Howell, 1981; Wolf and Pool, 1987). On the other hand, an extended period of evaporative cooling may have delayed bud development and increased freezing tolerance. Bud break was delayed in Chardonnay vines evaporatively cooled by intermittent sprinklers (Lipe et al., 1992). In contrast, emerging buds were injured in noncooled vines following a late spring freeze. When freezing occurred before either the cooled or noncooled Cabernet Sauvignon vines broke bud, only the noncooled vines were injured, with fewer buds breaking and lower yields at the end of the season (Lipe et al., 1992).

After bud break, the actively growing shoot is very sensitive to freezing injury and has limited ability to acclimate to lower temperatures until the end of the growing season. Freezes in late spring and late summer to early fall cause severe damage to actively growing shoots, leaves, and fruit. Freeze damage during bud break and shoot development may be compensated for by growth of the secondary buds. A long growing season may allow the vine to recover and produce fruit, but yields are generally lower. Actively growing shoots are severely injured or killed at temperatures below −2.5°C (Fuller and Telli, 1999; Probesting et al., 1980). Typically these freeze events are the result of radiation freezes, making site selection and other cultural practices critical for avoiding freezing injury during the growing season (Fuller and Telli, 1999; Proebsting et al., 1991; Reynolds, 1987; Stergios and Howell, 1977b).

EVALUATING FREEZING TOLERANCE AND INJURY IN GRAPEVINES

Detecting differences in freezing tolerance between cultivars and accurately appraising the influence of environmental conditions and management practices on vine freezing tolerance provides information necessary to match cultivars to production regions and identify or modify growing sites. Screening cultivars under field conditions over a series of years in several locations provides information on cultivar suitability to specific growing regions and environmental conditions that contribute to injury. Controlled testing of cultivars helps identify tissues most susceptible to injury, critical temperatures for injury, and influence of genotype by environment or management practices on freezing tolerance and injury. A combination of field and controlled testing provides the best information base for understanding acclimation and freezing tolerance in the vine.

Field Analysis of Freezing Tolerance

The water-soaked appearance of leaves and shoots, followed by browning as the injured tissue desiccates, is striking evidence of a damaging freeze event. Monitoring chance field freezing events in the spring and fall, vine survival over several winters, and damage after specific extreme low temperature events provides information on the types of damage incurred on vines. Analysis of freezing injury and vine performance after early winter and midwinter extreme freeze episodes readily identifies cultivar differences. Performance of *V. vinifera*, *V. labruscana*, and other hybrid cultivars in several locations indicated that following optimal acclimation conditions, significant midwinter freeze damage began to occur at about −20°C in *V. vinifera* cultivars and vine survival was limited at temperatures below −26°C (Ahmedullah, 1985; Bordelon et al., 1997; Hamman, 1993; Mullins, 1986; Wolf and Miller, 2001; Wolf and Warren, 2000). In contrast, some *V. riparia* hybrids have survived −38°C with less than 50% bud injury (Hemstead and Luby, 2000). Observations of grape cultivar performance over a ten-year period on the Tennessee Cumberland Plateau, a region with a short growing season, early winter freeze episodes, and fluctuating winter temperatures, indicated that early winter freezes were particularly devastating (Mullins, 1986). Although the cultivars Concord and Rougeon (*V.* spp.) were productive in the short growing season, repeated injury rather than a single low temperature winter eventually destroyed the Tennessee vineyard. A five-year evaluation of cultivars in Colorado with minimum midwinter temperatures reaching −22 to −25°C indicated that many *V. vinifera* and hybrid cultivars could survive with minimal damage (Hamman, 1993). However, severe trunk damage was apparent on some cultivars. Rougeon and Riesling were the most freezing tolerant in this planting, sustaining little damage at −29°C. In Washington, evaluations of *V. vinifera* and hybrid cultivars after temperatures as low as −26°C indicated that many of the cultivars could survive low temperatures with less than 50% bud injury (Ahmedullah, 1985; Clore et al., 1974). In Indiana and Ohio, *V. labruscana* and *V.* spp. cultivars survived temperatures of −31 to −32°C, but all *V. vinifera* cultivars except Riesling sustained severe damage (Bordelon et al., 1997). In Minnesota, severe damage was noted in all cultivars except Michurintz (*V. amurensis*), Beta (*V. riparia* × *V. labruscana*), and several hybrid selections (*V. riparia* × *V. vinifera*) after a field temperature of −38°C (Hemstead and Luby, 2000). These studies illustrate genotypic differences in freezing tolerance but also emphasize the importance of early acclimation, acclimation conditions and temperature conditions just prior to a severe freeze event as some cultivars performed differently in the different regions.

Controlled Freeze Testing and Analysis of Freezing Tolerance

Since spring, fall, and winter conditions vary widely over the grape production regions and potential test winters can not be predicted, controlled freezing tests are frequently used to determine freezing tolerance in grapevines. Controlled freeze testing, with accurate tissue temperatures and viability assays, provides an early assessment of new genotypes and aids in identifying critical environmental and cultural influences on established vines. In grapevines, controlled freezing tests are typically characterized by a dose response test or thermal analysis, and the impact of a particular test is evaluated using one or more viability tests. Reviews of freeze test methodology can be found elsewhere, only studies that use grapevine tissues are addressed here.

Tissue Viability Analysis

Accurately evaluating injury after a freezing event requires a method of assaying tissue viability that is relatively quick and objective. Bud break and shoot growth and yield components provide the best indicators of vine viability after a freeze episode; however, these methods require weeks to months to complete. A more rapid test to determine damage prior to bud break would allow a grower to adjust pruning levels in accordance with bud injury and train new trunk systems as needed.

Electrolyte leakage or specific conductivity is frequently used to assess cellular injury after a freezing event. Specific conductivity of apoplastic water can be measured after thawing of frozen tissues. Comparisons of electrolyte leakage from freeze injured and uninjured tissues provide a rapid method of detecting injury during dormancy; however, caution must be used with this technique during induction of dormancy and bud break periods, as transition to active bud growth alone in the absence of a freezing event will increase specific conductivity of the apoplastic water (Barka and Audran, 1996; Jones et al., 1999; Stergios and Howell, 1973). Similarly, triphenyl tetrazolium chloride (TTC) can be applied to tissues, and a high level of reduction of TTC is correlated with viability of the tissue (Stergios and Howell, 1973). Both TTC and specific conductivity testing can be done immediately following a freeze event and are objective when used with proper controls and calibrated with bud break, growth, or tissue browning assays; however, they are not practical for screening large numbers of samples.

Chlorophyll fluorescence using the ratio of the variable fluorescence to maximum fluorescence (Fv/Fm) provided a measure of phloem and cambial injury in the cane when compared with tissue browning or growth response after a freezing event (Jiang et al., 1999). Decreased Fv/Fm ratios in buds subjected to controlled freezes corresponded to rate of injury determined from by oxidative browning (Düring et al., 1990). This provides an objective method for assessing bud and cambial tissue after calibration curves are established; however, specialized equipment is needed to process samples.

Oxidative browning of cane and bud tissues provides a visual indication of freezing injury. A strong correlation exists between browning and freezing of supercooled buds and lack of bud break and growth (Hemstead and Luby, 2000; Jones et al., 1999 Stergios and Howell, 1973). Uninjured buds and phloem of the canes have a bright green color, but after cellular damage and release of cellular content the tissue turns brown. Buds and cane tissue are typically examined two to seven days after thawing to detect tissue browning. Although the test is subjective and a short waiting period after the freeze episode is required, a large number of samples can be rapidly screened. Tissue browning can be used reliably in the field, making it possible to quickly check for damage prior to pruning.

Analysis of Freezing Events

Ice nucleation and propagation in grapevine tissues are determined by an interaction of the temperature and ice nucleating agents, ionic and colloidal components of cellular water, tissue organization, and the presence or absence of surface moisture. Determining the actual temperature when a damaging freeze event takes place in a tissue requires the ability to simultaneously monitor tissue temperature and ice formation and correlate this with tissue injury. In grapevines, determining whether a freezing event will cause damage is further complicated by the fact that different tissues tolerate freezing temperatures by either tolerating the presence of ice or avoiding ice formation. Conducting MRI over a series of subzero temperatures can indicate the temperature at which water freezes in different tissues (Figure 3); however, it is not possible to view the initiation and progression of the freezing event with MRI.

The recent advent of infrared thermography provides an opportunity to monitor freezing events in real time. Freezing exotherms are visualized using an infrared camera and video thermal imaging equipment (Fuller and Wisniewski, 1998; Wisniewski, 1998). Infrared thermography of dormant grape canes indicated that the cane froze before the supercooled bud promordia (Wisniewski, 1998). Examination of freezing events during bud break and early shoot growth also indicated that water in the cane froze before water in the bud (Hamed et al., 2000). Ice developed from the point of ice nucleation due to internal ice nucleators, and progresses rapidly throughout the cane tissue. The rate of ice spread in canes during bud break suggested that ice was propagated through the vascular system and most likely through the xylem (Hamed et al., 2000). Buds on the cane froze only after the cane had completely frozen and then the buds froze randomly. At higher freezing temperatures, such as those that occur in radiation freezes, many buds remained unfrozen (Fuller and Telli, 1999; Hamed et al., 2000). It was not possible to determine from these thermography studies whether buds remained unfrozen due to a barrier to ice propagation from the cane. It was also not possible to determine whether random bud freezing occured due to internal nucleation or ice being propagated from the cane as the barrier/resistance to ice propagation decreased with bud development (Hamed et al., 2000; Jones et al., 2000).

Controlled Freeze Testing

Freezing tolerance of grapevine tissues can be determined by subjecting ice-nucleated plant samples to decreasing temperatures and removing samples at specific temperature intervals to observe viability. Combining controlled freezing at 2 to 5°C/hour with tissue browning or electrolyte leakage is a frequently used method for evaluating freezing tolerance of grape buds and cane tissues. The lethal temperature for 50% of the buds (LT_{50}), calculated using the Spearman-Karber method, has been related to bud survival and productivity in the field (Bittbender and Howell, 1974; Clore et al., 1974; Howell et al., 1978; Stergios and Howell, 1977a; Wolf and Cook, 1994). Care must be taken, however, to uniformly sample the vine, using well exposed canes that might otherwise be kept during pruning (Howell and Shaulis, 1980). This technique has been particularly useful for monitoring cane and trunk samples. Much of the cane water freezes extracellularly and although supercooling was observed in the canes and related to cane xylem injury, thermal analysis has not routinely been used to determine damage in cane tissues (Pierquet and Stushnoff, 1980; Pierquet et al., 1977; Hamman et al., 1990). Dose response freeze test and monitoring electrolyte leakage or oxidative browning have been used to identify phloem and xylem damage (Slater et al., 1991; Jones et al., 1999). These techniques have also been useful for determining freezing tolerance during bud development transition phases at the end of the growing season or during bud break when bud supercooling is limited (Anderson et al., 1980; Fennell and Hoover, 1991; Johnson and Howell, 1981; Schnabel and Wample, 1987; Wolpert and Howell, 1986a).

The ability to supercool and the lethal intracellular freezing that usually takes place with ice nucleation of supercooled water make it possible to readily detect damaging freeze events in grape buds (Andrews et al., 1984; Pierquet and Stushnoff, 1980; Pierquet et al., 1977 Quamme, 1986; Wolf and Pool, 1987). Thermal analysis detects freezing events in tissues from the heat of fusion released (exotherm) upon freezing. A grape node section typically has a high temperature exotherm (HTE) as water in the cane and bud scales freeze and one or more low temperature exotherms (LTE) as supercooled water in the bud primordia freezes (Andrews et al., 1984; Pierquet and Stushnoff, 1980; Pierquet et al., 1977; Quamme, 1986; Wolf and Pool, 1987). Two exotherms were identified in the cane internode of Concord, and samples that were thawed and refrozen showed one exotherm, suggesting that bud tissues were killed (Stergios and Howell, 1973). The relationship between LTE and bud death was verified in *V. riparia* grape buds by removing internode samples at temperatures higher than the predicted LTE or immediately after the LTE occurred and observing tissue viability (Pierquet and Stushnoff, 1980). Grape bud supercooling and bud death upon freezing were further characterized by comparing thermal analysis and LTE with LT_{50} determined by dose response freezing and indicated by tissue browning. Supercooling was observed in 20 genotypes (*V. labruscana*, *V. vinifera*, and hybrid cultivars) and the LTE and LT_{50} were within 1°C of each other, verifying the usefulness of thermal analysis for predicting freezing tolerance (Andrews et al., 1984).

Subsequent studies emphasized the correct methods for using thermal analysis to accurately predict freezing tolerance. Cooling rates of

1.5 to 10°C/hour produced consistent LTE temperatures; however, very fast cooling rates (40°C/hour) produced inconsistent LTEs (Quamme, 1986). Interpretation of the LTE can be confounded by the fact that the compound bud can produce multiple LTEs; however, the primary bud typically has a larger LTE at a higher temperature than the secondary and tertiary bud (Andrews et al., 1984; Pierquet et al., 1977; Wolf and Cook, 1994; Wolf and Pool, 1987). In the early fall, the HTE was difficult to separate from LTE when buds, excised with 2 mm of subjacent tissue, were frozen without ice nucleation. Placing samples on moistened filter paper and ice nucleating provided a consistent HTE that was more easily distinguished from the lethal LTE (Wolf and Pool, 1987). It was also noted that buds excised from the cane have a higher LTE temperature or no LTE than buds excised with an intact cane or with a portion of the cane still attached. This indicated that the tissue subjacent to the bud provided a barrier or resistance to ice propagation that was essential for the bud to supercool (Quamme, 1986; Wolf and Pool, 1987).

LTEs can occur over a range of temperatures (up to 7°C difference), even in samples taken from the same planting and from canes of similar diameter and position in the canopy (Bourne and Moore, 1991; Gu, 1999; Wolf and Pool, 1987). A mean LTE or median LTE of several samples is typically reported as the freezing injury temperature (Andrews et al., 1984; Bourne and Moore, 1991; Jones et al., 1999; Kang et al., 1998; Quamme, 1986; Wolf and Pool, 1987; Wolf and Cook, 1994). Examination of the LTEs in multiple samples and cultivars indicated that some cultivars have a narrower range of bud kill than do others and that median lethal temperature combined with a lethal temperature coefficient describing the rate of increased injury for each degree temperature decrease could provide a better interpretation of cultivar responses (Gu, 1999). However, the relationship between LTEs and field survival and productivity has been difficult to accurately establish. Thermal analysis studies of *V. rotundifolia* Michx. cultivars indicated that the buds were more freezing tolerant than what was usually considered the minimum temperature for their survival (Clark and Watson, 1998). Freezing tests that are conducted just prior to a natural extreme low temperature event provide the best estimate the relevance of thermal analysis for predicting grape bud hardiness. In *V. rotundifolia* injury occured in the field at higher temperatures than was predicted with thermal analysis (Clark and Watson, 1998; Clark et al., 1996). However, relative differences detected between cultivars with thermal analysis did correspond to relative differences in winter survival observed between cultivars in the field (Bourne et al., 1991). Studies of *V. vinifera*, *V. labruscana*, and hybrid cultivars showed that there was a strong relationship between the mean LTE and killing temperatures under field conditions (Wolf and Cook, 1994; Wolf and Warren, 2000). These studies indicate that appropriately performed thermal analysis provides a rapid and consistent method to evaluate grape bud freezing tolerance.

FACTORS INFLUENCING FREEZING TOLERANCE

Freezing tolerance varies with phenology; however, climatic conditions and management practices can alter a vine's tolerance (Table 2). Genotypic differences and interactions between genotype and environment also influence freezing tolerance. Identifying the genetic potential of different cultivars and the critical environmental and cultural factors influencing the expression of maximum freezing tolerance provides information needed for successfully managing vineyards and developing new cultivars with a greater freezing tolerance.

Genetic Influences

Wide distribution of the different species of grapes indicates a considerable genetic variability for cold hardiness within *Vitis*. *V. rotundifolia* (Muscadine grape) from the southeastern United States is recommended for regions with winter temperatures above −12°C, with death of the vine possible at −18°C (Ahmedullah and Himelrick, 1990; Clark and Watson, 1998). *V. vinifera* cultivars are most frequently grown in regions with midwinter temperatures above −22°C, although some cultivars survive lower temperatures (Alleweldt et al., 1990; Wolf and Miller, 2001; Wolf and Warren, 2000). Some species native to the northern regions of North America and Asia can tolerate much lower temperatures than *V. vinifera*. *V. riparia*, and *V. amurensis* have been reported to survive −30 to −40°C (Ahmedullah, 1985; Alleweldt and Possingham, 1988; Alleweldt et al., 1990; Hemstead and Luby, 2000; Pierquet and Stushnoff, 1980; Seyedbagheri and Fallahi, 1994). Hybrid cultivars that had a high percentage of *V. labruscana* were more freeze tolerant than genotypes with a high percentage of *V. vinifera* (Bourne et al., 1991). Breeding programs that use accessions from the northern limits of the native range of *V. riparia* or *V. amurensis* in crosses with *V. vinifera* have produced cultivars that have superior winter survival to many *V. vinifera* cultivars (Bordelon et al., 1997; Hemstead and Luby, 2000;

Luby, 1991; Pierquet and Stushnoff, 1978; Seyedbagheri and Fallahi, 1994).

Cultivars that cease vegetative growth and mature late can experience bud, cane, and trunk damage by low temperatures in fall and early winter (Ahmedullah, 1985; Clore et al., 1974; Meiering et al., 1980; Mullins, 1986; Paroschy et al., 1980). In hybrid cultivars of *V. vinifera* and *V. labruscana*, the cultivars with a higher percentage *V. vinifera* acclimated more slowly than those with a greater component of *V. labruscana* in their genotype (Bourne et al., 1991). In midwinter, Cabernet Franc (*V. vinifera*) buds were more freezing tolerant (1 to 2°C) than Cabernet Sauvignon, but Cabernet Franc deacclimated more rapidly than Cabernet Sauvignon (Wolf and Cook, 1991). Observations of rootstocks Kober 5BB (5BB), Couderc 3309 (C3309), and Selection Oppenheimer No. 4 (SO4, *V.* spp.) under natural conditions indicated that C3309 acclimated more rapidly than 5BB and SO4 (Miller et al., 1988). In contrast, 5BB deacclimated more rapidly than C3309 and SO4. Buds of *V. rotundifolia* cultivars survived freezing temperatures of −20 to −23°C in laboratory tests but were killed by −13°C freeze in the field after a warm period in the midwinter (Clark and Watson, 1998). *V. amurensis* can deacclimate rapidly when subjected to high midwinter temperatures and can be damaged in regions with fluctuating winter temperatures (Reisch and Pratt, 1996). A review of ten hybrid cultivars showed considerable variation in timing of bud break and rate of shoot growth (Anderson et al., 1980). Cultivars that resumed growth early in the spring or developed more rapidly after bud break may be more susceptible to spring freezes (Fuller and Telli, 1999). Thus, genetic differences in rate of response to changing environmental conditions also contribute to the freezing tolerance of a genotype and should be considered relative to the common environmental conditions of a particular growing region when selecting cultivars.

Environmental Influences

Ambient temperature is a major factor determining grapevine tissue freezing tolerance. Upon exposure to low temperatures in the fall vines begin acclimating and tolerate lower freezing temperatures. Bud and cane freezing tolerances increased rapidly in Concord when fall temperatures dropped below 10°C (Wolpert and Howell, 1985b). Microclimates within the same planting location provide direct evidence of the influence of small temperature differences on vine freezing tolerance. Low elevation field sites usually have air temperatures lower than higher elevation sites. In Concord, buds and canes acclimated more rapidly, had a greater freezing tolerance through the winter, and deacclimated more slowly in lower elevation sites than in higher elevation sites (Stergios and Howell, 1977b). However, low elevation sites are also more prone to late spring and early fall freezes and radiation freeze events during the growing season.

In dormant vines, decreasing ambient temperatures and exposure to subzero temperatures enhance freezing tolerance. Concord, Riesling, and Cabernet Sauvignon gained maximum freezing tolerance when exposed to temperatures below −5°C (Wolf and Cook, 1992). Seyval Blanc (*V.* spp.) buds exposed to −5°C were more freezing tolerant than those that were not (Slater et al., 1991). Chardonnay reached its maximum level of freezing tolerance when average temperatures were below 0°C (Jones et al., 1999). It is well established that the ambient temperatures the vines have been exposed to prior to an extreme low temperature influence survival (Hubáčková, 1996; Proebsting et al., 1980). The mean daily temperature for the five day period prior to a freezing event were more strongly correlated with the freezing tolerance of grape buds than other time periods (Wolf and Cook, 1992).

Deacclimation occurs in response to seasonal increases in temperature or a warming period during the midwinter. Merlot, Riesling, and Muller-Thurgau cultivars (*V. vinifera*) began deacclimating when mean ambient temperatures rose above 4°C (Hamman et al., 1990; Hubáčková, 1996). An increase in the LTE in *V. riparia* occurred after 24 hours at 4.5°C (Pierquet et al., 1977). Seyval Blanc buds and canes deacclimated more quickly upon exposure to 16°C in February than in January. The buds began deacclimating within 12 to 24 hours after warming and lost 3°C freezing tolerance in 48 hours (Slater et al., 1991). Concord, Riesling, and Cabernet Sauvignon buds deacclimated more rapidly at 23°C as the winter progressed, with Concord deacclimating most rapidly and Cabernet Sauvignon deacclimating most slowly (Wolf and Cook, 1992). Under natural conditions in Prosser, Washington, Chelois (*V.* spp.) deacclimated more rapidly than Concord, and Riesling (*V. vinifera*) deacclimated more slowly than Chelois and Concord (Andrews et al., 1984). Therefore, resistance to deacclimation during transient temperature increases could be a critical factor in regions where temperatures fluctuate during the winter, so differences in genotype should be considered.

Buds have the ability to reacclimate upon exposure to low temperatures following a brief warm period; however, once bud break begins vines can not reacclimate (Fuller and Telli, 1999; Schnabel and Wample,

1987; Stergios and Howell, 1977b). Genotypes with an early growth response are therefore most susceptible to spring freezes (Anderson et al., 1980; Fuller and Telli, 1999; Lipe et al., 1992).

The distribution of light within the canopy influences maturation and carbohydrate levels in canes and buds. High light levels promote a greater freezing tolerance, which is most apparent in the early acclimation period and can be easily seen in the brown canes in the exterior of the canopy, as opposed to green and yellow canes in shaded areas (Wolpert and Howell, 1986a). High light levels within the canopy also enhance the midwinter freezing tolerance of the canes and buds, most likely due to the increased level of carbohydrate storage (Hamman et al., 1996; Jones et al., 1999; Stergios and Howell, 1977a).

Decreasing daylength contributes to early acclimation response in some genotypes. In Washington, acclimation of grape canes can begin by the third week of August (Ahmedullah, 1985). In Concord, use of night interruption during natural fall daylengths maintained more actively growing shoots than found in the natural short days (Wolpert and Howell, 1986b). No increase in freezing tolerance was detected between the night interrupted vines (simulating long day) and vines receiving a decreasing photoperiod. In Riesling no acclimation was observed in response to short day lengths (Schnabel and Wample, 1987). In *V. riparia* and Concord, short photoperiods promoted growth cessation, the onset of dormancy and increased freezing tolerance (Fennell and Hoover, 1991; Fennell and Mathiason, 2002). Short days did not promote maturation and dormancy in Seyval Blanc; however, an F1 (*V. riparia* x Seyval Blanc) did mature and increase freezing tolerance in response to short photoperiods, although more slowly than *V. riparia* (Fennell and Mathiason, 2002; Wake and Fennell, 2000). Similarly, short photoperiods promoted increased periderm development, freezing point depression, and acclimation in rootstock C-1613 (*V. solonis* × Othello [*V. labruscana* × *V. riparia* × *V. vinifera*]). It is apparent that some of the northern species are responsive to changes in daylength and that this, coupled with decreasing temperatures, may promote a more rapid acclimation response than found in species that respond predominantly to decreasing temperature.

Vine Management Practices

Choosing an optimal site and matching cultivars to the climatic conditions of the region are the first considerations for limiting vine freezing injury. Other management decisions, such as type of training system, rootstock selection, fertilization and irrigation practices, and cropping level impact the overall vine health and may influence freezing tolerance (Ahmedullah and Himelrick, 1990; Reynolds, 1987; Seyedbagheri and Fallahi, 1994; Wample and Wolf, 1996).

Training systems that ensure maximum light exposure and balance vegetative and crop load maximize wood maturity and carbohydrate levels in canes. Matching a training system with cultivar and climatic conditions is discussed in detail in other publications, therefore only a few relevant factors impacting freezing tolerance will be included here. Timing of dormant pruning has a small influence on freezing tolerance. Concord vines pruned in the fall had more bud winter kill than unpruned vines (Wolpert and Howell, 1984). Midwinter pruning of Cabernet Sauvignon and Chardonnay had no effect on freezing tolerance and did not promote earlier bud break (Wample, 1994). In Merlot, pruning had no effect on the bud's ability to supercool and did not accelerate deacclimation, indicating that late winter pruning had no impact on vine freezing tolerance (Hamman et al., 1990).

Early defoliation and excessive crop loads can delay acclimation (Howell et al., 1978; Stergios and Howell, 1977a). Chardonnay vines with a low, moderate, or high crop load (7, 14, 21 MT/ha) all showed similar wood maturity; however, there was a greater freezing tolerance in the low crop load vines than in the high crop load vines during the fall acclimation period. There were no differences in freezing tolerance due to crop load during the midwinter and deacclimation period (Wample and Wolf, 1996). Early, late, or no harvest had only a small influence on cane carbohydrate levels and did not affect freezing tolerance of Chardonnay, Riesling, and Cabernet Sauvignon, indicating that timing of harvest in well maintained vines does not impact freezing tolerance (Hamman et al., 1996, Wample and Bary, 1992).

Rootstocks may have a direct influence on scion biochemistry and freezing tolerance or indirect impact on freezing tolerance by influencing vine size and canopy shading (Striegler and Howell, 1991). C3309 vines acclimated more rapidly and were more freezing tolerant than 5BB and SO4 vines (Miller et al., 1988). There was no difference in freezing tolerance of Seyval Blanc canes when they were grafted on 5BB or C3309 rootstocks (Striegler and Howell, 1991). Similarly, Chardonnay on C3309 rootstock had greater canopy shading than Chardonnay on Elvira rootstock; however, there was no difference in freezing tolerance among canes of good wood maturity that would normally be retained for fruiting (Wolf and Pool, 1988). A study using the hybrid St. Pepin on MN Rip 64 (*V. riparia*), MN 1065 (a hybrid containing *V. riparia*) and

C3309 rootstocks indicated that vines had a greater survival on the MN rootstocks than on C3309 after six winters (Hoover et al., 2002). The MN rootstocks also promoted an earlier cane maturation in the fall than did C3309 (Hoover personal communication). It is possible that rootstocks have a greater impact during acclimation and deacclimation phases rather than midwinter freezing tolerance, but this has not been addressed.

Fertilization and irrigation practices impact vine size and health and have been suggested to influence freezing tolerance. Three levels of nitrogen fertilization applied before bloom had no influence on cane maturation or freezing tolerance in Chardonnay vines on two different rootstocks (Wolf and Pool, 1988). Three levels of nitrogen fertilization in Riesling also indicated that recommended rates of nitrogen applied in the spring did not influence freezing tolerance (Wample and Wolf, 1996; Wample et al., 1993). High levels of irrigation can delay acclimation of vines. A mild water stress in Concord at the end of the season had no impact on primary bud freezing tolerance but did increase secondary and tertiary bud survival under field conditions (Wample and Wolf, 1996). In contrast, varying the irrigation level during the growing season did not influence soluble carbohydrate content or freezing tolerance in Cabernet Sauvignon (Hamman and Dami, 2000). In general, these studies suggest that standard management practices that maintain the health of the plant and balance vine vigor and productivity will maximize the inherent freezing tolerance of a cultivar.

CONCLUSIONS

Recent progress in describing the mechanisms involved in freezing tolerance and the use of controlled freeze testing coupled with field testing have promoted a greater understanding of freezing tolerance in grapevines. Freezing tolerance in grapevines is dynamic, changing with stage of development and differences in ambient temperature. Different tissues within an individual vine avoid injury upon exposure to freezing temperatures by either supercooling or tolerating the presence of ice. Freezing tolerance is also impacted by genotype by environment interactions which may result in differences in rate of acclimation and deacclimation between cultivars, thereby impacting expression of their inherent freezing tolerance. However, identification of the seasonal periods critical for damage in a given region and characterization of cultivar interaction with environment provide information needed for matching cultivars with growing sites and aiding the breeding or selection of cultivars with greater freezing tolerance.

REFERENCES

Ahmedullah, M. (1985). An analysis of winter injury to grapevines as a result of two severe winters in Washington. Fruit Var. J. 39:29-34.

Ahmedullah, M. and D.G. Himelrick. (1990). Grape management. Pp. 383-471. In: *Small Fruit Crop Management*. Prentice Hall, Englewood Clilffs, NJ.

Alleweldt, G. and J.V. Possingham. (1988). Progress in grapevine breeding. Theor. Appl. Genet. 75:669-673.

Alleweldt, G., P. Spiegel-Roy and B. Reisch. (1990). Grapes (*Vitis*). Acta Hort. 290:291-327.

Anderson, K.A., G.S. Howell and J.A. Wolpert. (1980). Phenological development of different *Vitis* (grape) cultivars (Relation to freeze damage). Fruit Varieties Journal 34:5-7.

Andrews, P.K., C.R. Sandidge III and T.K. Toyama. (1984). Deep supercooling of dormant and deacclimating *Vitis* buds. Am. J. Enol. Vitic. 35:175-177.

Barka, E.A. and J.C. Audran. (1996). Utilisation de la conductivite specificque comme critere d'estimation de la viabilite au niveau de l'appareil aerien des vignes champenoises soumises aux temperatures neagtives. Can. J. Bot. 74:413-418.

Barka, E.A. and J.C. Audran. (1997). Response of champenoise grapevine to low temperatures: Changes of shoot and bud proline concentrations in response to low temperatures and correlations with freezing tolerance. J. Hort. Sci. 72:577-582.

Barka, E.A. J.C. Audran, O. Brun, C. Leddet and J. Dereuddre. (1995). Formation et distribution de glace dans les bourgeons de *Vitis vinifera* avant et au cours de leur debourrement. Can. J. Bot. 73:1878-1888.

Bittenbender, H.C. and G.S. Howell. (1974). Adaptation of the Spearman-Karber method for estimating the T50 of cold-stressed flower buds. J. Amer. Soc. Hortic. 99:187-189.

Boone, D.A. and R.E. Durham. (1996). Photoperiod response of three *Vitis* rootstocks and Cabernet Sauvignon. Pp. 54-56. In *Proc. 4th Intl. Symp. Cool Climate Viticulture and Enology*, T. Henick-Kling, T.E. Wolf and E.M. Harkness, (eds.), New York State Agricultural Experiment Station, Geneva, New York.

Bordelon, B.P., D.C. Ferree and T.J. Zabadal. (1997). Grape bud survival in the Midwest following the winter of 1993-1994. Fruit Varieties Journal 51:53-59.

Bourne, T.F. and J.N. Moore. (1991). Cold hardiness in grape cultivar development. Fruit Var. J. 45:26-28.

Bourne, T.F., J.N. Moore and M.F. George. (1991). Primary bud hardiness of four genotypes of grapes in Arkansas. J. Amer. Soc. Hort. Sci. 116:835-837.

Clark, J.R. and P. Watson. (1998). Evaluation of dormant primary bud hardiness of muscadine grape cultivars. Fruit Varieties J. 52:47-50.

Clark, J.R., T.K. Wolf and M.K. Warren. (1996). Thermal analysis of dormant buds of two muscadine grape cultivars and of *Vitis labrusca* L. 'Mars'. HortScience 31:79-81.

Clore, W.J., M.A. Wallace and R.D. Fay. (1974). Bud survival of grape varieties at subzero temperatures in Washington. Amer. J. Enol. Vitic. 25:24-29.

Düring, H., T.V. Ortoidze and B. Bushnell. (1990). Effects of subzero temperatures on chlorophyll fluorescence of grapevine buds. J. Plant Physiol. 136:758-760.

FAO Production Yearbook. (1999). FAO Statistics Series No. 156, Vol. 53:65-80. Chief, Publishing and Multimedia Service, Information Division, FAO, Rome, Italy.

Fennell, A. and E. Hoover. (1991). Photoperiod influences growth, bud dormancy, and cold acclimation in *Vitis labruscana* and *V. riparia*. J. Amer. Soc. Hort. Sci. 116: 270-273.

Fennell, A. and M.J. Line. (2001). Identifying tissue response in grapes during induction of endodormancy using magnetic resonance imaging. J. Amer. Soc. Hort. Sci. 126:681-688.

Fennell, A. and K. Mathiason. (2002). Early acclimation response in grapes. Pp. 93-108. In: P.H. Li. and E. Tapio Palva (eds.), *Plant Cold Hardiness: Gene Regulation and Genetic Engineering*. Plenum Press, New York.

Fennell, A., C. Wake and P. Molitor. (1996). Use of ^{1}H-NMR to monitor changes in bud water status during the onset of dormancy. J. Amer. Soc. Hort. Sci. 121: 1112-1116.

Fuller, M.P. and G. Telli. (1999). An investigation of the frost hardiness of grapevine (*Vitis vinifera*) during bud break. Ann. Appl. Biol. 135:589-595.

Fuller, M.P. and M. Wisniewski. (1998). The use of infrared thermal imaging in the study of ice nucleation and freezing of plants. J. Therm. Biol.23:81-89.

Gu, S. (1999). Lethal temperature coefficient–a new parameter for interpretation of cold hardiness. J. Hort. Sci. Biotech. 74:53-59.

Hamman Jr., RA. (1993). Wine grape performance of 32 cultivars in western Colorado 1982-1986. Fruit Varieties Journal 47:59-63.

Hamman Jr., R.A. and I.E. Dami. (2000). Effects of irrigation on wine grape growth and fruit quality. HortTechnology 10:162-168.

Hamman Jr., R.A., I.-E. Dami, T.M. Walsh and C. Stushnoff. (1996). Seasonal carbohydrate changes and cold hardiness of Chardonnay and Riesling grapevines. Am. J. Enol. Vitic. 47:31-36.

Hamman Jr., R.A., A.R. Renquist and H.G. Hughes. (1990). Pruning effects on cold hardiness and water content during deacclimation of Merlot bud and cane tissues. Am.. J. Enol. Vitic. 41:251-260.

Hamed, F., M.P. Fuller and G. Telli. (2000). The pattern of freezing of grapevine shoots during early bud growth. CryoLetters 21:255-260.

Hemstead, P.R. and J.J. Luby. (2000). Utilization of *Vitis riparia* for the development of new wine varieties with resistance to disease and extreme cold. Acta Hort. 528:487-490.

Himelrick, D.G., R.M. Pool and P.J. McInnis. (1991). Cryoprotectants influence freezing resistance of grapevine bud and leaf tissue. HortScience 26:406-407.

Hoover, E., P. Hemstead, J. MacKenzie and F. Propsom. (2002). Rootstock influence on scion vigor, hardiness, yield and fruit quality of 'St. Pepin' grape. Abstract #770. XXVI International Horticultural Congress, August 11-17, 2002. Toronto, Canada.

Howell, G.S. and N. Shaulis. (1980). Factors influencing within-vine variation in the cold resistance of cane and primary bud tissues. Amer. J. Enol. Vitic. 131:158-161.

Howell, G.S., B.G. Stergios and S.S. Stackhouse. (1978). Interrelation of productivity and cold hardiness of Concord grapevines. Am. J. Enol. Vitic. 29:187-191.

Hubácková, M. (1996). Dependence of grapevine bud cold hardiness on fluctuations in winter temperatures. Am. J. Enol. Vitic. 47:100-102.

Jiang, H., G.S. Howell and J.A. Flore. (1999). Efficacy of chlorophyll fluorescence as a viability test for freeze-stressed woody grape tissues. Can. J. Plant Sci. 79:401-409.

Johnson, D.E. and G.S. Howell. (1981). Factors influencing critical temperatures for spring freeze damage to developing primary shoots on Concord grapevines. Am. J. Enol. Vitic. 32:144-149.

Jones, K.S., J. Paroschy, B.D. McKersie and S.R. Bowley. (1999). Carbohydrate composition and freezing tolerance of canes and buds in *Vitis vinifera*. J. Plant Physiol. 155:101-106.

Jones, K.S., B.D. McKersie and J. Paroschy. (2000). Prevention of ice propagation by permeability barriers in bud axes of *Vitis vinifera*. Can. J. Bot. 78:3-9.

Kang, S.K., H. Motosugi, K. Yonemori and A. Sugiura. (1998). Supercooling characteristics of some deciduous fruit trees as related to water movement within the bud. J. Hort. Sci. and Biotechnology 73:165-172.

Lipe, W.N., L. Baumhardt, C.W. Wendt and D.J. Rayburn. (1992). Differential thermal analysis of deacclimating Chardonnay and Cabernet Sauvignon grape buds as affected by evaporative cooling. Am. J. Enol. Vitic. 43:355-361.

Luby, J.J. 1991. Breeding cold-hardy fruit crops in Minnesota. HortSci. 26:507-512.

Meiering, A.G., J.H. Paroschy, R.L. Peterson, G. Hostetter and A. Neff. (1980). Mechanical freezing injury in grapevine trunks. Am. J. Enol. Vitic. 31:81-89.

Millard, M.M., O.B. Veisz, D.T. Krizek and M. Line. 1995. Magnetic resonance imaging (MRI) of water during cold acclimation and freezing in winter wheat. Plant, Cell and Environment 18:534-544.

Miller, D.P., G.S. Howell and R.K. Striegler. (1988). Cane and bud hardiness of selected grapevine rootstocks. Am. J. Enol. Vitic. 39:55-59.

Moncur, M.W., K. Rattigan, D.H. Mackenzie and G.N. McIntyre. 1989. Base temperatures for bud break and leaf appearance of grapevines. Am. J. Enol. Vitic. 40:21-26.

Mullins, C.A. (1986). Grape cultivar performance on the Tennessee Cumberland Plateau. Fruit Varieties J. 40:18-23.

Mullins, M.G., A. Bouquet and L.E. Williams. (1992). *Biology of the Grapevine*. Cambridge University Press, Cambridge, Cambridge, UK.

Okamoto, G., S. Wang and K. Hirano. (2000). Cold resistance in root and cane of own-root 'Kyoho' grapevines. Scientific Reports of the Faculty of Agriculture Okayama University 89:23-29.

Paroschy, J.H., A.G. Meiring, R.L. Peterson, G. Hostetter and A. Neff. (1980). Mechanical winter injury in grapevine trunks. Am. J. Enol. Vitic. 31:227-232.

Pierquet, P. and C. Stushnoff. (1978). Variation and breeding potential of some northern clones of *Vitis riparia* Michx. Fruit Var. J. 32:74-84.

Pierquet, P. and C. Stushnoff. (1980). Relationship of low temperature exotherms to cold injury in *Vitis riparia* Michx. Am. J. Enol. Vitic. 31:1-6.

Pierquet, P., C. Stushnoff and M.J. Burke. (1977). Low temperature exotherms in stem and bud tissues of *Vitis riparia* Michx. J. Amer. Hort. Sci. 102:54-55.

Pool, R.T. (a) *http://www.nyaes.cornell.edu/hort/faculty/Pool/trunkinjury/tihtml/trninjkinds.html* In: *Dealing with Winter Cold Injury to Grapevine Canes and Trunks*. Cornell Viticulture, http://www.nysaes.cornell.edu/hort/faculty/pool/trunkinjury/tihtml/trkinjtablecontents.html. Accessed May 15, 2002.

Pool, R.T. (b) *http://www.nyaes.cornell.edu/hort/faculty/Pool/trunkinjury/tihtml/trninjbudsurv.html* In: *Dealing with Winter Cold Injury to Grapevine Canes and Trunks*. Cornell Viticulture, http://www.nysaes.cornell.edu/hort/faculty/pool/trunkinjury/tihtml/trkinjtablecontents.html. Accessed May 15, 2002.

Pool, R., T. Wolf, M.J. Welser and M.D. Goffinet. (1992). Environmental factors affecting dormant bud cold acclimation of three Vitis cultivars. Pp. 611-616. In: *Proceedings of the IV International Symposium on Grapevine Physiology*. Fondazone Giovani Dalmasso, Torino, Italia.

Proebsting, E.L., M. Ahmedullah and V.P. Brummund. (1980). Seasonal changes in low temperature resistance of grape buds. Am. J. Enol. Vitic. 31: 329-336.

Proebsting, E.L., V.P. Brummund and W.J. Clore. (1991). Critical temperatures for Concord grapes. Washington State University Cooperative Extension Bulletin 1615.

Quamme, H.A. (1986). Use of thermal analysis to measure freezing resistance of grape buds. Can. J. Plant Sci. 66:945-952.

Quamme, H.A. (1995). Deep supercooling in buds of woody plants. Pp. 183-200 In: R.E. Lee Jr., G.J. Warren and L.V. Gusta (eds.), *Biological Ice Nucleation and Its Applications*. APS St. Paul, MN.

Rajashekar, C.B. and M.J. Burke. (1996). Freezing characteristics of rigid plant tissues. Development of cell tension during extracellular freezing. Plant Physiol. 111:597-603.

Rajashekar, C.B. and A.C. Lafta. (1996). Cell-wall changes and cell tension in response to cold acclimation and exogenous abscisic acid in leaves and cell cultures. Plant Physiol. 111:605-612.

Reisch, B.I. and C. Pratt. (1996). Grapes. Pp. 297-369. In: J. Janick and J.N. Moore (eds.), *Fruit Breeding VolumeII: Vine and Small Fruit Crops*. John Wiley and Sons. New York.

Reynolds, A.G. (1987). Cultural practices affecting vine hardiness in British Columbia. The Northwest Grape Grower November/December:29-34.

Salzman, R.A., R.A. Bressan, P.M. Hasegawa, E.N. Ashworth and B.P. Bordelon. (1996). Programmed accumulation of LEA-like proteins during desiccation and cold acclimation of overwintering grape buds. Plant, Cell and Environment 19: 713-720.

Schnabel, B.J. and R.L. Wample. (1987). Dormancy and cold hardiness in *Vitis vinifera* L. cv. White Riesling as influenced by photoperiod and temperature. Am. J. Enol. Vitic. 38:265-272.

Seyedbagheri, M.M. and E. Fallahi. (1994). Physiological and environmental factors and horticultural practices influencing cold hardiness of grapevines. J. Small Fruit Vitic. 2:3-38.

Slater, J.V., M.R. Warmund, M.R. George and M.R. Ellersieck. (1991). Deacclimation of winter hardy 'Seyval Blanc' grape tissue after exposure to 16 C. Scientia Horticulturae 45:273-285.

Stergios, B.G. and G.S. Howell. (1973). Evaluation of viability tests for cold stressed plants. J. Amer. Soc. Hort. Sci. 98:325-330.

Stergios, B.G. and G.S. Howell. (1977a). Effects of defoliation, trellis height and cropping stress on the cold hardiness of Concord Grapevines. Am. J. Enol. Vitic. 28:34-42.

Stergios, B.G. and G.S. Howell. (1977b). Effect of site on cold acclimation and deacclimation of Concord grapevines. Am. J. Enol. Vitic. 28:43-48.

Striegler, R.K. and G.S. Howell. (1991). The influence of rootstock on the cold hardiness of Seyval grapevines. I. Primary and secondary effects on growth, canopy development, yield, fruit quality and cold hardiness. Vitis 30:1-10.

Stushnoff, C., R.L. Remmele, Jr., V. Esensee and M. McNeil. (1993). Low temperature induced biochemical mechanisms: implications for cold acclimation and de-acclimation. Pp. 647-657. In: M.B. Jackson and C.R. Black (eds.), *Interacting Stresses on Plants in a Changing Climate, Vol. 116.* NATO ASI Series. Springer-Verlag, Berlin-Heidelberg.

Wake, C.F.M. and A. Fennell. (2000). Morphological, physiological and dormancy responses of three *Vitis* genotypes to short photoperiod. Physiol. Plant. 109:203-210.

Wample, R.L. (1994). A comparison of short- and long-term effects of mid-winter pruning on cold hardiness of Cabernet Sauvignon and Chardonnay buds. Am. J. Enol. Vitic. 45:388-392.

Wample, R.L. and A. Bary. (1992). Harvest date as a factor in carbohydrate storage and cold hardiness of Cabernet Sauvignon grapevines. J. Amer. Soc. Hort. Sci. 117: 32-36.

Wample, R.L., S.E. Spayd, R.G. Evans and R.G. Stevens. (1993). Nitrogen fertilization of white Riesling grapes in Washington: Nitrogen seasonal effects on bud cold hardiness and carbohydrate reserves. Am. J. Enol. Vitic. 44:159-167.

Wample, R.L. and T.K. Wolf. (1996). Practical considerations that impact vine cold hardiness. Pp. 23-28. In: T. Henick-Kling, T.E. Wolf and E.M. Harkness (eds.), *Proc. 4th Intl. Symp.Cool Climate Viticulture and Enology*. New York State Agricultural Experiment Station, Geneva, New York.

Warmund, M.R., C.J. Starbuck and L. Lockshin. (1986). Growth, cold hardiness and carbohydrate content of Vidal Blanc grapevines propagated by hardwood vs. softwood cuttings. Am. J. Enol. Vitic. 37:215-219.

Wisniewski, M. (1998). The use of infrared video thermography to study freezing in plants. Pp. 311-316. In: P.H. Li and T.H.H. Chen (eds.), *Plant Cold Hardiness: Molecular Biology, Biochemistry, and Physiology*. Plenum Press, New York.

Wisniewski, M., T. Wolf and L. Fuchigami. (1996). Biochemical and biophysical mechanisms of cold hardiness in woody plants. Pp. 14-22. In: T. Henick-Kling, T.E. Wolf and E.M. Harkness, (eds.), *Proc. 4th Intl. Symp.Cool Climate Viticulture and Enology*, New York State Agricultural Experiment Station, Geneva, New York.

Wolf, T.K. and M.K. Cook. (1991). Comparison of 'Cabernet Sauvignon' and 'Cabernet Franc' grapevine dormant bud cold hardiness. Fruit Varieties Journal 45:17-21.

Wolf, T.K. and M.K. Cook. (1992). Seasonal deacclimation patterns of three grape cultivars at constant, warm temperature. Am. J. Enol. Vitic. 43:171-179.

Wolf, T.K. and M.K. Cook. (1994). Cold hardiness of dormant buds of grape cultivars: Comparison of thermal analysis and field survival. HortScience 29:1453-1455.

Wolf, T.K. and M.K. Miller. (2001). Crop yield, fruit quality, and winter injury of 12 red-fruited grape cultivars in Northern Virginia. J. Amer. Pomological Soc. 55: 241-250.

Wolf, T.K. and R.M. Pool. (1987). Factors affecting exotherm detection in the differential thermal analysis of grapevine dormant buds. J. Amer. Soc. Hort. Sci. 112: 520-525.

Wolf, T.K. and R.M. Pool. (1988). Nitrogen fertilization and rootstock effects on wood maturation and dormant bud cold hardiness of cv. Chardonnay grapevines. Am. J. Enol. Vitic. 39:308-312.

Wolf, T.K. and M.K. Warren. (2000). Crop yield, grape quality and winter injury of eight wine grape cultivars in Northern Virginia. J. Amer. Pomological Soc. 54:34-43.

Wolpert, J.A. and G.S. Howell. (1984). Effects of cane length and dormant season pruning date on cold hardiness and water content of concord bud and cane tissues. Am. J. Enol. Vitic. 35:237-241.

Wolpert, J.A. and G.S. Howell. (1985a). Cold acclimation of Concord grapevines. I. Variation in cold hardiness within the canopy. Am. J. Enol. Vitic. 36:185-188.

Wolpert, J.A. and G.S. Howell. (1985b). Cold acclimation of Concord grapevines. II. Natural acclimation pattern and tissue moisture decline in canes and primary buds of bearing vines. Am. J. Enol. Vitic. 36:189-194.

Wolpert, J.A. and G.S. Howell. (1986a). Cold acclimation of Concord grapevines. III. Relationship between cold hardiness, tissue water content, and maturation. Vitis 25:151-159.

Wolpert, J.A. and G.S. Howell. (1986b). Effect of night interruption on cold acclimation of potted 'Concord' grapevines. J. Amer. Soc. Hort. Sci. 111:16-20.

Physiology and Regulation of Seasonal Nitrogen Cycling in Woody Plants

Gary D. Coleman

SUMMARY. Seasonal nitrogen cycling in trees is a mechanism for conserving and reusing nitrogen that may contribute to nitrogen use efficiency. In this review the physiological and molecular regulation of seasonal nitrogen storage is examined. Research using the 32 kDa bark storage protein gene, *bspA*, of poplar (*Populus*) as a model system has revealed important facets regulating gene expression and subsequent nitrogen storage. Recent progress has demonstrated that *bspA* expression is a photoperiod response mediated by phytochrome and secondary messengers including Ca^{2+} and protein phosphorylation are involved in signaling pathways. Furthermore, evidence indicates that photoperiod modulated expression likely involves metabolic signaling mediated by glutamine and sucrose. A model is proposed that integrates the various facets of poplar *bspa* expression.

Gary D. Coleman is affiliated with the Department of Natural Resource Sciences and Landscape Architecture and Program in Molecular and Cellular Biology, University of Maryland, College Park, MD 20742-4452 (E-mail: gc76@umail.umd.edu).

Research in the author's laboratory has been made possible by the generous support of the National Science Foundation and by the Plant Responses to the Environment Program of the National Research Initiative (NRI) Competitive Grants Program, USDA.

[Haworth co-indexing entry note]: "Physiology and Regulation of Seasonal Nitrogen Cycling in Woody Plants." Coleman, Gary D. Co-published simultaneously in *Journal of Crop Improvement* (Food Products Press, an imprint of The Haworth Press, Inc.) Vol. 10, No. 1/2 (#19/20), 2004, pp. 237-259; and: *Adaptations and Responses of Woody Plants to Environmental Stresses* (ed: Rajeev Arora) Food Products Press, an imprint of The Haworth Press, Inc., 2004, pp. 237-259. Single or multiple copies of this article are available for a fee from The Haworth Document Delivery Service [1-800-HAWORTH, 9:00 a.m. - 5:00 p.m. (EST). E-mail address: docdelivery@haworthpress.com].

http://www.haworthpress.com/web/JCRIP

Digital Object Identifer: 10.1300/J411v10n01_10

KEYWORDS. Seasonal nitrogen storage, vegetative storage proteins, bark storage proteins, photoperiod, gene regulation

INTRODUCTION

In most environments one or more resources will be limiting during plant growth. Limited resources will influence physiological and metabolic activities and plant growth and productivity (Sheriff et al. 1995). Thus, productivity is associated with the efficiency that a particular plant, species or ecosystem uses a limiting resource. Understanding how plants use resources and the components that contribute to economy of use are basic to understanding plant growth, plant competition and ecosystem function (Sheriff et al..1995). Nitrogen is typically a limiting resource in natural forest systems (Sheriff et al. 1995) and a number of processes may contribute to economy of nitrogen use including increasing the time that a given unit of nitrogen is held in the plant, increasing the amount of nitrogen absorbed per unit of root mass, and increasing the function of nitrogen at lower internal concentrations by fixing more carbon per unit of nitrogen (Vitousek 1982; Birk and Vitousek 1986). Nitrogen storage appears to be an important process for the retention of nitrogen and may contribute to overall use efficiency. In addition, accumulation of nitrogen for subsequent re-use provides trees with some independence from external nitrogen supplies that may reduce annual variations in growth (Ryan and Bormann 1985). Stored nitrogen may also reduce potential loss from mineralization and subsequent leaching (Ryan and Bormann 1982).

Nitrogen can be considered stored if mobilization and reuse occurs. Nitrogen storage is independent of growth rate and the ability to store nitrogen appears to be independent of plant nitrogen status (Millard 1988). Chapin et al. (1990) recognized three types of storage: accumulation, reserve formation and recycling. Nitrogen compounds accumulate when supply exceeds plant demand (Millard 1988). Reserve formation is a regulated process involving synthesis and compartmentation that directly competes with other plant functions (Chapin et al. 1990). Recycling involves the reuse of physiological metabolites that might otherwise be lost as litter (Chapin et al. 1990). Storage occurs on a number of time scales including daily, weekly, seasonal and lifetime storage (Chapin et al. 1990). The metabolic costs of storage can be reduced if the storage compound has additional physiological functions (Millard 1988; Chapin et al. 1990).

Nitrogen cycling in forests is a complex interaction involving nitrogen deposition, leaching, plant and soil uptake and return, and internal redistribution within the plant (see Attiwill and Adams 1993, for review). Internal redistribution of nitrogen in woody perennials occurs on two levels. Nitrogen is reallocated during growth according to sink demands and is mobilized from source to sink tissues. In addition, redistribution involves seasonal cycling where nitrogen is resorbed from leaves during the fall, stored in perennial tissues, and remobilized during spring to support new growth. Results from ^{15}N labeling experiments indicate that stored nitrogen is rapidly partitioned to new growth and can account for 30%-75% of the nitrogen transported to new shoots during spring growth (Deng et al. 1989; Millard and Neilsen 1989; Millard and Proe 1991).

Seasonal nitrogen resorption is a mechanism for retaining and conserving nutrients and appears to be an important component of nitrogen use efficiency (Ostman and Weaver 1982; Ryan and Bormann 1982; Vitousek 1982; Chapin and Kedrowski 1983; Boerner 1984; Kost and Boerner 1985; Birk and Vitousek 1986; Aerts, 1990). Studies indicate that from 50%-90% of total leaf nitrogen is translocated from senescing leaves during autumn (Taylor and May 1967; Kang and Titus 1980; Luxmoore et al. 1981; Ostman and Weaver 1982; Boerner 1984; Chapin and Kedrowski 1993). The majority of resorbed leaf nitrogen is stored as protein (Taylor and May 1967; Chapin and Kedrowski 1983; Oaks et al. 1991) Major storage sites during overwintering include shoot and root bark, and xylem ray cells (Kang and Titus 1980; Sauter et al. 1989; Wetzel et al. 1989; Coleman et al. 1991). Seasonal nitrogen is a common mechanism used to conserve nitrogen in trees and seasonal storage proteins have been reported for a number of angiosperm and gymnosperm tree species (for review see Stepien et al. 1994).

Seasonal nitrogen storage has been studied in a number of tree species with most research focused on describing and characterizing nitrogen and protein storage. In most tree species, the molecular basis of seasonal nitrogen storage and the physiological and molecular factors regulating storage has not been investigated. The majority of molecular studies have been conducted in the genus *Populus* (poplars), which has contributed significantly to our understanding of seasonal nitrogen storage. Because poplars have been extensively used as a model for studying the molecular biology of seasonal nitrogen storage, this review will focus on the molecular biology of poplar seasonal nitrogen cycling.

SEASONAL NITROGEN CYCLING IN POPLARS

Poplars, like other deciduous hardwood species, resorb nitrogen from senescing leaves and store this nitrogen in perennial tissues (McColl 1980; Pastor and Bockheim 1984; Alban 1985; Sauter et al. 1989; Killingbeck et al. 1990; Pregitzer et al. 1990). From 40 to 60% of foliar nitrogen is resorbed during fall leaf senescence (Pastor and Bockheim 1984) and the rate of resorption is dependent on leaf senescence (Killingbeck et al. 1990). Reductions in leaf nitrogen content during fall leaf senescence correlate with a localized increase in leaf free amino acid concentrations and an increase in bark nitrogen content (Côté and Dawson 1986; Côté et al. 1989). During growth about 75% of total tree nitrogen is contained in leaves and after leaf senescence almost 80% of tree nitrogen is still contained in the tree (Pregitzer et al. 1990). The majority (about 90%) of the nitrogen contained within the stem is protein (Hollwarth 1976).

POPLAR BARK STORAGE PROTEINS ARE A VEGETATIVE STORAGE PROTEIN FOUND IN XYLEM RAYS AND INNER BARK PARENCHYMA

Vegetative storage proteins (VSP) are a heterogeneous class of proteins characterized by a role in storage rather than an enzymatic function. Because VSP temporarily store nitrogen these proteins are characterized by accumulation during periods of excess N availability, including periods when nitrogen is resorbed. VSP function by supplying stored carbon and nitrogen for use by the plant through VSP catabolism (see Staswick 1994 for review). In addition VSP may also play a regulatory role in the allocation of nitrogen to various tissues (Coleman, unpublished data). VSP have been characterized in a wide range of both herbaceous and perennial species (reviewed in Staswick 1994). In some species, such as potato and sweet potato, VSP are localized in specialized storage organs (Paiva et al. 1983; Maeshima et al. 1985). In soybean and *Arabidopsis thaliana*, VSP can be found in different tissues including seedlings, leaves, stems, flowers and developing fruit (Franceschi and Giaquinta 1983; Mason and Mullet 1990; Staswick 1990; Berger et al. 1995; Utsugi et al. 1998). VSP in woody perennial plants are thought to be involved in seasonal nitrogen storage and are characterized by accumulation during fall and winter followed by a decline in abundance during spring growth (Wetzel et al. 1989). Because most seasonal VSP of perennial woody plants accumulate in bark tissues they have been termed bark storage proteins (BSP). Seasonal nitrogen storage utilizing BSP appears to be a conserved storage mechanism among trees since BSP have been identified in a wide range of tree species (see Stepien et al. 1994 and Coleman 1997 for reviews).

Initial studies of protein accumulation in the wood of dormant poplars identified a 32-kDa protein that was localized in protein storage vacuoles of xylem ray parenchyma (Sauter et al. 1989). This protein accumulated in xylem rays during the dormant season and disappeared during spring bud growth (van Cleve et al. 1988; Sauter and van Cleve 1990). Wetzel et al. (1989) showed that a protein of similar mass (32 kDa) accumulated in protein storage vacuoles of inner bark parenchyma and demonstrated a similar pattern of accumulation with highest levels during the fall and winter and declining in spring and summer. In a study of the structural and immunological properties of the 32-kDa proteins localized in the bark and xylem rays revealed that these proteins are similar and were probably the same protein (Stepien et al. 1992).

Although both wood and bark serve as storage sites, the protein content of these tissues vary with age (Sauter et al. 1989). The highest protein concentration was detected in the bark (31.9 μg mg^{-1} DW) and wood (11.3-13.5 μg mg^{-1} DW) of 1-year-old shoots (Sauter et al. 1989). Protein content declined basipetally by 70% and then increased in roots (Sauter et al. 1989). Thus, both wood and bark are used for protein storage and serve as storage tissue for seasonal nitrogen cycling. The differences in protein storage levels in different aged shoots may have an adaptive role. Storage in tissues adjacent to sites of shoot and leaf growth (both vegetative and reproductive) would reduce the metabolic costs associated with translocation of nitrogen and carbon during storage protein synthesis and also during nitrogen and carbon remobilization.

Poplars appear to have three isoforms of BSP with molecular masses of 32, 36, and 38 kDa (Stepien and Martin 1992). All of the isoforms react with periodic acid-Schiff's reagent and lectin-digoxigenin conjugates, indicating that all three polypeptides are glycoproteins (Stepien and Martin 1992). Furthermore, treatment with endo-β-N-acetylglucosaminidase H reduced the mass of all three polypeptides to 29 kDa and it was estimated that BSP in the bark of *P. x euroamericana* consists of 10-22% covalently attached carbohydrates (Stepein and Martin 1992).

In addition to the 32 kDa BSP, a second VSP termed WIN4 may also play a role in vegetative N storage (Lawrence et al. 1997). WIN4 was initially identified as a RNA that was systemically wound induced (Parsons et al. 1989). BSP and WIN4 share about 60% sequence identity (both nucleotide and amino acid) and the respective genes are differentially regulated (Davis et al. 1993; Coleman et al. 1994). Although the evidence is not conclusive, BSP appears to be involved primarily in seasonal nitrogen storage, whereas WIN4 may be involved in short-term nitrogen storage during growth (Davis et al. 1993; Coleman et al. 1994). The WIN4 protein is found in youngest leaves of growing poplars and accumulation increased in plants fertilized with 50 mM NH_4NO_3 (Lawrence et al. 1997). Furthermore, WIN4 is localized to immature leaf epidermis, stipules and in the cells adjacent to the vasculature (Lawrence et al. 1997). Based upon the pattern of induction and sites of accumulation it was postulated the WIN4 might play a role in temporary nitrogen storage in immature tissues (Lawrence et al. 1997).

The poplar 32 kDa BSP accumulates primarily in xylem rays and bark parenchyma and gene expression is induced by a number of environmental factors including photoperiod and increased nitrogen availability (reviewed in Stepien et al. 1994; Coleman 1997). Since poplar BSP is encoded by a small multigene family (Coleman et al. 1992), these different patterns of *bsp* expression and BSP accumulation could either involve the expression of the same gene in the various tissues and environmental conditions or different genes. Either case could account for the various patterns of expression that have been reported. Poplar *bspA* has been cloned (Coleman and Chen 1993) and it was found that the promoter from this gene is sufficient to confer both short-day photoperiod and nitrogen-inducibility to the GUS reporter gene (Zhu and Coleman 2001b). In addition, it was also shown that the *bspA* promoter was active in both bark parenchyma and xylem rays (Zhu and Coleman 2001b). Therefore, one gene family member, *bspA*, can account for the various expression patterns observed by short-day photoperiods and elevated nitrogen availability in both bark parenchyma and xylem rays.

POPLAR BSP AND RELATED POPLAR VSP SHARE SIMILARITY WITH NUCLEOSIDE PHOSPHORYLASES

Although a number of BSP have been identified in trees, few of the genes encoding these proteins have been cloned and characterized. In poplar, the gene encoding the 32 kDa BSP (*bspA*) has been cloned (Coleman and Chen 1993). *BspA* is a member of a small multigene family (Coleman and Chen 1993) and is linked to a related gene family of wound-inducible genes (*WIN4*) (Davis et al. 1993). Based upon results from ClustalW analysis (Thompson et al. 1994) the poplar 32 kDa BSPA peptide is closely related to poplar WIN4 and also shares similarity to a poplar VSP cDNA PNI288, an *Arabidopsis* unknown protein and a WIN-4 like protein from rice (Figure 1). Therefore, the poplar BSP appears to be related to proteins from at least two herbaceous species but differs from the other known VSP of herbaceous plants. The ClustalW alignment of BSP, WIN4, PNI288, rice WIN4-like, and the *Arabidopsis* putative protein is shown in Figure 2.

FIGURE 1. Relationship dendrogram of VSP. VSP peptide sequence relationships were determined using ClustalW.

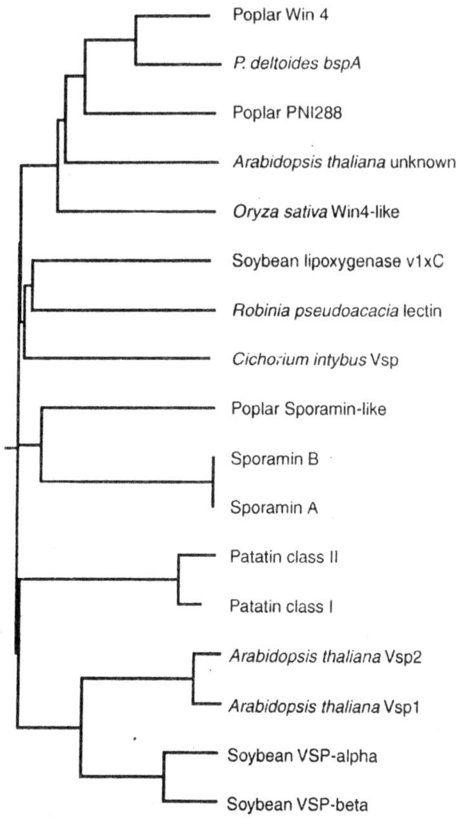

FIGURE 2. BSP-like sequence alignments. Peptide sequence similarities were determined using ClustalW. Sequence identities are indicated by black-boxed regions while sequence similarities are indicated by grey-boxed regions.

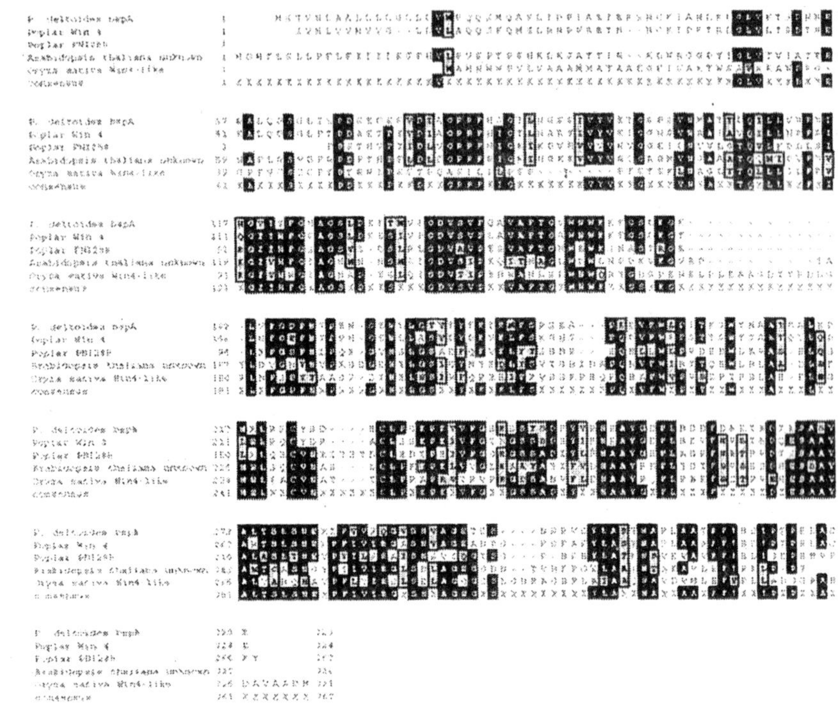

In addition to storing carbon and nitrogen, some VSP have also been found to exhibit enzyme activity. For example, patatin from potato is a lipid acyl hydrolase (Andrews et al. 1988) while soybean VSPα and VSPβ are acid phosphatases (DeWald et al. 1992). Although the BSP of *Sophora japonica* (Hankins et al. 1987), *Robina pseudoacacia* (Yoshida et al. 1994), and *Sambucus nigra* (Nsimba-Lubaki and Peumans, 1986) do not appear to have enzyme activity but are carbohydrate-binding lectins. An enzymatic role for the poplar BSPA peptide has not been shown but sequence analysis suggests that this protein may belong to a family of nucleoside phosphorylases, or alternatively, evolved from this family of enzymes (Mushegian and Koonin 1994; Cornell and Riscoe 1998). These enzymes catalyze phosphorolysis and hydrolysis of the *N*-glycosidic bond in nucleotides and nucleosides (Neuhard and Nygaard

1987). Mushegian and Koonin (1994) proposed three conserved motifs of nucleosidases which also occur in the poplar BSPA peptide. The first motif includes amino acids 82-111 and is postulated to be involved in phosphate binding (Mushegian and Koonin 1994). The second motif includes amino acids 118-143 and appears to be the nucleoside-binding site, while a function for the third motif is not known (Mushegian and Koonin 1994). Interestingly these same regions show sequence similarity among poplar BSPA, poplar WIN4, the rice WIN4-like peptide, and the *Arabidopsis* unknown protein. As shown in Table 1, BLAST analysis revealed that BSPA, WIN4, rice WIN4-like, and the *Arabidopsis* protein; all contain peptide sequences that share similarity with purine nucleoside phosphorylase conserved domains (Mao et al. 1997). However, confirmation of nucleoside phosphorylase activity has not been demonstrated experimentally and it is not clear if such enzymatic activity would have a role in nutrient cycling and storage.

POPLAR BSP AND SEED STORAGE PROTEINS (SSP) ARE RELATED

In addition to storage of carbon and nitrogen in vegetative tissue, plants also store these nutrients in seeds as seed storage proteins (SSPs). Usually VSP and SSP are different unrelated storage proteins (Staswick 1990). For example, in soybean VSPα and VSPβ are unrelated to the soybean SSPs both immunologically (Wittenbach 1983) and at the DNA sequence level (Staswick 1989). The situation in poplar appears to be more complex. Beardmore et al. (1996) demonstrated that the pop-

TABLE 1. Conserved Domain Analysis of BSPA and BSPA-Related Proteins. Conserved Domains Were Identified Using Basic Local Alignment Search Tool (BLAST).

Protein	Accession #	% aligned	Region of alignment (amino acid position)	E-value
Poplar BSPA	CAA49669	38	75-160	1×10^{-5}
Poplar WIN4	AAA16342	79	69-293	3×10^{-9}
Rice Win4-like	BAA94780	61	76-279	9×10^{-5}
Arabidopsis unknown protein	AAK44093	25	6-141	1×10^{-7}
		35	251-328	1×10^{-5}

lar 36 kDa SSP and the 32 kDa BSPA share homology. This homology was based on antibody cross-reactivity, chemical characteristics, and proteolytic digestion products (Beardmore et al. 1996). Although poplar BSPA and SSP share homology, SSP appear to possess disulfide bonds while BSPA is hydrogen bonded (Beardmore et al. 1996).

Because of the long generation interval in poplars, Zhu and Coleman (2001b) studied the likelihood that poplar BSPA also functions as a SSP by using tobacco transformed with a chimeric gene consisting of the *bspA* promoter fused to β-glucuronidase (GUS). In tobacco, it was shown that the *bspA* promoter was active in a variety of tissues including flowers, developing seeds and germinating seeds (Zhu and Coleman 2001b). Although the *bspA* promoter was active in these tissues the pattern of GUS activity was not consistent with that predicted for a SSP. For example, during seed formation GUS activity was greatest after about 5 DPA (day post anthesis) and then declined during continued seed development (Zhu and Coleman 2001b). During seed germination, GUS activity was low until the emergence of the seed radicle (Zhu and Coleman 2001b). Histochemical assays showed that during seed development, GUS staining was localized to developing flowers, ovary vascular tissues and placental tissue and little staining was observed in developing seeds, mature seeds or fruit (Zhu and Coleman 2001b). Furthermore, during seed germination GUS staining was located to cotyledons and root tips (Zhu and Coleman 2001b). This pattern of gene expression is consistent with a VSP rather than a SSP.

As discussed by Beardmore et al. (1996), the homology between poplar BSP and SSP could result from SSP and BSP being encoded by the same gene but processed differently depending upon the tissue. Alternatively, SSP and VSP genes may have evolved from a common ancestral gene (Beardmore et al. 1996). Based upon the results with transgenic tobacco carrying the *bspA* promoter fused to GUS, it is likely that a common ancestral gene appears to account for the homology between BSP and SSP as opposed to the same gene and protein functioning in both vegetative and seed storage.

POPLAR BSPA EXPRESSION IS A SHORT-DAY PHOTOPERIOD RESPONSE MEDIATED BY PHYTOCHROME

Plants can be classified according to their photoperiodic responses. Photoperiod classifications are usually related to the flowering response, however, similar categories exist for responses such a tuberisation and dormancy (Thomas and Vince-Prue 1997). This classification is based on the type of photoperiod required for induction of the response and includes short-day plants (SDP), long-day plants (LDP), and day-neutral plants (DNP) (reviewed in Thomas and Vince-Prue 1997). SD responses in woody plants include a number of adaptive traits ranging from cessation of stem growth (Nitsch 1957), terminal bud formation, and initiation of bud dormancy (Perry 1971), and cold acclimation (Weiser 1970).

Photoperiodic responses are related to the light and dark periods of a 24-hour cycle. SDP's photoperception involves the photoreceptor phytochrome and is characterized by inhibition of the inductive response by a night-break with red (R) light which is reversible with far-red (FR) light (Borthwick et al. 1952). Two interconvertible forms of phytochrome exist. Pr, the physiologically inactive form, is converted to Pfr, the active form, by R light while Pfr is converted to Pr by FR light (Smith and Whitelam 1990). A hallmark of SD photoperiodic responses is the inhibition by R light and the reversibility by FR (reviewed in Thomas and Vince-Prue 1997). This reversibility has been used to identify phytochrome-mediated responses including phytochrome mediated gene expression (Quail et al. 1995).

Accumulation of the 32-kDa poplar BSPA follows a seasonal pattern (Wetzel et al. 1989). The seasonal accumulation of BSP is correlated with a corresponding change in the steady state levels of BSPA mRNA (Clausen and Apel 1991). Using controlled environment studies it was shown that accumulation of poplar BSPA was induced by SD (Coleman et al. 1991; Langheinrich and Tischner 1991). Subsequently, it was also shown that the SD increase in BSPA accumulation was correlated with an increase in the steady state levels of *bspA* mRNA (Coleman et al. 1992). Using poplars transformed with a gene consisting of the *bspA* promoter fused to GUS it was demonstrated that the promoter from *bspA* could confer both SD and N inducibility to GUS (Zhu and Coleman 2001b). Furthermore, it was also shown that the SD inducibility conferred to GUS by the *bspA* promoter was inhibited with a 15-minute night interruption with incandescent light (Zhu and Coleman 2001b). In addition to inhibiting SD GUS induction, night interruptions also inhibited SD increases in the steady state levels of *bspA* mRNA (Zhu and Coleman 2001b). Although these results were consistent with *bspA* gene expression being a SD response, definitive proof required confirmation of the role of phytochrome in this response.

Poplars transformed with a chimeric gene consisting of the *bspA* promoter fused to the β-glucuronidase gene were used to define the role of photoperiod and phytochrome mediated photoperception in regulating the activity of the *bspA* promoter (Zhu and Coleman 2001a). By using the *bspA* promoter fused to GUS, it was possible to monitor transcriptional activation of *bspA*, as opposed to measuring changes in steady state levels of *bspA* mRNA or BSPA protein levels. Therefore, post-transcriptional aspects of BSPA accumulation are essentially eliminated. As shown in Figure 3, SD induced activation of the *bspA* promoter resulted in a significant increase in bark GUS activity and this induction was inhibited by interruption of the night cycle with R light (Zhu and Coleman 2001a). Furthermore, the R inhibition was FR reversible, a characteristic of phytochrome-mediated responses (Zhu and Coleman 2001a). In addition to the low fluence response (LFR) characterized by R-FR reversibility, it also appears that transcriptional activation of *bspA* also involves a very low fluence response (VLFR) (Zhu and Coleman 2001a). This is based upon the observation that FR alone partially inhibited GUS induction and the R followed by FR treatments were comparable to the FR treatment (Zhu and Coleman 2001a). Since FR alone partially

FIGURE 3. Phytochrome meditated photoperiod perception of poplar *bspA* induction. Poplars transformed with the *bspA* promoter GUS chimeric gene were treated with either 8 hr light, 16 hr darkness with a 15 min night interruption with incandescent light during the middle of the dark cycle (SD/NB), 8 hr light, 16 hr darkness (SD), SD with a 5 min night interruption with R light (R), SD with a 5 min interruption with FR light (FR), SD with a 5 min night interruption with FR followed by 5 min of R (FR/R), SD with a 5 min night interruption with R followed by 5 min of FR (R/FR). Treatments were conducted for 4 wks after which time bark GUS activity was measured (adapted from Zhu and Coleman 2001a).

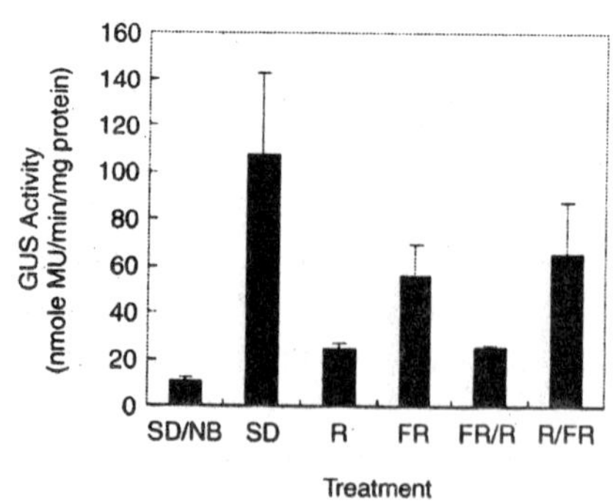

inhibited induction of GUS, complete reversibility by the R-FR treatment would not be expected (Zhu and Coleman 2001a). Therefore, the inhibition observed by FR most likely represents the VLFR since these responses occur with any light and are not photoreversible (Mancinelli 1994). These results conclusively show that SD induction of *bspA* expression is a photoperiod response that involves a dark dominant mechanism that is mediated by phytochrome (Thomas and Vince-Prue 1997). However, the components of the signal transduction pathway between photoperiod perception and *bspA* gene expression are not known, although calcium and protein phosphorylation appear to be involved (Zhu and Coleman 2001a).

Photoperiod activation of the *bspA* promoter involves *cis*-acting sequences located in the distal 1.6 kb of the 2.8 kb *bspA* promoter (Zhu and Coleman 2001b). As shown in Figure 4, the 2.8 kb *bspA* promoter resulted in a 33-fold increase in GUS activity in plants exposed to SD. In contrast, when the distal 1.6 kb of the *bspA* promoter was deleted SD activation of the *bspA* promoter was eliminated (Figure 4). Although N activation of the *bspA* promoter was also reduced when the distal 1.6 kb sequences were deleted, N treatment of plants transformed with the truncated promoter still retained about a 3-fold increase in GUS activity compared to control plants (Figure 4) (Zhu and Coleman 2001b). The

FIGURE 4. Poplar *BspA* promoter activation of the full-length promoter and the truncated promoter. Poplars transformed with either the 2.8 kb full length *bspA* promoter or the 1.2 kb truncated promoter (1.6 kb of distal DNA sequences were deleted) fused to GUS were treated either with SD photoperiods or elevated levels of N. Bark GUS activity was measured after 4 weeks of treatment (adapted from Zhu and Coleman 2001b).

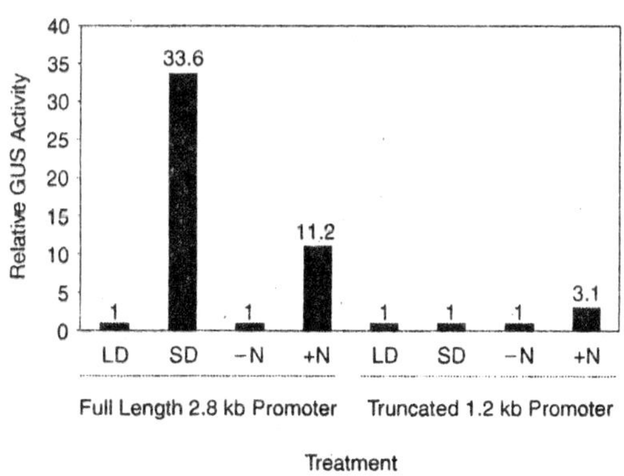

distal 1.6 kb region of the *bspA* promoter contains three regions with sequences that resemble known enhancer sequences (Figure 5) (Zhu and Coleman 2001b; for review see Guilfoyle 1997). The significance of these sequences is not known. Research is currently underway using a promoter deletion strategy to define the sequences involved in SD induction of *bspA* (Coleman, unpublished results).

SIMILAR TO OTHER VSPS POPLAR BSPA EXPRESSION IS MODULATED BY SOURCE-SINK RELATIONS

Altered source-sink relations related to environmental and developmental signals influence soybean VSP gene expression (Staswick 1994; Bunker et al. 1995). Since SD alters the patterns of growth and development, regulation of nitrogen storage by photoperiod must take into

FIGURE 5. Potential *cis*-acting DNA elements contained in the 1.6 kb distal poplar *bspA* promoter. The locations of the potential enhancer elements are shown with the number of nucleotides between each element indicated (see Guilfoyle 1997 for a description of enhancer elements).

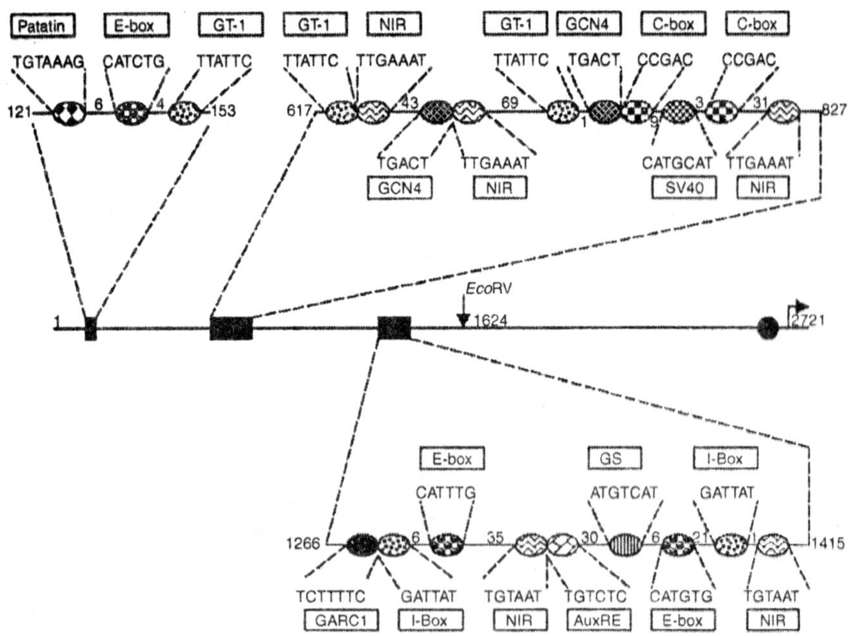

account changes in source-sink relations. Short-term assimilate partitioning between storage compounds is known to be influenced by photoperiod (for review see Thomas and Vince-Prue 1997). However, the influence of photoperiod on assimilate partitioning is poorly understood in woody plants. Changes in carbohydrate pools in tree stems vary on a seasonal basis (for review see Ceulemans and Isebrands 1996), but because most of these studies were not performed under controlled environmental conditions it is difficult assess the role of photoperiod or temperature in altering assimilate partitioning. Dickson and Nelson (1982) studied the influence of photoperiod on carbon fixation and partitioning in poplar. During growth the majority of fixed carbon was incorporated into leaf proteins or translocated from source leaves (Dickson and Nelson 1982). SD treatment resulted in decreased carbon fixation in source leaves and after SD induced bud formation the majority of fixed carbon was found in the sugar fraction (Dickson and Nelson 1982). These studies indicate that photoperiod influences carbon assimilation and partitioning in poplar.

In addition to photoperiod, *bspA* expression is influenced by source-sink relations (Zhu and Coleman 2001a). When shoot growth was stimulated under SD conditions by treatment with GA, activation of the *bspA* promoter was inhibited (Zhu and Coleman 2001a). Conversely, when growth has inhibited under growth promoting conditions the *bspA* promoter was activated (Zhu and Coleman 2001a). Such results are consistent with SD photoperiods altering source-sink relations that in turn activate *bspA* expression. If photoperiod regulates *bspA* expression by altering source-sink relations, then the changes in source-sink relations must somehow be perceived. Upon perception of altered source-sink relations, the signal must be transduced, eventually resulting in induction of *bspA*. Changes in carbon and nitrogen metabolites may serve as metabolic signals for altered source-sink relations.

Gln is the major form of nitrogen transported from leaves to stems during the fall in poplars (Sagisaka 1974; Sauter 1981; Sauter et al. 1992). Gln was shown to induce *bspA* expression and this induction was synergistically enhanced by the presence of sucrose while sucrose alone had no effect (Zhu and Coleman 2001a). In addition, the synergistic activity of sucrose was specific to Gln or ammonium nitrate compared to other amino acids (Zhu and Coleman 2001a). These results are consistent with photoperiod altering source-sinks relations whereby carbon and nitrogen metabolites are transported from source leaves to bark. It is likely that Gln and sucrose serve as both transport molecules and metabolic signals since activation of *bspA* was specific to Gln rather than a general change in carbon/nitrogen balance. Whether these metabolites directly signal *bspA* induction is not clear, however it appears that metabolic signaling involving Gln and sucrose is part of a signaling pathway that involves protein phosphorylation and Ca^{2+} (Zhu and Coleman 2001a).

The direct synergistic induction of *bspA* by Gln and sucrose is in contrast to the induction of *vsp* expression in herbaceous crops. For example, although soybean *vsp* expression is related to source-sink relations no direct nitrogen effect was found (Staswick et al. 1991). Furthermore, the soybean VSP, sweet potato storage protein (sporamin) and the potato storage protein (patatin) are all sucrose inducible (Wenzler et al. 1989; Hattori et al. 1991; Mason et al. 1992) while sucrose alone had no effect on induction of poplar *bspA* expression (Zhu and Coleman 2001a). Thus the regulation of poplar *bspA* expression appears directly related to both carbon and nitrogen metabolism with sucrose and Gln acting as a metabolic signals.

BSPA INDUCTION INVOLVES SECONDARY MESSENGERS

As discussed in the previous sections, poplar *bspA* induction is a photoperiod response involving photoperception by phytochrome. Photoperiod induction is likely coupled to altered source-sink relations leading to changes in cellular carbon and nitrogen status. The change in carbon and nitrogen status appears to involve changes in the cellular concentrations of sucrose and Gln (Zhu and Coleman 2001a). In addition to providing the carbon and nitrogen needed for BSP synthesis it also seems that Gln and sucrose act as specific metabolic signals involved in *bspA* induction as opposed to induction from altered carbon/nitrogen balance since the response is specific to Gln (Zhu and Coleman 2001a). A growing body of evidence indicates and carbon and nitrogen metabolites are involved in complex signaling cascades in plants (Coruzzi and Bush 2001).

The involvement of secondary messengers including Ca^{2+} and protein phosphorylation/dephosphorylation have also been shown to be involved in Gln plus sucrose induction of *bspA* (Zhu and Coleman 2001a). The Ca ionophore, ionomycin, can activate the *bspA* promoter and Gln plus sucrose activation of the *bspA* promoter was inhibited by the calcium chelator EGTA (Zhu and Coleman 2001a). Inhibition of *bspA* induction by EGTA was relieved by addition of Ca^{2+} and the response was specific for Ca^{2+} (Zhu and Coleman 2001a). Furthermore, Gln plus sucrose activation of the *bspA* promoter was inhibited by inhibitors of protein phosphorylation (K-252a) and dephosphorylation (okadaic acid) (Zhu and Coleman 2001a). Based on these studies, it was concluded that Ca^{2+} and phosphorylation are both involved in the Gln plus sucrose activation of the *bspA* promoter, however it is not known if these components are part of the same pathway, different pathways, or convergent pathways (Zhu and Coleman 2001a).

PROPOSED MODEL FOR REGULATING BSPA EXPRESSION

The study of seasonal nitrogen storage in poplars has served as a valuable model for understanding how this process is regulated. Figure 6 presents a proposed model for the activation of poplar *bspA* expression based upon our current knowledge. In this model, photoperiod is the primary environmental signal that initiates subsequent steps leading to *bspA* induction. Photoperiod perception involves phytochrome and results in subsequent changes in growth and source-sink relations. The changes in growth and source-sink relations lead to a mobilization and repartitioning N and C metabolites which involve the transport of Gln and sucrose to stem storage sites. In addition to serving as a source of carbon and nitrogen for BSP synthesis, Gln and sucrose also serve as metabolic signals for *bspA* induction. The localized increase in the cellular concentration of Gln and sucrose is sensed by some unknown mechanism. Secondary messengers, including protein phosphorylation and Ca^{2+}, are involved in the signaling process of Gln and sucrose. These secondary messengers could either be involved in the sensing of these metabolites, in subsequent signaling steps or both.

This model represents a potential mechanism for regulating BSP storage in poplar. It must be recognized that this is a simplistic model and in reality the steps involved from photoperiod perception to BSP accumulation are going to be complex and interrelated. However, by constructing such a model experiments can be designed to understand the various steps in this physiological process. In addition to the practical value of understanding how seasonal N storage is regulated, research related to *bspA* induction may prove useful in understanding environmental and metabolic signaling in trees.

FIGURE 6. Proposed model of poplar *bspA* regulation.

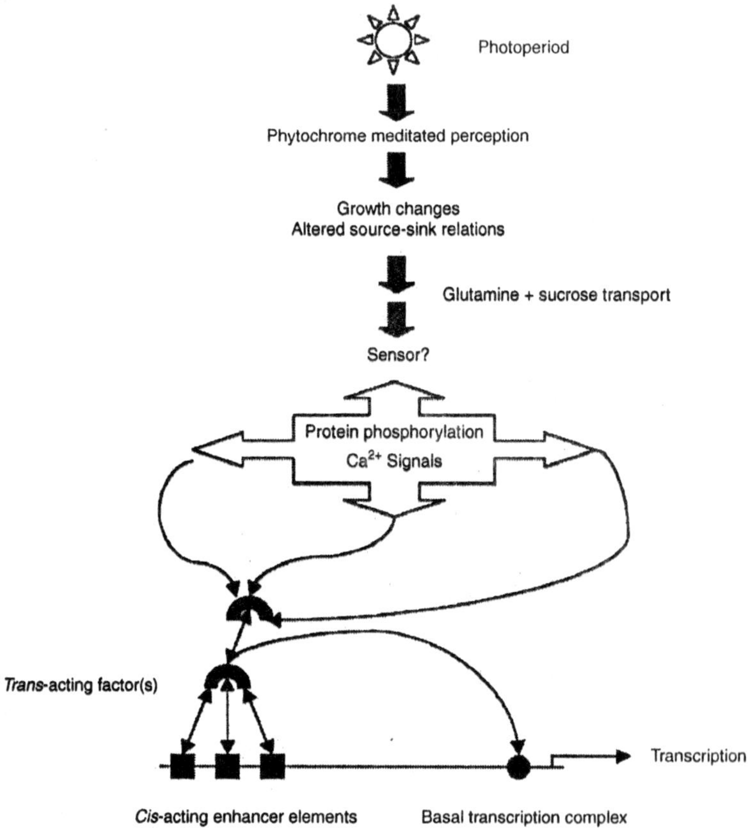

REFERENCES

Aerts, R. (1990). Nutrient use efficiency in evergreen and deciduous species from heathlands. Oecologia. 84:391-397.

Alban, H. (1985). Seasonal changes in nutrient concentration and content of aspen suckers in Minnesota. For. Sci. 31:785-794.

Andrews, D.L., Deames, B., Summers, M.D. and W.D. Park. (1988). Characterization of the lipid acyl hydrolase activity of the major potato (*Solanum tuberosum*) tuber protein, patatin, by cloning and abundant expression in a baculovirus vector. Biochem. J. 252:199-206.

Attiwill, P.M., and M.A. Adams (1993). Nutrient cycling in forests. New Phytol. 124:561-582.

Beardmore, T., Wetzel, S., Burgess, D. and P.J. Charest. (1996). Characterization of seed storage proteins in *Populus* and their homology with *Populus* vegetative storage proteins. Tree Physiol. 16:833-840.

Berger, S., Bell, E., Sadka, A. and J.E. Mullet. (1995). *Arabidopsis thaliana Atvsp* is homologous to soybean *VspA* and *VspB*, genes encoding vegetative storage protein acid phosphatases, and is regulated similarly by methyl jasmonate, wounding, sugars, light and phosphate. Plant Molecular Biology. 27:933-942.

Birk, E.M. and P.M. Vitousek. (1986). Nitrogen availability and nitrogen use efficiency in loblolly pine stands. Ecology. 67:69-79.

Borthwick, H.A., Hendricks, S.B. and M.W. Parker. (1952). The reaction controlling floral initiation. Proc. Nat. Acad. Sci. USA 38:929-934.

Boerner, R.E. (1984). Foliar nutrient dynamics and nutrient use efficiency of four deciduous tree species in relation to site fertility. J. Appl. Ecol. 21:1029-1040.

Bunker, T.W., Koetje, D.S., Stephenson, L.C., Creelman, R.A., Mullet, J.E. and H.D. Grimes. (1995). Sink limitation induces the expression of multiple soybean vegetative lipoxygenase mRNAs while the endogenous jasmonic acid level remains low. Plant Cell. 7:1319-1331.

Ceulemans, R. and J.G. Isebrands. (1996). Carbon acquisition and allocation. *In* Biology of *Populus* and Its Implications for Management and Conservation. R.F. Stettler, H.D. Bradshaw, P.E. Heilman, T.M. Hinckley (eds.), NRC Research Press. pp. 355-354.

Chapin, F.S. and R.A. Kedrowski. (1983). Seasonal changes in nitrogen and phosphorus fractions and autumn retranslocation in evergreen and deciduous taiga trees. Ecology. 64:376-391.

Chapin, F.S., Schulze, E.-D. and H.H. Mooney. (1990). The ecology and economics of storage in plants. Annu. Rev. Ecol. Syst. 21:423-447.

Clausen, S. and K. Apel. (1991). Seasonal changes in the concentration of the major storage protein and its mRNA in xylem ray cells of poplar trees. Plant Molecular Biology. 17:669-678.

Coleman, G.D. (1997). Seasonal vegetative storage proteins of poplar. *In* Micropropagation, Genetic Engineering, and Molecular Biology of *Populus*. Gen Tech. Rep. RM-GTR-297. Fort Collins, CO. U.S. United State Department of Agriculture, Rocky Mountain Forest and Range Experiment Station. Klopfenstein, N.B., Chun, Y.W., Kim, M.S., Ahuja, M.R., eds. pp. 124-130.

Coleman, G.D., Chen, T.H.H., Ernst, S.G. and L.H. Fuchigami. (1991). Photoperiod control of poplar bark storage protein accumulation. Plant Physiol. 96:686-692.

Coleman, G.D., Chen, T.H.H. and L.H. Fuchigami. (1992). Complementary DNA cloning of poplar bark storage protein and control of its expression by photoperiod. Plant Physiol. 98:687-693.

Coleman, G.D. and T.H.H. Chen. (1993). Sequence of a poplar bark storage protein gene. Plant Physiol. 102:1347-1348.

Coleman, G.D., Banados, M.P. and T.H.H. Chen. (1994). Poplar bark storage protein and a related wound-induced gene are differentially induced by nitrogen. Plant Physiol. 106:211-215.

Cornell, K.A. and M.K. Riscoe. (1998). Cloning and expression of *Escherichia coli* 5′-methylthioadenosine/S-adenosylhomocysteine nucleosidase: Indentification of the *pfs* gene product. Biochemica et Biophysica Acta 1396:8-14.

Coruzzi, G. and D.R. Bush. (2001). Nitrogen and carbon nutrient and metabolite signaling in plants. Plant Physiol. 125:61-64.

Côté, B. and J.O. Dawson. (1986). Autumnal changes in total nitrogen, salt-extractable proteins and amino acids in leaves and adjacent bark of black alder, eastern cottonwood and white basswood. Physiol. Plant. 67:102-108.

Côté, B., Vogel, C.S. and J.O. Dawson. (1989). Autumnal changes in tissue nitrogen of autumn olive, black alder and eastern cottonwood. Plant and Soil 118:23-32.

Davis, J.M., Egelkrout. E.E., Coleman, G.D., Chen, T.H.H., Haissig, B.E., Riemenschneider, D.E. and M.P. Gordon. (1993). A family of wound-induced genes in *Populus* shares common features with genes encoding vegetative storage proteins. Plant Mol. Biol. 23:135-143.

Deng, X., Weinbaum, S.A. and T.M. DeJong. (1989). Use of labeled nitrogen to monitor the transition in nitrogen dependence form storage to current-year uptake in mature walnut trees. Trees 3:11-16.

DeWald, D., Mason, H.S. and J.E. Mullet. (1992). The soybean vegetative storage protein *Vsp*α and *Vsp*β are acid phosphatases active on polyphosphates. J. Biol. Chem. 267:15958-15964.

Dickson, R.E. and E.A. Nelson. (1982). Fixation and distribution of ^{14}C in *Populus deltoides* during dormancy induction. Physiol. Plant. 54:393-401.

Franceschi, V.R. and R.T. Giaquinta (1983). The paraveinal mesophyll of soybean leaves in relation to assimilate transfer and compartmentation. Planta. 157:411-421.

Guilfoyle, T.J. (1997). The structure of plant gene promoters. Genetic Engineering 19:15-47.

Hankins, C.N., Kindinger, J. and L.M. Shannon. (1987). The lectins of *Sophora japonica 1.* purification, properties, and n-terminal amino acid sequences of two lectins from leaves. Plant Physiol. 83:825-829.

Hattori, T., Fukumoto, H., Nakagawa, S. and K. Nakamura. (1991). Sucrose-induced expression of genes coding for the tuberous root storage protein, sporamin, or sweet potato in leaves and petioles. Plant Cell Physiol. 32:79-86.

Hollwarth, M. (1976). Seasonal changes in nitrogen metabolism and its relation to temperature in the bark of popiars. Z. Pflanzenphysiol. 80:215-224.

Kang, S.M. and J.S. Titus. (1980). Qualitative and quantitative changes in nitrogenous compounds in senescing leaf and bark tissues of the apple. Physiol. Plant. 50:285-290.

Killingbeck, K.T., May, J.D. and S. Nyman. (1990). Foliar senescence in an aspen (*Populus tremuloides*) clone: The response of element resorption to interramet variation and timing of abscission. Can. J. For. Res. 20:1156-1164.

Kost, J.A. and R.E.J. Boerner. (1985). Foliar nutrient dynamics and nutrient use efficiency in *Cornus florida*. Oecologia 66:602-606.

Langherinrich, Y. and R. Tischner. (1991). Vegetative storage proteins in poplar: Induction and characterization of a 32- and a 36-kilodalton polypeptide. Plant Physiol. 97:1017-1025.

Lawrence, S.D., Greenwood, J.S., Korhnak, T.E. and J.M. Davis. (1997). A vegetative storage protein homolog is expressed in the growing shoot apex of hybrid poplar. Planta. 203:237-244.

Luxmoore, R.J., Grizzard, T. and R.H. Strand. (1981). Nutrient translocation in the outer canopy and understory of an eastern deciduous forest. Forest Sci. 27:505-518.

Maeshima, M., Sasaki, T. and Y. Asahi. (1985). Characterization of major proteins in sweet potato tuberous roots. Phytochem. 24:1899-1902.

Mancinelli, A.L. (1994). The physiology of phytochrome action. *In* R.E. Kindrick, G.H.M. Kronberg, eds., Photomorphogenesis in Plants. 2nd Ed. Kluwer Academic Publishers, Dordrecht, The Netherlands, pp. 211-269.

Mao, C., Cook, W.J., Zhou, M., Koszalka, G.W., Krenitsky, T.A., and S.E. Ealick (1997). The crystal structure of *Escherichia coli* purine nucleoside phosphorylase: A comparison with the human enzyme reveals a conserved topology. Structure. 5:1373-1383.

Mason, H.S. and J.E. Mullet. (1990). Expression of two vegetative storage protein genes during development and in response to water deficit, wounding and jasmonic acid. Plant Cell. 2:569-579.

Mason, H.S., DeWald, D.B., Creelman, R.A. and J.E. Mullet. (1992). Coregulation of soybean vegetative storage protein gene expression by methyljasmonate and soluble sugars. Plant Physiol. 98: 859-867.

McColl, J.G. (1980). Seasonal nutrient variation in trembling aspen. Plant and Soil. 54:323-328.

Millard, P. (1988). The accumulation and storage of nitrogen in herbaceous plants. Plant, Cell, and Environ. 11:1-8.

Millard, P. and G.H. Neilsen. (1989). The influence of nitrogen supply on the uptake and remobilization of stored N for the seasonal growth of apple trees. Annals Bot. 63:301-309.

Millard, P. and M.F. Proe. (1991). Leaf demography and seasonal internal cycling of nitrogen in sycamore (*Acer pseudoplatanous* L.) seedlings in relation to nitrogen supply. New Phytol. 117:587-596.

Mushegian, A.R. and E.V. Koonin. (1994). Unexpected sequence similarity between nucleosidases and phosphoribosyltransferases of different specificity. Protein Science. 3:1081-1088.

Neuhard, J. and P. Nygaard. (1987). Purines and pyrimidines. *In* Ingraham, J.L., Low, K.B., Magasanik, B., Schaechter, M., Umbarger, H.E., eds. *Escherichia coli* and *Salmonells typhinurium*. Cellular and Molecular Biology. Washington, DC: American Society for Microbiology. pp. 445-473.

Nitsch, J.P. (1957). Photoperiodism in woody plants. Proc. Am. Soc. Hort. Sci. 70:526-544.

Nsimba-Lubaki, M. and W.J. Peumans. (1986). Seasonal fluctuations of lectins in barks of elderberry (*Sambucus nigra*) and Black Locust (*Robinia pseudoacacia*). Plant Physiol. 80:747-751.

Oaks, A., Clark, C.J. and J.S. Greenwood. (1991). Nitrogen assimilation if higher plants: Strategies for annual and perennial plant species. *In* International Symposium on Nitrogen in Grapes and Wine, 1991, pp. 43-51.

Ostman, N.L. and G.T. Weaver. (1982). Autumnal nutrient transfers by retranslocation, leaching, and litter fall in a chestnut oak forest in southern Illinois. Can. J. For. Res. 12:40-51.

Paiva, E., Lister, R.M. and W.D. Park. (1983). Induction and accumulation of major tuber proteins in stems and petioles. Plant Physiol. 71:161-168.

Parsons, T.J., Bradshaw, H.D. and M.P. Gordon. (1989). Systemic accumulation of specific mRNAs in response to wounding in poplar trees. Proc. Natl. Acad. Sci. USA 86:7895-7899.

Pastor, J. and J.G. Bockheim. (1984). Distribution and cycling of nutrients in an aspen-mixed-hardwood-spodosol ecosystem in northern Wisconsin. Ecology. 65:339-353.

Perry, T.O. (1971). Dormancy of trees in winter. Science. 171:29-36.

Pregitzer, K.S., Dickmann, D.J., Hendrick, R. and P.U. Nguyen. (1990). Whole-tree carbon and nitrogen partitioning in young hybrid poplars. Tree Physiology. 7:79-93.

Quail, P.H., Boylan, M.T., Parks, B.M. Short, T.W., Xu, Y. and D. Wanger. (1995). Phytochromes: Photosensory perception and signal transduction. Science. 268: 675-680.

Ryan, D.F. and F.H. Bormann. (1982). Nutrient resorption in northern hardwood forests. BioScience 32:29-32.

Sagisaka, S. (1974). Effect of low temperature on amino acid metabolism in wintering poplar. Plant Physiology. 53:319-322.

Sauter, J.J. (1981). Seasonal variation of amino acids in the xylem sap of *Salix*. Z. Pflanzenphysiol. 101:399-411.

Sauter, J.J. and B. van Cleve. (1990). Biochemical, immunochemical and ultrastructural studies of protein storage in poplar (*Populus x canadensis* 'robusta') wood. Planta 183:92-100.

Sauter, J.J., van Cleve, B. and S. Wellenkamp. (1989). Ultrastructural and biochemical results on the localization and distribution of storage proteins in a poplar tree and in twigs of other tree species. Holzforschung 43:1-6.

Sauter, J.J., van Cleve, B. and S. Wellenkamp. (1992). Seasonal variation of amino acids in the xylem sap of *Populus* x *canadensis* and its relation to protein body mobilization. Trees. 7:26-32.

Sheriff, D.W., Margolis, H.A., Kaufmann and P. B. Reich. (1995). Resource use efficiency. *In* Resource Physiology of Conifers. W.K. Smith and T.M. Hinckley (eds.). Academic Press, San Diego. pp. 143-178.

Smith, H. and G.C. Whitelam. (1990). Phytóchrome, a family of photoreceptors with multiple physiological roles. Plant Cell Envrion. 13:695-707.

Staswick, P.E. (1989). Preferential loss of an abundant storage protein from soybean pods during seed development. Plant Phyisol. 90:1252-1255.

Staswick, P.E. (1990). Novel regulation of vegetative storage protein genes. Plant Cell. 2:1-6.

Staswick, P.E. (1994) Storage proteins of vegetative plant tissues. Annu. Rev. Plant Physiol. Plant Mol. Biol. 45:303-322.

Staswick, P.E., Haung, J. and Y. Rhee. (1991) Nitrogen and methyl jasomonate induction of soybean vegetative storage protein genes. Plant Physiol. 96:130-136.

Stepien, V. and F. Martin. (1992). Purification, characterization and localization of the bark storage proteins of poplar. Plant Physiol. Biochem. 30:399-407.

Stepien, V., Sauter, J. and F. Martin. (1994). Vegetative storage proteins in woody plants. Plant Physiol. Biochem. 32:185-192.

Taylor, B.K. and L.H. May. (1967). The nitrogen nutrition of the peach tree 11. Storage and mobilization of nitrogen in young trees. Aust. J. Biol. Sci. 20:389-411.

Thomas B. and D. Vince-Prue. (1997). Photoperiodism in Plants. Academic Press. London.

Thompson, J.D., Higgins, D.G. and T.J. Gibson. (1994). Improving the sensitivity of progressive multiple sequence alignment throughout sequence weighting, positions-specific gap penalties and weight matrix choice. Nuc. Acids Res. 22:4673-4680.

Utsugi, S., Sakamoto, W., Murata, M. and F. Motoyoshi. (1998). *Arabidopsis thaliana* vegetative storage protein (VSP) genes: Gene organization and tissue specific expression. Plant Mol. Biol. 38:565-576.

van Cleve, B., Clausen, S. and J.J. Sauter (1988). Immunochemical localization of a storage protein in poplar wood. J. Plant Physiol. 133: 371-374.

Vitousek, P. (1982). Nutrient cycling and nutrient use efficiency. Am. Nat. 119: 553-572.

Weiser, C.J. (1970). Cold resistance and injury in woody plants. Science. 169:269-278.

Wenzler, H., Mignery, G., Fisher, L. and W. Park. (1989). Sucrose-regulated expression of a chimeric potato tuber gene in leaves of transgenic tobacco plants. Plant Mol. Biol. 13: 347-54.

Wetzel, S., Demmers, C. and J.S. Greenwood. (1989). Seasonally fluctuating bark proteins are a potential form of nitrogen storage in three temperate hardwoods. Planta 178:275-281.

Wetzel, S. and J.S. Greenwood. (1989). Proteins as a potential nitrogen compound in bark and leaves of several softwoods. Trees 3:149-153.

Wetzel, S. and J.S. Greenwood. (1991). A survey of seasonal bark proteins in eight temperate hardwoods. Trees 5:153-157.

Wittenback, V.A. (1983). Purification and characterization of a soybean leaf storage glycoprotein. Plant Physiol. 73:125-129.

Yoshida, L., Baba, K., Yamamoto, N. and K. Tazaki. (1994). Coning of a lectin cDNA and seasonal changes in levels of the lectin and its mRNA in the innter bark of *Robinia pseudoacacia*. Plant Mol. Biol. 25: 845-853.

Zhu, B. and G.D. Coleman. (2001a). Phytochrome-mediated photoperiod perception, shoot growth, glutamine, calcium, and protein phosphorylation influence the activity of the poplar bark storage protein gene promoter (bspA). Plant Physiol. 126: 342-351.

Zhu, B. and G.D. Coleman. (2001b). The poplar bark storage protein gene (*Bspa*) promoter is responsive to photoperiod and nitrogen in transgenic poplar and active in floral tissues, immature seeds and germinating seeds of transgenic tobacco. Plant Mol. Biol. 46:383-394.

Physiology and Molecular Biology of a Family of Pathogenesis-Related PR-10 Proteins in Conifers

Abul K. M. Ekramoddoullah

SUMMARY. PR-10 is a group of pathognesis-related proteins that are induced by biotic and abiotic stresses. In conifers, PR-10 proteins and their genes were isolated and characterized. In some species, there is a multi-gene family encoding PR-10 proteins. These PR-10 proteins were shown to be up-regulated by pathogens in two conifer pathosystems and were associated with overwintering of conifers. In western white pine, the level of PR-10 protein is associated with frost hardines of white pine

---

Abul K. M. Ekramoddoullah is Senior Research Scientist, Forest Biotechnology, Pacific Forestry Centre, Canadian Forest Service, Natural Resources Canada.

Address correspondence to: Abul K. M. Ekramoddoullah, Pacific Forestry Centre, 506 West Burnside Road, Victoria, British Columbia, Canada, V8Z 1M5 (E-mail: aekramoddoul@pfc.forestry.ca).

The research of the author is supported by the Canadian Forest Service Biotechnology Network, the Canadian Forest Service Science and Technology Opportunity Fund, the Canadian Biotechnology Strategy Fund, the Canada-BC Forest Resources Development Agreement and Forest Renewal British Columbia.

The author wishes to thank the following members of his laboratory for their contribution to this work: Ms. Joanne Davidson, Dr. Jun-Jun Liu, Ms. Nina Piggott, Mr. Doug Taylor, Dr. Xeushi Yu and Ms. Arezoo Zamani.

[Haworth co-indexing entry note]: "Physiology and Molecular Biology of a Family of Pathogenesis-Related PR-10 Proteins in Conifers." Ekramoddoullah, Abul K. M. Co-published simultaneously in *Journal of Crop Improvement* (Food Products Press, an imprint of The Haworth Press, Inc.) Vol. 10, No. 1/2 (#19/20), 2004, pp. 261-280; and: *Adaptations and Responses of Woody Plants to Environmental Stresses* (ed: Rajeev Arora) Food Products Press, an imprint of The Haworth Press, Inc., 2004, pp. 261-280. Single or multiple copies of this article are available for a fee from The Haworth Document Delivery Service [1-800-HAWORTH, 9:00 a.m. - 5:00 p.m. (EST). E-mail address: docdelivery@haworthpress.com].

http://www.haworthpress.com/web/JCRIP
Digital Object Identifer: 10.1300/J411v10n01_11

needles. Ultrastructural localization revealed that western white pine PR-10 protein is an intracellular protein and is found in chloroplasts of mesophyll and endodermis, and in the stroma. In infected pine needles, the PR-10 protein was bound to the cell surface of the blister rust pathogen. PR-10 genes are organized along two exons intervened by an intron. Homology of PR-10 proteins is higher among five-needle pines than in either Douglas-fir or white spruce. Cloning, sequencing, and functional analysis of PR-10 promoters are underway to identify cis-acting regulatory elements responsive to fungal infection and cold stress. Future research and possible applications are discussed.

KEYWORDS. Pathogenesis-related proteins, conifers, western white pine, sugar pine, Douglas-fir, white pine blister rust

INTRODUCTION

Plants, including conifers, respond to biotic and abiotic stress by expressing an array of genes. Understanding the expression of these genes is important in managing genetic improvement of the adaptive traits of long-lived plants such as forest trees. Pathogenesis-related (PR) proteins are among the most widely studied components of plant defense responses (Bol et al., 1990; Linthorst, 1991; Karenlampi et al., 1994). Eleven groups of PR proteins have now been delimited (van Loon et al., 1994). The PR-10 family has now been discovered in a wide variety of species (Fristensky et al., 1988; Matton and Brisson, 1989; Breiteneder et al., 1989, 1995; Walter et al., 1990, 1996; Crowell et al., 1992; Warner et al., 1992; Moiseyev et al., 1994; Breda et al., 1996; Ekramoddoullah et al., 1995, 1997, 1998; Lo et al., 1999). PR-10 proteins have been shown to be induced by elicitors (Walter et al., 1990), pathogens (Swoboda et al., 1995; Ekramoddoullah et al., 1998), wounding (Warner et al., 1992), and metal stress (Utriainen et al., 1998). Unlike most PR proteins, the PR-10 family are intracellular proteins that may have ribonuclease-like activity (Moiseyev et al., 1994). Although normally induced in diseased plants, there is increasing evidence that some PR proteins may also be induced during cold acclimation (Ekramoddoullah et al., 1995; Hon et al., 1995). Recently, it has been suggested that cold hardening may increase plants resistance to snow mold fungi. This may indicate an interaction between cold hardening and fungal induced expression of PR genes (Ergon et al., 1998).

DISCOVERY OF *Pin l I*, A PR-10 PROTEIN IN WHITE PINE BLISTER RUST PATHOSYSTEM

White pine blister rust is a disease of five-needle pines, e.g., western white pine (*Pinus monticola* D.) and sugar pine (*Pinus lambertiana* Doug.), caused by the blister rust fungus, *Cronartium ribicola*. The fungus was inadvertently introduced into western North America in Vancouver, British Columbia in 1910. The fungus has five different spore stages on two unrelated hosts, white pine and *Ribes*. The basidiospores are produced on *Ribes* in the fall and are transported by wind to pine foliage, where they germinate. The mycelium of the fungus then grows down the needle until it reaches the bark, causing a canker which eventually kills the tree. Although some large and older trees can survive, infected younger trees almost inevitably die.

To identify genes involved in this host-pathogen interaction, we began to analyze this pathosystem (Ekramoddoullah and Hunt, 1993). One of our approaches was to examine the induction of proteins in white pine following inoculation with the rust fungus. We discovered that conifer proteins were extremely difficult to extract (Ekramoddoullah, 1991). We then had to develop methods to extract conifer proteins (Ekramoddoullah, 1991) and develop protocols for 2-D electrophoresis of conifer proteins (Ekramoddoullah and Tan, 1998). The new method produced protein extracts that contained 4% sodium dodecyl sulphate (SDS) and 5% mercaptoethanol. These large amounts of detergent and reducing agent used in the extraction and solubilization of proteins interfere with the best known spectrophotometric assay for protein (Bradford, 1976; Lowry et al., 1951). We therefore developed a specialized membrane-based protein dot blot assay (Ekramoddoullah and Davidson, 1995) whereby the interference by SDS and mercaptoethanol was eliminated. During a 2-D gel electrophoretic analysis of sugar pine proteins (Ekramoddoullah and Hunt, 1993), we observed two intensely stained protein spots in the low molecular weight and acidic region of the gel. These proteins were detected in large amounts in the foliage of control and infected seedlings only in samples taken in late fall (Ekramoddoullah et al., 1995). Since the amount detected exceeded the quantification limit of the 2-D gel software used, no quantitative difference could be ascertained at the time. When the control seedlings were sampled in the summer, we could not detect this protein. This was an initial indication that this protein may be seasonally regulated.

We obtained a partial N-terminal sequence of this protein (Gene Bank Accession # A40451) from both 1-D and 2-D gels which gave no homology with known proteins in the database and named this *Pin l* I. Although several PR-10 proteins were known at that time (Somssich et al., 1988; Fristensky et al., 1988; Breiteneder et al., 1989; Matton and Brisson, 1989), it was surprising that the N-terminal sequence did not show homology with these PR-10 proteins. Only recently, the deduced amino acid sequences from *Pin l* I cDNA revealed that this protein belongs to the PR-10 family (Ekramoddoullah et al., 2000a).

PREPARATION OF SPECIFIC ANTI-*Pin l* I ANTIBODY

To further study the possible role of this protein in western white pine, we developed an antibody probe to *Pin l* I utilizing the following strategy. We first synthesized a synthetic peptide according to the N-terminal sequence information. During the synthesis, we added an amino acid, cysteine, at the carboxy terminal of the peptide. The sulphohydryl group of cysteine allowed us to conjugate the peptide to a carrier molecule, Keyhole Limphet Hemocyanine (KLH). The peptide, being a small molecule, is by itself poorly immunogenic but when coupled to a large protein molecule such as KLH it becomes immunogenic. Peptide-KLH conjugate was injected into rabbits. Since rabbits do not discriminate between peptide and KLH, they produced antibodies to both molecules. This crude rabbit anti-sera was then passed onto an immuno-affinity column which was prepared by covalently linking the peptide to a solid support, e.g., agarose. The bound antibody (i.e., anti-peptide antibody) was then eluted off the column. This affinity purified antibody was termed anti-*Pin l* I antibody. The specificity of the antibody was demonstrated using western immunoblot analysis (Ekramoddoullah et al., 1995) by its ability to bind only *Pin l* I protein present in the extract of sugar pine foliage.

DETECTION AND SEASONAL VARIATION OF *Pin m* III IN WESTERN WHITE PINE

Having demonstrated the specificity of anti-*Pin l* I antibody, this antibody was used to detect a homologous protein of similar size in western white pine. The homologous protein was named *Pin m* III. In an experiment, 4-month-old western white pine seedlings were kept in a shelter house under natural daylight and temperature for a year. Foliar samples were collected monthly for a year. Proteins were extracted and *Pin m* III content was quantified by immunoblot using anti-*Pin l* I antibody probe. *Pin m* III content increased as fall progressed, reached a maximum in winter months, and decreased to low levels in summer months (Ekramoddoullah et al., 1995). Interestingly, the level of a 23-kDa western white pine photosystem II protein slowly decreased in the fall and reached a minimum in December and then slowly increased as spring approached, reaching a maximum in early summer (Ekramoddoullah and Taylor, 1996). While the variation in the level of photosystem II may be a reflection of photosynthetic activity in winter months, *Pin m* III is associated with over-wintering of pine seedlings. When we compared the total protein profiles of winter and summer tissues by 2-D electrophoresis, *Pin m* III remained the predominant protein increasing in winter months (Ekramoddoullah and Taylor, 1996).

ASSOCIATION OF Pin m III WITH FROST HARDINESS OF WESTERN WHITE PINE

The timing and accumulation of *Pin m* III is reminiscent of the accumulation of vegetative storage proteins. Storage proteins of molecular mass in the range of 27 kD to 32 kD were detected in increasing amounts in other species during fall (Coleman et al., 1991; Roberts et al., 1991; Wetzel and Greenwood 1989, 1991; Wetzel et al., 1989). These storage proteins are usually basic and rich in glutamine and asparagine. Some of the storage proteins store nitrogen over winter that will then be mobilized for new growth in the spring. Amino acid composition from the hydrolysate of *Pin m* III (Ekramoddoullah et al., 1995) or the deduced amino acid sequences from *Pin m* III cDNA (Yu et al., 2000a) did not show a preponderance of asparagine and glutamine. Although the size, charge and amino acid composition of *Pin m* III do not resemble those of vegetative storage proteins, the possibility of this protein being a storage protein cannot be ruled out. The high percentage of *Pin* m III in winter samples of white pine and its low level in summer samples suggests a particular functional role for *Pin m* III.

An increase in soluble proteins has been found to be associated with frost hardiness in a *Pinus* species (Pomeroy and Simonovitch, 1970). Similarly, increase in protein content with the onset of cold hardening has been documented in *Picea abies* (Eriksen, 1997; Żeigler and Kandler, 1980). The role of proteins in cold hardening is not clear. These proteins may act as cryoprotectants by binding to the cell walls and membranes and by acting as a barrier to ice propagation (Griffith et al., 1977). Cold acclimation closely parallels an accumulation of membrane-bound proteins in the leaves of Korean boxwood (Gusta and Wesser, 1972). Proteins isolated from cold acclimated spinach has been shown to protect isolated thylakoid membranes against freeze-thaw damage (Hincha et al., 1989). Several cold regulated genes some of which encode either late embryogenesis abundant (LEA) proteins or other polypeptides (Thomashow 1998) have been described. One polypeptide, COR15a may act as a cryoprotectant by decreasing the propensity of lipid bi-layers which form deleterious hexagonal II phase lipids during freeze-induced membrane lesions (Steponkus et al., 1998). The process of cold-acclimation may involve multiple genes (Thomashow, 1990). In support of this, it has been possible to induce several cold-regulated genes by overexpressing a transcription factor, a regulatory gene CBF1, in transgenic *Arabidopsis thaliana.* These transgenic plants were shown to have enhanced freezing tolerance (Jaglo-Ottosen et al., 1998) and were associated with multiple biochemical changes (Gilmour et al., 2000). In woody plants, dehydrin-like proteins accumulate in the cytoplasm and the nucleus in winter months (Wisniewski et al., 1996). Dehydrins may also have cryoprotectant properties by protecting cells against dehydration (Close 1997). They may stabilize protiens and cell membranes against freeze damage (Thomashow, 1999). Since some dehydrins have been identified in conifers (Jarvis et al., 1996; Close, 1993), it was of interest to see whether *Pin m* III could be a dehydrin-like protein particularly since its size was similar to that of a dehydrin. An antibody raised against dehydrins did not react with *Pin m* III (Ekramoddoullah, unpublished) and the deduced protein sequence of *Pin m* III cDNA has no homology with dehydrin.

Over 730 million forest tree seedlings are planted annually in Canada (Canadian Council of Forest Ministers, 1993) to maintain a sustainable timber harvest. As a planting strategy, it is common practice to develop cold hardiness seedlings by inducing dormancy in them. At present in British Columbia, the dormancy condition of conifer seedling stock prior to lifting and cold storage is assessed using the −18°C freezing test. In this procedure, the seedlings are frozen at −18°C, then placed in a controlled environment chamber for 7 days at which time the foliage is visually assessed for damage. Considering the narrow window available for lifting, this method is time consuming and also imprecise. Thus, it was of interest for us to establish an association of *Pin m* III with frost hardiness that will lead to the development of quicker and reliable biochemical methods to detect dormancy induction. We have carried out a series of experiments that clearly established that *Pin m* III is significantly associated with frost hardiness of western white pine needles (Ekramoddoullah et al., 1995). More recently, we have detected a protein cross-reacting with anti-*Pin m* III antibody that was associated with induction of dormancy in white spruce (unpublished). Although the biological role of *Pin m* III in frost hardiness is not known, a detailed study at the molecular level into *Pin m* III gene expression may help to identify conifer seedlots well suited for planting in areas with specific frost problems, e.g., late spring frost, or summer frost, and to select seedlots that are suitable for planting in a particular site within a biogeoclimatic zone.

FUNGAL INDUCTION OF Pin m III AND Pin l I IN WHITE PINE BLISTER RUST PATHOSYSTEM

We analyzed bark proteins of mature resistant western white pines (i.e., free of cankers) and susceptible trees with several cankers (Davidson and Ekramoddoullah, 1997). We decided to look at the level of *Pin m* III protein in this system. To our surprise, we found that the level of this protein is significantly lower in resistant trees than in susceptible trees. However, in resistant and susceptible trees, the level of *Pin m* III was higher in winter than in summer (Ekramoddoullah et al., 1998). Importantly, trees with cankers maintained a high level of *Pin m* III in summer months and all times of the year; particularly in the spring, the level of *Pin m* III was higher in susceptible trees than in resistant trees. In a series of experiments, the level of *Pin m* III in forty-three 7-year-old previously inoculated white pine seedlings was examined. In all susceptible seedlings tested, cankered tissues had high levels of *Pin m* III, and samples collected from the outside edge of the canker margin had low levels of *Pin m* III; this protein was also detected in healthy bark of cankered trees. Since the level of *Pin m* III in healthy white pine trees is normally lowest in summer months, the high level of *Pin m* III in summer samples of infected tissues is thought to be a consequence of fungal infection. The level of *Pin l* I, on the other hand, was significantly higher in the foliage of inoculated sugar pine seedling carrying R gene for resistance than in the foliage of inoculated susceptible sugar pine seedlings

(Ekramoddoullah and Tan, 1998). Stem samples of both resistant and susceptible sugar pine seedlings had comparable levels of *Pin l* I.

ULTRASTRUCTURAL LOCALIZATION OF Pin m III PROTEIN

We examined the localization of *Pin m* III in an infected needle spot of susceptible western white pine immunocytochemically with anti-*Pin l* I antibody using both light and electron microscopy (unpublished). Results (Figure 1) illustrate that the *Pin m* III protein is bound to the blister rust fungus regardless of whether the fungus is inside or outside the host cell. Similar results were found under light microscopy utiliz-

FIGURE 1. An electron micrograph of a portion of a host cell from infection spot of inoculated susceptible western white pine taken in summer months. Sections were probed rabbit with anti-*Pin l* I antibody followed by gold-labeled goat anti-rabbit IgG antibody. The fungal cell is seen inside the host cell. The fungal cell wall is clearly labeled with gold particles.

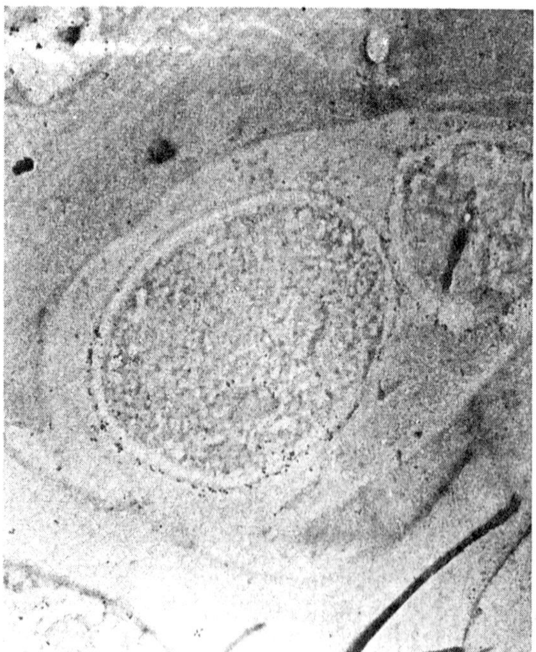

ing a silver enhancement technique. This is the first time that a PR-10 protein has been shown to bind to the pathogen.

The hypersensitive response is a manifestation of a dominant resistance gene (R) whereby the host cells surrounding the pathogen undergo immediate cell death (Keen, 1992). In the sugar pine system, we attempted to detect *Pin m* III at a hypersensitive reaction spot. The reaction spot was visible only in winter months following inoculation with the fungus in late summer. *Pin m* III protein was found in this infection spot as well as on healthy needles; however, the fungus itself could not be detected in this hypersensitive spot. On examination of the hypersensitive infection spot by light microscopy it was revealed that there was a formation of an extra layer of endothelial cells in the vascular bundle. It is of interest to note that, during an incompatible interaction between *Medicago sativa* and *Pseudomonas* pv. *Pisi*, a MsPR-10-1 gene transcript was observed in the vascular bundle (Breda et al., 1996). Because *Pin m* III was also up-regulated in winter months, there was too much background noise to delineate the role of *Pin m* III in this hypersensitive response. In fact, in winter samples of primary western white pine needles from healthy seedlings, the highest levels of *Pin m* III are found in the chloroplasts of the mesophyll and endodermis distributed in the stroma (Figure 2). Any starch deposits are devoid of *Pin m* III. Cytoplasm, nuclei, and mitochondria have the next highest amount, about half the amount of *Pin m* III in the chloroplasts. Vacuoles show about one-sixth the amount as chloroplasts and cell walls are essentially devoid of *Pin m* III. None was found in the intercellular spaces, indicating that *Pin m* III is an intracellular protein.

CLONING AND CHARACTERIZATION OF cDNA ENCODING Pin m III

To further characterize the *Pin m* III protein, we prepared an expression cDNA library from poly(A)+ mRNA isolated from needles of healthy mature western white pine trees collected in winter months. The library was screened with anti-*Pin l* I antibody and subsequently a full length cDNA was cloned (Yu et al., 2000a). An open reading frame of 486 base pairs encodes a protein of 161 amino acid residues with a molecular mass of 18 kDa and a predicted isoelectric point of 5.5 which is consistent with observed values for *Pin m* III protein. Moreover, the deduced N-terminal amino acid sequence is identical to that determined by protein sequence analysis. BLAST search of the protein database revealed similarities among the sequence of *Pin m* III and other PR-10

FIGURE 2. An electron micrograph of a portion of a mesophyll cell from a primary western white pine needle collected in February, 1999. Sections were probed with rabbit anti-*Pin l* I antibody followed by gold-labeled goat anti-rabbit IgG antibody. A chloroplast (C) is closely pressed against the cell wall (W) and there are two small mitochondria (M) associated it. The wall is devoid of gold particles, while the chloroplast, mitochondria, and to a much lesser extent, the vacuole (V) have gold particles overlaying them.

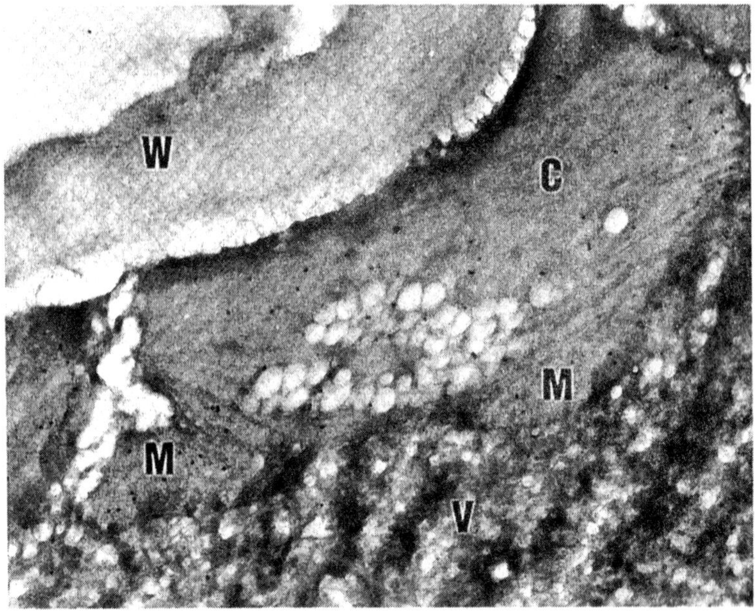

proteins. *Pin m* III shares 22 to 36% amino acid homology with representative PR-10 proteins. From northern analysis, the amount of *Pin m* III transcript changed seasonally, preceding the parallel changes in the amount of *Pin m* III protein. There was a larger accumulation of *Pin m* III transcript and the protein in root systems than other tissues. The amount of transcript was also shown to be correlated with the severity of fungal infection in white pine.

GENE STRUCTURE OF *Pin m III* AND ITS HOMOLOGUES IN OTHER PINACEAE

To study the gene structure of *Pin m* III and its homologues in other Pinaceae, the genomic sequences were cloned by PCR. PCR primers were designed to amplify the genomic sequence for the coding region of the *Pin m* III gene and its homologues from the following pines: western white, eastern white, sugar, Korean, Himalayan, limber, ponderosa, and loblolly. Only a 650-bp band was amplified and the fragment sequences showed high homology (90%) among the nine Pinaceae (Yu et al., 2000b). There are two exons interrupted by an intron. The size of intron is in the range of 150 to 180 bp. The intron position deduced by comparison of the genomic sequence and the cDNA of *Pin m* III revealed a regular intron-exon junction.

CLONING AND CHARACTERIZATION OF SUGAR PINE *Pin l* I cDNA

To further characterize *Pin l* I, an expression cDNA library from poly(A)+mRNA extracted from a pooled collection of resistant sugar pine needles 2 days to 6 weeks after inoculation with blister rust was screened with a digoxigenin-labelled *Pin m* III cDNA probe. From this cDNA library, a 3′ end partial cDNA was cloned and sequenced (Ekramoddoullah et al., 2000a). Using primers based on the sequence of *Pin m* III cDNA, a RT-PCR fragment for the full length coding region was cloned and characterized. Sequence analysis indicated an open reading frame of 486 bases encoding a protein of 161 amino acids with a molecular mass of 18 kD and a predicted isoelectric point of 5.53. The deduced amino acid sequence showed highest homology (~95%) to *Pin m* III.

CLONING OF WHITE SPRUCE PR-10 GENE

A cDNA library was constructed from cold hardened white spruce (*Picea glauca*) seedling roots (Mattheus et al., 2000) which was screened with labelled *Pin m* III cDNA. Three highly related sequences termed as Picg1, Picg2, and Picg4 were isolated. Deduced amino acid sequences showed that Picg1 has 140 amino acids with pI of 5.0, Picg2 has 160 amino acids with pI of 5.5 and Picg4 has 161 amino acids with pI of 5.3. As illustrated in Figure 3, multiple sequence alignment showed that similarities at amino acid level of Picg1, Picg2, Picg4 with *Pin m* III are 67, 87, and 76%, respectively. As expected, all three white spruce putative proteins were highly homologous to intracellular pathogenesis-re-

FIGURE 3. Alignment of the cDNA deduced amino acid sequences of a western white pine PR10 gene *Pin m* III and its homologous picg1 (AF197344), picg2 (AF197343), and picg4 (AF197342) from white spruce. The identities between *Pin m* III and its white spruce homologues picg1, picg2, and picg4 are 67%, 87% and 76%, respectively.

```
picg2      MVSGTPVTEDVCQVEARKLWNPMAKDSHNLLPKVLPEI-FASVTLLQGDGGVGTIKQLNF 59
Picg4      MVSRTAITEDVCQVEAKRLWNALSKDSHNLLPKILPEI-FASVTLLQGDGGVGTIKQLNF 59
Pinm III   MVSGTSSTEEVVQVEARRLWNATTKDSHNFLPKVLPEV-FTSVTLLQGDGGVGTVKQLNF 59
picg1      ----------------------MVKDGHNLYPKALHEFNISSVTVLQGDGGVGTVRQLNF 38
                                 **.**: ** * *. ::***:*********::****

picg2      TPANKDFSFVKERVDEIDEEKMVYKYTTIEGGSLGKKLSAASFEVKISPRKEGGCVATWV 119
Picg4      TPANKDFSFIKERVDEIDEEKMVYKYTTIEGGSLGEKLSAASFEVKLVPRKEGGCVASWI 119
Pinm III   TPGKKDFSFIKERVDELDQENFVYKYTAIEGGPLGKQLSSACFEVKLIPRKEGGCVARWT 119
picg1      TPANKDFSYVKERLDVIDEDNMVHKYAAIEGGSLGKKLSALNFELKFVHREEGGCALTWI 98
            **.:****::***:* :*::::*:**::****.**::**:  **:*:  *:****.  *

picg2      CNYETLPGGQLDEGRVKEIKENSVAMLKKIEQYLLSNPTLY- 160
Picg4      CNYETLPGAQLDEGKTKEIKENSIAMLKKIEQYLLSNPNLYC 161
Pinm III   CNYETLPGVQPDEGKLKEIKEDSFGMLKKVEQYLLSNPNLYC 161
picg1      CNYETLPGAPDGETRVEEIKKMDDAMFKKIEEYLISNPDLYC 140
            ********    .* : :***: . .*:**:*:**:*** **
```

* – homologous
: – conserved substitution
. – semi-conserved
---- – gap

lated PR-10 proteins. Northern analysis revealed that the amount of PR-10 transcript is elevated in white spruce during cold-hardening.

STRATEGY FOR DETECTION OF PR-10 PROTEIN IN DOUGLAS-FIR AND WHITE SPRUCE

To gain more insight into the up-regulation of PR-10 proteins in other conifers, we extended our research to the pathosystem of Douglas-fir, *Pheliinus weirii* (Murr.) Gilb. The anti-*Pin l* I antibody can detect PR-10 proteins in several five-needle pine, e.g., eastern white pine, western white pine, sugar pine, Korean pine, Himalyan pine, limber pine. However, this antibody failed to detect the corresponding PR-10 proteins in other conifers, such as Douglas-fir and white spruce (Ekramoddoullah, unpublished).

We have devised a strategy to develop probes to identify and characterize PR-10 proteins in Douglas-fir. This strategy involves polymerase chain reaction (PCR) cloning of Douglas-fir (*Pseudotsuga menziesii*) genomic DNA utilizing primers based on a white pine *Pin m* III cDNA sequence. The deduced amino acid sequence of Douglas-fir PR-10 genomic sequence was compared with the white pine PR-10 protein which showed about 80% homology. Since the primer sequence of white pine PR-10 protein was reflected in N- and C-terminal sequence, the homology was overestimated. We synthesized a peptide based on the sequence information away from the N-terminal but well conserved between white pine and Douglas-fir PR-10 protein. This would allow us to use white pine samples as positive controls for checking the antibody reactivity. An antibody was made to this synthetic peptide, which also contains a p-loop structure (Saraste et al., 1990). With this antibody, PR-10 protein was detected in Douglas-fir tissues and it showed a seasonal variation and was up-regulated following infection with laminated root-rot (Ekramoddoullah et al., 2000b). The antibody could also detect a PR-10 protein in white spruce, the level of which also varied with season (Ekramoddoullah, unpublished data). When compared with five needle pines, the amount of protein expressed in both Douglas-fir and white spruce is extremely low, suggesting differences in the regulation of PR-10 gene expression in these conifers.

PR-10 GENE PROMOTER

Since our study clearly showed that the expression of PR-10 genes is induced by both biotic and abiotic stress, we are investigating whether cis-acting regulatory elements in the gene promoter are responsive to different stresses. In particular, we are interested in delineating cold and fungal responsive regulatory elements in the promoter which may shed light on the cross-communication between these two different stimuli.

Few PR-10 gene promoters have been isolated and characterized (Korfhage et al., 1994; Rushton et al., 1996; Euglem et al., 1999; and Desveaux et al., 2000). Some regulatory elements responsive to pathogens have tentatively been characterized (Eulgen et al., 1999). These studies revealed that the process of transcriptional activation by elicitation differs among various PR-10 genes within the same species. DNA-binding proteins, WRKY, have been identified which bind to an elicitor-response element (ERE) of parsley PR-10 gene promoter (Rushton et al., 1996). An elicitor-response element in the potato PR-10 promoter has also been identified (Matton et al., 1993). An accumulation of the PR-10 mRNA was correlated with the binding of two transcription factors to these EREs, indicating elicitor-dependent activation of PR-10 gene (Despres et al., 1995). Hajela et al. (1990) reported evidence that a cold-regulated gene, *corl* 5a of *Arabidopsis*, has a cold-inducible promoter, and subsequently a cis-acting regulatory element that can impart cold-regulated gene expression was identified (Baker et al., 1994).

Since no other PR-10 genes, other than those in conifers, have been shown to be induced by cold stress, studies on conifer PR-10 gene promoters would be of value in understanding the mechanisms of cross-communication between different external stimuli such as cold and pathogens. Thus, we started a systematic analysis of PR-10 gene promoters. We have recently isolated, cloned, and sequenced three PR-10 gene promoters, each one having between 85% to 100% homology with *Pin m* III cDNA in an open reading frame (Liu, J.-J. and Ekramoddoullah, A.K.M., unpublished data). It would appear that there are different PR-10 genes in western white pine, which is consistent with three PR-10 genes being isolated from white spruce (Mattheus et al., 2000). At the protein level, we have clearly shown that both *Pin m* III and *Pin l* I have at least two predominant isoforms with the same size (Ekramoddoullah et al., 1995; Ekramoddoullah and Taylor, 1996). More recent data indicates that there may be as many as seven isoforms of *Pin m* III (Figure 4). Work is now underway to isolate and purify these isoforms in an attempt to study their respective roles in defense and in cold stress. Preliminary data indicates that only one isoform of *Pin m* III protein is induced by mechanical wounding of the western white pine needles (Liu, J.-J. and Ekramoddoullah, A.K.M., unpublished data).

Sequence analysis of the promoters revealed that they have characteristic regulatory elements responsive to various stresses. Deletion analysis of western white pine PR-10 gene promoter utilizing a transient

FIGURE 4. Two-dimensional western blot analysis of "slow canker growth" resistant western white pine foliage collected February 1996 (sample 3117B 31-1 second year's growth). Sixty micrograms total protein was loaded and probed with anti-*Pin l* I antibody at 1/500 dilution. The antibody detected seven isoforms all at approximately 18 kDa and ranging in pI from 5.2 to 7.0.

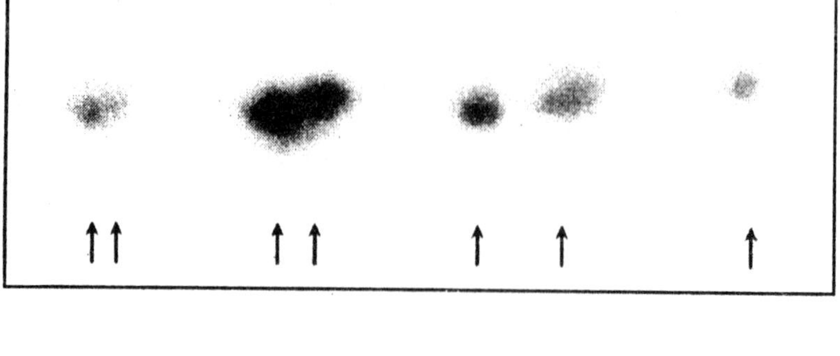

expression system showed that the promoter activity could be increased three-fold (Liu, J.-J. and Ekramoddoullah, A.K.M., unpublished data). These results suggest the presence of a suppressive regulatory element in the white pine PR-10 promoter.

CONCLUDING REMARKS

The discovery of a family of PR-10 genes in conifers with highly homologous protein sequences offers an opportunity to examine how long-living and slow-growing species adapt to different environmental stresses. It has been suggested that the evolution of conifers and angiosperms has proceeded along different paths (Kinlaw and Neale, 1997). Does the presence of several PR-10 genes in a given conifer species offer any special advantage? A systematic study of the biochemical separation of these proteins either in natural or recombinant form would be essential to analyze the functional role of these proteins in various stresses. The functional analysis of the promoters of the various conifer PR-10 genes may help us to understand how these genes respond to various stimuli. Comparative analysis of PR-10 gene regulation in different conifer pathosystems would also enhance our understanding of how the external stimuli are perceived and processed. In terms of tree improvement or crop development, there are potential applications of this knowledge. One potential application is the use of PR-10 protein as a marker for frost hardiness. If a pathogen-specific promoter of a PR-10 gene could be established, then one could engineer resistance in trees mediated by a PR-10 promoter driven anti-microbial peptide (or specific anti-microbial antibody). The advantage is that trees would need to produce these compounds only when there is a threat of the pathogen attack. Conifer gene promoters might also be applied in other crop systems for the study of frost tolerance and disease resistance.

REFERENCES

Baker S.S., Wilhelm K.S., and Thomashow M.F. (1994). The 5′-region of *Arabidopsis thaliana* cor15a has cis-acting elements that confer cold-, drought- and ABA-regulated gene expression. Plant Mol. Biol. 24: 701-713.

Bol J.F., Linthorst H.J.M., Cornelissen B.J.C. (1990). Plant pathogenesis-related proteins induced by virus infection. Ann. Rev. Phytopathol. 28: 113-138.

Bradford M.M. (1976). A rapid and sensitive method for the quantification of microgram quantities of protein utilizing the principle of protein-dye binding. Anal. Biochem. 72:248-254.

Breda C., Sallaud C., ElTurk J., Buffard D., deKozak I., Esnault R., and Kondorosi A. (1996). Defense reaction in Medicago sativa: A gene encoding a class 10 PR protein is expressed in vascular bundles. Mol. Plant Micr. Interact. 9: 713-719.

Breiteneder H., Pettenburger K., Bito A., Valenta R., Kraft D., Rumpold H., Scheiner O., and Breitenbach M. (1989). The gene coding for the major birch pollen allergen *Bet v* I is highly homologous to a pea disease resistance response gene. EMBO J. 8: 1035-1938

Breiteneder H., Hoffmann-Sommerguber K., O'Riordain G.O., Susani M., Ahorn H., Ebner C., Kraft D., and Scheiner O. (1995). Molecular characterization of *Api g* 1, the major allergen of celery (*Apium graveolens*), and its immunological and structural relationships to a group of 17-kDa tree pollen allergens. Eur. J. Biochem. 233: 484-489.

Close T.J., Fenton R.D., and Moonan F. (1993). A view of plant dehydrins using antibodies specific to the carboxy terminal peptide. Plant Mol. Biol. 23:279-286.

Close T.J. (1997). Dehydrins: A commonality in the response of plants to dehydration and low temperature. Physiol. Plant. 100:291-296.

Coleman G.D., Chen T.H.H., Ernst S.G., and Fuchigami L. (1991). Photoperiod control of poplar storage protein accumulation. Plant Physiol. 96: 686-692.

Compendium of Canadian Forestry Statistics. (1992). National Forestry Database. Ottawa: Canadian Council of Forest Ministers, 1993.

Crowell D.N., John M.E., Russell D., and Amasino R.M. (1992). Characterization of a stress-induced, developmentally regulated gene family from soybean. Plant Mol. Biol. 18: 459-466.

Davidson J.J. and Ekramoddoullah A.K.M. (1997). Analysis of bark protein in blister rust-resistant and susceptible western white pine (*Pinus monticola*). Tree Physiol. 17:663-669.

Despres C., Subramanium R., Matton D.P., and Brisson N. (1995). The activation of the potato PR-10a gene requires the phosphorylation of nuclear factor PBF-1. The Plant Cell. 7: 589-598.

Desveaux D., Despres C., Joyeux A., Subramaniam R., and Brisson N. (2000). PBF-2 is a novel single-stranded DNA binding factor implicated in PR-10a gene activation in potato. Plant Cell. 12: 1477-1490.

Ekramoddoullah A.K.M. (1991). Analyses of proteins of western white pine (*Pinus monticola* D.) needles. In: Hiratsuka Y., Blenis P.V., Crane P.E., Laishley B.L. (eds.) Rusts of Pine. Proc. 3rd IUFRO Rusts of Pine Working Party Conference, September 18-22, 1989 Banff, Alberta. For. Can, Northwest Reg., North. For. Cent. Edmonton, Alberta. Inf. Rep. NOR-X-317, pp. 102-108.

Ekramoddoullah A.K.M. (1993). Analysis of needle proteins and N-terminal amino acid sequence of two photosystem II proteins of western white pine (*Pinus monticola* D. Don). Tree Physiol. 12: 101-106.

Ekramoddoullah A.K.M., and Hunt R.S. (1993). Changes in protein profile of susceptible and resistant sugar pine foliage infected with the white pine blister rust fungus *Cronartium ribicola*. Can. J. Plant Pathol. 15:259-264.

Ekramoddoullah A.K.M., and Davidson J.J. (1995). A method for the determination of conifer foliage protein extracted using sodium dodecyl sulfate and mercaptoethanol. Phytochem. Anal. 6: 20-24

Ekramoddoullah A.K.M., Taylor D., and Hawkins B.J. (1995). Characterization of a fall protein of sugar pine and detection of its homologue associated with frost hardiness of western white pine needles. Can. J. For. Res. 25: 1137-1147.

Ekramoddoullah A.K.M., and Taylor D.W. (1996). Seasonal variation of western white pine (*Pinus monticola* D. Don) foliage proteins. Plant Cell Physiol. 37: 189-199.

Ekramoddoullah A.K.M., Taylor D.W., Yu X., and Misra S.M. (1997). Cloning and characterization of *Pin m* III gene in white pine inducible by cold and a fungal pathogen. U.S. patent application no. 08/955,629.

Ekramoddoullah A.K.M., Davidson J.J., and Taylor D. (1998). A protein associated with frost hardiness of western white pine is up-regulated by infection in the white pine blister rust pathosystem. Can. J. For. Res. 28: 412-417.

Ekramoddoullah A.K.M., and Tan Y. (1998). Differential accumulation of proteins in resistant and susceptible sugar pine (*Pinus lambertiana*) seedlings inoculated with white pine blister rust fungus (*Cronartium ribicola*). Can. J. Plant Pathol. 20:308-318.

Ekramoddoullah A.K.M, Yu X., Taylor D.W., and Piggott N. (2000a). Characterization of a defense responsive PR-10 gene in sugar pine. Year 2000 meeting of the western international forest disease work conference. Waikoloa, Big Island of Hawaii.

Ekramoddoullah A.K.M., Yu X., Sturrock R., Zamani A., and Taylor D.W. (2000b). Detection and seasonal expression pattern of a pathogenesis-related protein (PR-10) in Douglas-fir (*Pseudotsuga menziesii*) tissues. Physiol. Plant. 110: 240-247.

Ergon A., Klemsdal S.S., Tronsmo A.M. (1998). Interactions between cold hardening and Microdochium nivale infection on expression of pathogenesis-related genes in winter wheat. Physiol. Mol. Plant Pathol. 53: 301-310.

Eriksen A.B. (1997). Winter vigour in *Picea abis* (L.) Karst. V. Biochemical changes in buds of four-year old spruce plants during the autumn. 1972. Medd. Nor. Inst. Skogfosk. 34:1-22.

Eulgem T., Rushton P., Schmelzer J.T., Hahlbrock K., and Somssich I.E. (1999). Early nuclear events in plant defence signalling: Rapid gene activation by Wrky transcription factors. The EMBO Journal 18: 4689-4699.

Fristensky B., Horovitz D., and Hadwiger L.A. (1988). cDNA sequences for disease response genes. Plant Mol. Biol. 11: 713-715.

Gilmour S.J., Sebolt C.A.M., Salazar M.P., Everard J.D., and Thomashow M.F. (2000). Overexpression of the Arabidopsis CBF3 transcriptional activator mimics multiple biochemical changes associated with cold acclimation. Plant Physiol. 124: 1854-1865.

Griffith M., Antikainen M., Hon W.C., PihakaskiMaunsbach K., Yu X.M., Chun J.U., and Yang D.S.C. (1997). Antifreeze proteins in winter rye. Physiol. Plant. 100: 327-332.

Gusta L.V., and Wesser C.J. (1972). Nucleic acid and protein changes in relation to cold acclimation and freezing injury of Korean boxwood leaves. Plant Physiol. 49:91-96.

Hajela R., Horbath D.V., Gilmour S.J., and Thomashaow M.F. (1990). Molecular cloning and expression of *cor* (*cold-regulated*) genes in *Arabidopsis thaliana*. Plant Physiol. 93: 1246-1252.

Hincha D.K., Heber U., and Schmitt J.M. (1989). Freezing ruptures thylakoid membranes in leaves, and rupture can be prevented in vitro by cryoprotective proteins. Plant Physiol. Biochem. 27:795-801.

Hon W.C., Griffith M., Mlynarz A., Kwok Y.C., and Yang D.S.C. (1995). Antifreeze proteins in winter rye are similar to pathogenesis-related proteins. Plant Physiol. 109: 879-889.

Jaglo-Ottosen K.R., Gilmour S.J., Zarka D.G., Schabenberger O., and Thomashow M.F. (1998). Arabidopsis CBF1 overexpression induces COR genes and enhances freezing tolerance. Science. 280: 104-106.

Jarvis S.B., Taylor M.A., Macleod M.R., and Davies H.V. (1996). Cloning and characterization of the cDNA clone of three genes that are differentially expressed during dormancy-breakage in the seeds of Douglas-fir (*Pseudotsuga menziesii*). J. Plant Physiol. 147:559-566.

Korfhage U., Trezzini G.F., Meier I., Hahlbrock K., and Somssich I.E. (1994). Plant homeodomain protein involved in transcriptional regulation of a pathogen defense-related gene. Plant Cell. 6: 695-708.

Karenlampi S.O., Airaksinen K., Miettinen A.T.E., Kokko H.I., Holopainen J.K., Karenlampi L.V., and Karjalainen R.O. (1994). Pathogenesis-related proteins in ozone-exposed Norway spruce [*Picea abies* (Karst) L.]. New Phytol. 126:81-89.

Keen N.T. (1992). The molecular biology of disease resistance. Plant Mol. Biol. 19:109-122.

Kinlaw C.S., and Neale D.B. (1997). Complex gene families in pine genomes. Trends in Plant Science. 2: 356-359.

Linthorst H.J.M. (1991). Pathogenesis-related proteins of plants. Crit. Rev. Plant Sc. 10: 123-150.

Lo S.C.C., Hipskind J.D., and Nicholson R.L. (1999). cDNA cloning of a sorghum pathogenesis-related protein (Pr-10) and differential expression of defense-related genes following inoculation with *Cochliobolus heterostrophus* or *Colletotrichum sublineolum*. Mol. Plant-Micr. Interact. 12: 479-489.

Lowry O.H., Rosebrough N.J., Farr A.L., and Randall R. (1951). Protein measurement with the folin phenol reagent. J. Biol. Chem. 193:265-275.

Matheus Y.N., Ekramoddoullah A.K.M., and Lee S.P. (2000). Isolation and characterization of transcripts encoding putative intracellular pathogenesis-related protein from cold hardened white spruce tissues. Abs# S31-71. 6th International Congress of Plant Molecular Biology, June 18-24, 2000, Quebec, Canada.

Matton D.P., and Brisson N. (1989). Cloning, expression, and sequence conservation of pathogenesis-related gene transcripts of potato. Mol. Plant-Microbe Interact. 2: 325-331.

Matton D.P., Prescott, G., Bertrand, C., Camirand, A., and Brisson, N. (1993). Identification of cis-acting elements involved in the regulation of the pathogenesis-related gene STH-2 in potato. Plant Mol. Biol. 22: 279-291.

Moiseyev G.P., Beintena J.J., Fedoreyeva L.I., and Yakovlev G.I. (1994). High sequence similarity between a ribonuclease from ginseng calluses and fungus-elicited proteins from parsley indicates that intracellular pathogenesis-related proteins are ribonucleases. Planta. 193: 470-472.

Pomeroy M.K., and Siminovitch D. (1970). Seasonal biochemical changes in the living bark and needles of red pine (*Pinus resinosa*) in relation to adaptation to freezing. Can. J. Bot. 48:953-967.

Roberts D.R., Toivonen P., and Mcinnis S.M. (1991). Discrete proteins associated with overwintering of interior spruce and Douglas-fir seedlings. Can. J. Bot. 69: 437-441.

Rushton P.J., Torres J.T., Parniske M., Wernert P., Hahlbrock K., and Somssich I.E. (1996). Interaction of elicitor-induced DNA-binding proteins with elicitor response elements in the promoters of parsley PR1 genes. The EMBO Journal. 15: 5690-5700.

Saraste M., Sbbald P.R., and Wittinghofer A. (1990). The P-loop-a common motif in ATP-and GTP-binding proteins. TIBS 15: 430-434.

Somssich I.E., Schmeizer E., Kawalleck P., and Hahlbrock K. (1988). Gene structure and in situ transcript localization of pathogenesis-related protein in parsley. Mol. Gen. Genet. 213:93-98.

Steponkus P.L., Uemura M., Joseph R.A., Gilmor S.J., and Thomashow M. (1998). Mode of action of the COR15a gene on the freezing tolerance of *Arabidopsis thaliana*. Proc. Natl. Acad. Sci. USA 95: 14570-14575.

Swoboda I., Scheiner O., Heberlebors E., and Vicente O. (1995). cDNA cloning and characterization of three genes in the Bet v 1 gene family that encode pathogenesis-related proteins. Plant Cell Environ. 18: 865-874.

Thomashow M.F. (1990). Molecular genetics of cold acclimation in higher plants. Adv. Genet. 28: 99-125.

Thomashow M.F. (1998). Role of cold-responsive genes in plant freezing tolerance. Plant Physiol. 118: 1-7.

Thomashow M.F. (1999). Plant cold acclimation: Freezing tolerance genes and regulatory mechanisms. Annual Review of Plant Physiology and Plant Molecular Biology. 50: 571-599.

Utriainen M., Kokko H., Auriola S., Sarrazin O., and Karenlampi S. (1998). PR-10 protein is induced by copper stress in roots and leaves of a Cu/Zn tolerant clone of birch, *Betula pendula*. Plant Cell Environ. 21: 821-828.

van Loon L.C., Pierpoint W.S., Boller T., and Conejero V. (1994). Recommendations for naming plant pathogenesis-related proteins. Plant Mol. Biol. Rep. 12: 254-264.

Walter M.H., Liu J.-W., Grand C., Lamb C.J., and Hess D. (1990). Bean pathogenesis-related (PR) proteins deduced from elicitor-induced transcripts are members of a ubiquitous new class of conserved PR proteins including pollen allergens. Mol. Gen. Genet. 222: 353-360.

Walter M.H., Liu J.-W., Wunn J., and Hess D. (1996). Bean ribonulease-like pathogenesis-related protein genes (Ypr-10) display complex patterns of developmental, dark-induced and exogenous-stimulus-dependent expression. Eur. J. Biochem. 239: 281-293.

Warner S.A.J., Scott R., and Draper J. (1992). Characterization of a wound-induced trancript from the monocot asparagus that shares similarity with a class of intracellular pathogenesis-related (PR) proteins. Plant Mol. Biol. 19: 555-561.

Wetzel S., and Greenwood J.S. (1989). Proteins as potential nitrogen storage compund in bark and leaves of softwoods. Trees. 3: 149-153.

Wetzel S., Demmers C., and Greenwood J.S. (1989). Seasonally fluctuating bark proteins are potential form of nitrogen storage in three temperate hardwoods. Planta. 178: 275-281.

Wetzel S., and Greenwood J.S. (1991). The 32-kilodalton vegetative storage protein of Salix-microstachya Turz-characterization and immunolocalization. Plant Physiol. 97: 771-777.

Wisniewski M., Close T.J., Artlip T., and Arora R. (1996). Seasonal patterns of dehydrins and 70-kda heat-shock proteins in bark tissues of eight species of woody plants. Physiol. Plan. 96:496-505.

Yu X., Ekramoddoullah A.K.M., and Misra S.M. (2000a). Characterization of *Pin m* III cDNA in western white pine. Tree Physiol. 20:663-671.

Yu X., Ekramoddoullah A.K.M., Taylor D.W., and Piggott N. (2000b). Analysis of pathogenesis-related PR-10 gene structure in nine Pinaceae. Can. J. Plant Pathol. 22:196.

Ziegler P., and Kandler O. (1980). Tonoplast stability as a critical factor in frost injury and hardening of spruce (*Picea abies* L. Karst.) needles. Z. Planzenphysiol. Bf. 99:393-410.

Nitrogen Fixation as a Stress-Avoidance Strategy Among Actinorhizal (Non-Legume) Trees and Shrubs

Heidi A. Kratsch
William R. Graves

SUMMARY. Actinorhizal species of trees and shrubs are a diverse group of plants that may survive in poor soils by virtue of their associations with the actinomycete *Frankia*. These species include several important woody plants that are well suited for horticultural use in temperate climates. The symbiosis between *Frankia* and actinorhizal species shows some similarity to symbioses between rhizobia and woody legumes, and a common ancestor has been proposed for the predisposition to root-nodule symbiosis. Despite their probable common origin, important differences exist between actinorhizal and leguminous symbioses; characteristics of the microsymbiont, nodule architecture, and mechanisms controlling oxygen relations of the nodule are among the ways the two systems differ. If nitrogen fixation is sustained under unfavorable conditions, woody plants that associate with nitrogen-fixing organisms may show enhanced tolerance of environmental stress; species of plants

---

Heidi A. Kratsch is Plant Sciences Institute Fellow and William R. Graves is Professor, Department of Horticulture, Iowa State University, Ames, Iowa, USA.

Address correspondence to: William R. Graves, Department of Horticulture, Iowa State University, Ames, IA 50011-1100 USA (E-mail: graves@iastate.edu).

[Haworth co-indexing entry note]: "Nitrogen Fixation as a Stress-Avoidance Strategy Among Actinorhizal (Non-Legume) Trees and Shrubs." Kratsch, Heidi A., and William R. Graves. Co-published simultaneously in *Journal of Crop Improvement* (Food Products Press, an imprint of The Haworth Press, Inc.) Vol. 10, No. 1/2 (#19/20), 2004, pp. 281-304; and: *Adaptations and Responses of Woody Plants to Environmental Stresses* (ed: Rajeev Arora) Food Products Press, an imprint of The Haworth Press, Inc., 2004, pp. 281-304. Single or multiple copies of this article are available for a fee from The Haworth Document Delivery Service [1-800-HAWORTH, 9:00 a.m. - 5:00 p.m. (EST). E-mail address: docdelivery@haworthpress.com].

http://www.haworthpress.com/web/JCRIP

Digital Object Identifer: 10.1300/J411v10n01_12

capable of nitrogen-fixing symbioses are known to have comparatively strong resistance to invasion by pathogens. Expansion of the capacity to form nitrogen-fixing symbioses to novel species is a goal of those concerned with the economic and ecological impact of chemical fertilizers. Small in-roads have been made in this regard, but much remains to be discovered about introducing nitrogen fixation to additional species. Herein we review biological aspects of actinorhizal symbioses; consider the horticultural potential of temperate, woody species that form these symbioses; and discuss how nitrogen-fixing symbiosis may impact the stress resistance and use of actinorhizal species as horticultural crops.

KEYWORDS. Actinorhizal, nodulation, nitrogen fixation, stress tolerance, nursery crops

INTRODUCTION

Actinorhizal species are diverse dicotyledonous plants capable of developing root nodules that house nitrogen-fixing bacteria in the genus *Frankia* (Actinomycetales). These plants, almost all of which are woody, formerly were known as non-leguminous nitrogen-fixing plants to distinguish them from the subset of legumes that forms symbioses with nitrogen-fixing rhizobia. "Actinorhizal" was adopted in 1978 to reflect the symbiotic nature of the plant-microbial interaction (Tjepkema and Torrey, 1979). "Actino-" refers to the actinomycete *Frankia*, and "-rhizal" specifies the location of the symbiosis on the plant root. Because *Frankia* fix atmospheric dinitrogen, actinorhizal plants often occur as pioneer species on nitrogen-poor soils and thereby play particularly important roles in the course of ecological succession. Actinorhizal plants can add significantly to the nitrogen economy of the soil in which they grow (Morris et al., 1974; Lepper and Fleschner, 1977; Dawson, 1990; Paschke, 1997) and pave the way for subsequent establishment of non-nodulating species. In managed landscapes, where ornamental plants are introduced to soils that are or become low in nitrogen, actinorhizal species often are regarded as unusually resilient by arborists and horticulturists who may not be aware that nitrogen fixation contributes to the success of certain woody species.

Promoting greater recognition of actinorhizal plants that are or have the potential to become important in horticultural commerce is one of our goals for this paper. We will highlight selected actinorhizal taxa with unusual horticultural merit and will contrast actinorhizal plants and nitrogen-fixing legumes, focusing on the unusual features of actinorhizal symbioses. Recognizing actinorhizal taxa with commercial potential is important, in part, because of their prevalence. There appear to be more temperate, actinorhizal species with horticultural importance or promise than there are temperate woody legumes that are common in the nursery industry and that associate with rhizobial bacteria. For example, several genera of temperate, leguminous trees that often are produced in nurseries, e.g., *Gleditsia*, *Cercis*, *Gymnocladus*, *Styphnolobium*, and *Cladrastis*, are comprised only of species that do not develop nitrogen-fixing symbioses (Allen and Allen, 1981; Foster, Horner, and Graves, 1998). Our final goal is to consider whether actinorhizal associations may confer enhanced stress resistance to the plant partner in the symbiosis. Although relatively little evidence has been collected on this subject, our discussion will be based on the hypothesis that actinorhizal species directly avoid stresses due to low soil nitrogen and also may avoid adverse effects of other biotic and abiotic stressors indirectly via increases in nitrogen content.

Phylogenetics and Geographic Distribution

A feature that sets actinorhizal plants apart from legumes is the occurrence of actinorhizal symbioses within multiple taxonomically distinct groups. In contrast to legumes in which symbioses occur within one family, Leguminosae (Fabaceae), actinorhizal plants represent eight plant families and 24 genera (Benson and Silvester, 1993). According to this conventional classification system, which is based on comparative anatomy and morphology (e.g., Cronquist, 1981), the evolution of actinorhizal symbioses seems to have occurred a number of times in largely unrelated species (Sprent, 1979). However, more recent molecular systematic studies based on DNA sequence comparisons of the chloroplast gene *rbc*L imply a closer relationship among actinorhizal species than was previously thought (Swenson and Mullin, 1997). Studies of relationships among taxa at the molecular level have allowed for discrimination of the more subtle aspects of plant classification and provide an opportunity to make inferences about the evolutionary relatedness between such groups. In this light, a single origin for the predisposition for root nodule symbiosis (both legume and actinorhizal) has been proposed (Soltis et al., 1995), with evidence for the multiple occurrence of nitrogen-fixing symbiosis within each group (Swenson, 1996). Non-symbiotic relatives exist within taxonomically related genera, but the precise set of conditions that favors the development of novel symbioses within the actinorhizal plants has yet to be determined.

There are around 200 known plant species that interact with *Frankia* (Swenson and Mullin, 1997; Huss-Danell, 1997), all of them woody except two species within the genus *Datisca* that have herbaceous shoots (Paschke, 1997). The family Betulaceae, one of the earliest to appear in the pollen fossil records (Muller, 1981), is also the family that has been the most extensively investigated and is considered by some to be a "model" plant for the study of actinorhizal symbioses in temperate regions (Bousquet and Lalonde, 1990). It includes both non-symbiotic and symbiotic plants, including the nitrogen-fixing genus *Alnus* (alders). *Alnus* spp. are widely distributed in wet to wet-mesic soils throughout the temperate zone, with representatives occurring in both North and South American, Europe and Asia, and in Oceania (Australia, New Zealand), where it has been introduced (Baker and Schwintzer, 1990; Dawson, 1990). Members of the family Myricaceae (*Myrica* and *Comptonia* spp.) appear about the same time in the fossil records and are native to the above geographic regions, plus Africa and Australia, where they occur under a wide range of conditions, from sandy soils to woodlands and coastal bogs (Morris et al., 1974; Wheeler and Miller, 1990).

Next to appear in the fossil records, the Casuarinaceae family includes plants in the genera *Casuarina*, *Allocasuarina*, *Ceuthostoma*, and *Gymnostoma*. These species have a much narrower natural distribution range, native only to Australia, Malaysia, and Polynesia, but occur widely in both the northern and southern hemisphere as an economically important, introduced species (Diem and Dommergues, 1990). Members of this family are unique in that they are angiosperms yet possess conifer-like, reduced leaves and photosynthetic, deciduous branchlets (Torrey and Berg, 1988). They grow primarily in warm and humid environments, but selected species can be adapted to a wide variety of conditions (Diem and Dommergues, 1990).

Among actinorhizal plants for which pollen fossil records exist, the families Rosaceae, Elaeagnaceae, and Coriariaceae have appeared most recently. Symbiotic members of the Rosaceae family (*Cercocarpus*, *Chamaebatia*, *Cowania*, *Dryas*, and *Purshia* spp.) are native primarily to Europe (Wheeler and Miller, 1990) and North America (Baker and Schwintzer, 1990), where a number of species are ecologically important in the harsh chaparral and desert environments of the rangelands in western United States (Paschke, 1997), and other species (*Dryas* spp.) survive as far north as the arctic tundra (Benson and Silvester, 1993). The distribution range of the Elaeagnaceae family (*Elaeagnus*, *Hippophae*, *Shepherdia* spp.), like Casuarinaceae, has been expanded greatly by economic activity, the natural distribution being limited to the temperate areas of North America and Europe (Baker and Schwintzer, 1990). Coriariaceae (*Coriaria* spp.) exist in disjunct populations in dry woods, hedges, and rocky places on every continent except Africa and Antarctica (Silvester, 1977; Baker and Schwintzer, 1990; Wheeler and Miller, 1990).

Fossil records do not exist for the Rhamnaceae (*Ceanothus*, *Colletia*, *Discaria*, *Kentrothamnus*, *Retanilla*, *Talguenea*, *Trevoa* spp.) and Datiscaceae (*Datisca* spp.). The natural distribution of members of the family Rhamnaceae appears to be limited to North and South America, and Oceania, where they are often exposed to extreme temperature and drought stress (Baker and Schwintzer, 1990). A number of species can be found only in South America (*Colletia*, *Retanilla*, *Talguenea*, *Trevoa* spp.), where they are limited to the xeric environment of the Chilean matorral (Silvester, Balboa, and Martinez, 1985). *Datisca* occurs in North America and Eurasia, where it grows along the banks of streams (Baker and Schwintzer, 1990; Wheeler and Miller, 1990).

Another non-legume species *Parasponia* (family Ulmaceae) forms nitrogen-fixing nodules, but with the legume-colonizing *Rhizobium* rather than the actinomycete *Frankia* (Trinick, 1973). It is, therefore, not considered a true actinorhizal species, even though nodules of *Parasponia* are anatomically and ontogenetically similar to actinorhizal nodules (Trinick and Galbraith, 1976). There have been no other reports of rhizobial infection of non-legume species. Considering the proposed common origin for the predisposition to nodulation (Soltis et al., 1995; Doyle, 1998), that legume symbioses are thought to be more recent on an evolutionary scale than actinorhizal symbioses (Pawlowski and Bisseling, 1996), and that bacterial nodulation genes appear to be transmissible to diverse groups of bacteria (Sullivan and Ronson, 1998), is it possible that the *Rhizobium-Parasponia* symbiosis represents the actinorhizal missing link that connects the evolution of legume symbioses to the earlier appearance of actinorhizal associations? It is more likely, however, that root nodule symbioses arose separately in these two groups (Swenson, 1996), with the *Rhizobium-Parasponia* symbiosis representing a broadening of the host range of the microsymbiont. Such taxonomical paradoxes serve as a reminder that organisms often defy human classification systems. Further phylogenetic studies of nodulation genes and their expression in diverse species (Doyle, 1998) are needed to surmise the evolutionary history of root nodule symbioses.

Selected Actinorhizal Species with Horticultural Merit for Temperate Climates

Horticultural features of selected temperate, actinorhizal taxa found in commerce are summarized below.

Alnus spp. (Alders)

All of the approximately 30 species of alders are considered actinorhizal. Plants in this genus (family Betulaceae) grow at rapid rates and develop into shrubs or trees that typically occupy riparian or wetland niches in their native habitats in North America and elsewhere (Figure 1). A few species have been introduced to the horticultural trade. As would be expected based on their ecology in the wild, alders generally are used in managed landscapes with heavy or wet soils, although at least certain species also are resistant to drought (Graves, Kroggel, and Widrlechner, 2002; Hennessey, Bair, and McNew, 1985). The species are monoecious, and all but three bloom early in the season. Staminate flowers are in catkins that may be considered subtly ornamental. Pistillate flowers develop into woody strobili, a recognizable marker for the genus. Several rather obscure species have not been evaluated thoroughly for use in horticulture but appear promising (Graves, Kroggel, and Widrlechner, 2002).

Ceanothus spp. (e.g., Redroot, New Jersey Tea)

There are over 50 deciduous and evergreen species in this genus of shrubs in the Rhamnaceae family. Numerous hybrids and cultivars of North American species have been developed; many of these are widely used in European gardens, where they are valued for attractive foliage and colorful flowers. *Ceanothus americanus* (New Jersey tea) is a widely distributed native of the United States with a growing, favorable reputation among native-plant enthusiasts. It forms showy, white flowers in panicles after many other shrubs in gardens have finished blooming. Whereas seasonal low temperatures restrict the use of many taxa of *Ceanothus*, New Jersey tea is cold-hardy to USDA hardiness zone 4 and may be the most suitable member of its genus for cold climates (Dirr, 1998).

FIGURE 1. Evaluation of the potential to use a rare species of alder in managed landscapes. Seedlings of *Alnus maritima* (seaside alder) were grown in rows along a slope in Ames, Iowa, USA. Rows were at various distances from a constant source of water at the bottom of the slope. Objectives of this trial included testing the cold hardiness of this species when planted north of its natural distribution, and determining whether plants could survive on drained soils; plants in the wild are found exclusively on saturated or continuously flooded soils. Commercialization of this attractive species is being pursued because plants proved hardy throughout the northern United States and have been shown capable of growth in dry soils.

Hippophae Rhamnoides (Common Seabuckthorn)

Common seabuckthorn, an attractive member of the Elaeagnaceae family, is native to western China and is commonly used in European landscapes. Although it is less common in North America, the species is considered hardy to USDA zone 4 and could be used in poor, sandy soils or in landscapes where plants are exposed to salt spray (Dirr, 1998). The foliage and fruits of this small tree are prominent and unusual. Leaves are linear to lanceolate and are covered with silvery scales. The fine-textured and lightly colored foliage is a striking backdrop for the bright, glossy, orange drupes that form in large masses on female plants (the species is dioecious) and persist from the autumn through early spring. Although they are acidic, the fruits are edible and can be processed into juice or other products that may be unusually beneficial to human health.

Elaeagnus spp. (e.g., Russian Olive)

This genus of the Elaeagnaceae family contains about 40 species, most (Dirr, 1998) or all of which are actinorhizal. *Elaeagnus angustifolia* (Russian olive) is the most commonly planted species of this genus. It is native to Europe and Asia but has been widely used in North American landscapes with harsh conditions. Like common seabuckthorn, this species is considered salt tolerant and is frequently planted in coastal landscapes and along northern highways on which de-icing salts are applied. While its silvery leaves, fruits, and twigs provide an interesting color in the landscape, the irregular form of adult trees, the structural weakness of trees, and the susceptibility of the species to disease limit the value of Russian olive; the species is best planted in masses where trees will be viewed from a considerable distance. Other members of the Elaeagnaceae (e.g., *Sheperdia* spp.) also are actinorhizal but are plagued by many of same problems associated with *Elaeagnus* spp.

Myrica spp. (e.g., Bayberries, Wax Myrtles)

Three of the many species in this genus of the Myricaceae family have received particular attention from horticulturists. *Myrica pensylvanica* (bayberry) is an outstanding shrub species for cold climates. Plants grow 2 to 3 m tall and colonize soils of poor quality with upright twigs that bear pest-resistant, fragrant leaves and attractive fruit (female plants only of this dioecious species). *Myrica cerifera* (southern bayberry or wax myrtle) is taller than bayberry (up to about 11 m) and evergreen, but unfortunately, the species is difficult to grow in areas colder than USDA zone 7. Several cultivars of *Myrica cerifera* have been selected. They vary in foliar qualities, plant habit, and cold hardiness but as a group are rather intolerant of dry, alkaline soils. *Myrica gale* (sweet gale), a fragrant shrub of about 1 m in height, is at least as cold-hardy as *Myrica pensylvanica* but is best restricted to uniformly moist, acidic soils.

Comptonia peregrina (Sweet Fern)

This genus of the Myricaceae contains only one species, sweet fern. Unlike some other species in its family, sweet fern is deciduous and monoecious. But, like *Myrica* spp., its leaves are fragrant and attractively dissected (pinnatifid) in a manner similar to many true ferns. The primary use for this cold-hardy (USDA zone 2) species is as a large-scale ground cover. Plants colonize most rapidly on acidic soils and develop fine-textured stems that reach about 1 m tall.

Benefits that result from the use of these and other actinorhizal taxa often are most evident at planting sites where low nitrogen content in the soil limits plant growth or prevents survival of plants not associated with nitrogen-fixing microsymbionts. Establishment of highly effective actinorhizal symbioses, however, is dependent on many factors. If we are to capitalize on the potential of actinorhizal taxa, it is important to understand the biology of both symbiotic partners, the formation and structure of nodules, and how two prevalent components of the environment, oxygen and nitrogen, regulate symbiosis.

Characterization of the Microsymbiont

Essential to gaining insight into nodule evolution is consideration of the coevolutionary patterns between the actinorhizal plant and its *Frankia* microsymbiont. *Frankia* spp. are genetically diverse, slow-growing, gram-positive bacteria widely distributed in the soil rhizosphere and occurring in both symbiotic and free-living states (Benson, 1982; Lechevalier and Lechevalier, 1990). Research into actinorhizal associations has lagged behind that of legume symbioses because of the difficulty in isolating and maintaining *Frankia* in culture. Only recently have attempts to isolate *Frankia* from actinorhizal nodules been successful (Callaham, Del Tredici, and Torrey, 1978). To date, *Frankia* strains from 20 of the 24 actinorhizal genera have been isolated, but these isolates represent only a small number of actinorhizal species. In addition, strains from only four of the eight families are able to reinfect the source plant (Benson and Silvester, 1993), and ineffective (non-nitrogen fixing) nodules are common. The events leading to the formation of ineffective nodules are unknown, but may be related to host incompatibility and/or environmental factors (Benson and Silvester, 1993; Quispel, 1958), and the potential for competitive interactions between effective and ineffective *Frankia* strains exists (Van Dijk and Sluimer-Stolk, 1990). Nevertheless, the genus *Frankia* is morphologically distinct and has been described.

The genus *Frankia* is placed in the order Actinomycetales and the family Frankiaceae. Attempts to differentiate the genus further into species, either by phenotypic characteristics or by host infectivity groups, have been confounded by the lack of strain isolates from a number of known host genera and the apparent promiscuity of some actinorhizal species and *Frankia* strains for their symbiotic partners. Fortunately, with the availability of molecular phylogenetic techniques for comparing 16S rRNA gene sequences (Normand et al., 1996), relationships between *Frankia* strains, even those that have so far resisted cultivation, can be discerned (Clawson, Caru, and Benson, 1998). In this way, meaningful groupings of related strains may eventually lead to the establishment of taxonomically sound *Frankia* species. The clustering of *Frankia* strains based on DNA sequence homology is consistent with similar grouping for actinorhizal plants, which speaks to the fidelity of the available technologies for discerning such relationships and suggests the potential for coevolution between *Frankia* and their actinorhizal hosts (Normand et al., 1996; Swensen and Mullin, 1997; Clawson, Caru, and Benson, 1998).

In culture, *Frankia* spp. are branching, filamentous bacteria. They can be discriminated from other actinomycetes by the presence of mulitlocular sporangia, containing non-motile spores, at the terminus of or sometimes intercalary to branched septate hyphae. These hyphae eventually differentiate into variably thick-walled, lipid-encapsulated structures called vesicles. (See Lechevalier and Lechevalier (1990) and Benson and Silvester (1993) for an extensive review of *Frankia* systematics and morphology.) Other distinguishing taxonomic features of *Frankia* include their specificity for actinorhizal hosts, unique cell wall composition, presence of the polysaccharide 2-*O*-methyl-D-mannose, and their growth characteristics in culture (Lechevalier, 1994). The presence and shape of *Frankia* vesicles and the morphology of symbiotic nodules are largely determined by the host plant (Huss-Danell, 1997).

Nodule Development and Morphology

Interactions between symbiotic partners begin with an exchange of chemical signals that triggers novel gene expression and allows the bacterial microsymbiont to over-ride the pathogen defense response normally elicited in the plant host (reviewed by Berry and Sunell, 1990; Pawlowski and Bisseling, 1996; Hull-Danell, 1997; Franche et al. 1998; Wall, 2000). In some hosts (Betulaceae, Casuarinaceae, Myricaceae), *Frankia* invades host cells intracellularly by way of deformed root hairs, and invading hyphae are encapsulated by a host-derived membrane. In other actinorhizal hosts (Elaeagnaceae, Rhamnaceae, Rosaceae, Coriariaceae, and Datiscaceae), infection occurs intercellularly. Hyphae of *Frankia* are not encapsulated within a host membrane during intercellular infection. But electron-dense materials rich in pectins and proteins are secreted by neighboring epidermal and cortical cells into the intercellular space as *Frankia* invades. *Frankia* strains capable of infecting multiple hosts may infect via the intercellular or the intracellular route, depending on the host plant (Miller and Baker, 1986). Concurrent with the infection process, early chemical signals induce divisions of cells of the pericycle, resulting in the formation of a nodule primordium, which ultimately is infected by *Frankia* hyphae and grows outward toward the root epidermis to form a mature nodule.

The morphology of actinorhizal root nodules varies widely among species (Table 1) and reflects the prevailing environment in which the plant evolved (Silvester, Harris, and Tjepkema, 1990). In general, actinorhizal nodules are perennial, coralloid structures comprised of one to multiple lobes, depending upon the age of the nodule (Figure 2). Each nodule lobe is discrete and develops in a way that is similar to lateral root development, the vascular cylinder of the nodule continuous with the vascular tissue of the root. An area of *Frankia*-infected cortical tissue interspersed with non-infected tissue surrounds the central vascular bundle, although in *Coriaria* and *Datisca*, the infected tissue is asymmetrically localized to one side of the vascular bundle (Berg, Langenstein, and Silvester, 1999). A periderm with one (*Coriaria* spp.) to many (*Alnus* spp.) lenticels surrounds each nodule lobe. The periderm in *Coriaria* and *Datisca* encloses the nodule lobe but also follows closely and nearly surrounds the localized area of infected cortical cells; the thickness and degree of suberinization of the periderm are sensitive to the oxygen environment in the root zone (Silvester and Harris, 1989). In *Casuarina*, *Datisca*, *Myrica*, *Comptonia*, and *Gymnostoma* spp., a nodule root arises from the nodule lobe apex and grows up towards the soil surface, substituting for the lenticel in facilitating gas diffusion into the nodule. Consistent with their lateral root-like development, actinorhizal nodule lobes are perennial structures having an indeterminate growth pattern, although active nitrogen fixation has only been demonstrated within the current season's growth (Schwintzer, Berry, and Tjepkema, 1980). Each nodule lobe exhibits a gradient of cell development, ranging from the uninfected, actively dividing meristematic cells at the tip and the infected cells of the nitrogen-fixation zone, to the senescence zone toward the base of the nodule, in which infected cells and their contents are being actively destroyed (described in Ribeiro et al., 1995).

In contrast to nodules of legumes and *Rhizobium*-infected *Parasponia*, in which infected cells are bounded by a diffusion-limiting cell layer that acts as a barrier to oxygen, most actinorhizal nodules are relatively well-aerated, allowing oxygen to diffuse freely throughout *Frankia*-in-

TABLE 1. Nodulation characteristics of various actinorhizal plant taxa, which are listed in order of their appearance in the pollen fossil record. A clade refers to a lineage branch that results from splitting in an earlier lineage, producing two distinct taxa. This information is based on reports by Newcomb and Heisey (1984), Miller and Baker (1986), Balboa, Avila, and Arce (1988), Silvester et al. (1990), Normand et al. (1996), Huss-Danell (1997), Swensen and Mullin (1997), and Wall (2000).

Plant host	Nodule anatomy	Infection route	Microsymbiont
Betulaceae (*Alnus*)	Infected cells surround central vascular bundle; oxygen diffusion via multiple periderm lenticels	Intracellular	*Frankia* isolates Clade I
Myricaceae	Infected cells surround central vascular bundle; oxygen perception/diffusion via nodule root	Intracellular	*Frankia* isolates Clades I & II
Casuarinaceae	Infected cells surround central vascular bundle; oxygen perception/diffusion via nodule root	Intracellular	*Frankia* isolates Clade I; (Gymnostoma–Clades I & II)
Elaeagnaceae	Infected cells surround central vascular bundle; route of oxygen diffusion not investigated; no nodule root	Intercellular	*Frankia* isolates Clade II
Rhamnaceae	Infected cells surround central vascular bundle; route of oxygen diffusion not investigated; no nodule root	Intercellular	*Frankia* isolates Clades II & III
Rosaceae	Infected cells surround central vascular bundle; route of oxygen diffusion not investigated; no nodule root	Intercellular	*Frankia* isolates Clade III
Coriariaceae	Asymmetric localization of infected cells nearly enclosed by diffusion-limiting cell layer and surrounding vascular bundle; oxygen diffusion via one periderm lenticel	Intercellular	*Frankia* isolates Clade III
Datiscaceae	Asymmetric localization of infected cells nearly enclosed by diffusion-limiting cell layer and surrounding vascular bundle; oxygen perception/diffusion via nodule root	Intercellular	*Frankia* isolates Clade III
Parasponia	Infected cells entirely enclosed by diffusion-limiting cell layer and surrounding vascular bundle; oxygen diffusion via leghemoglobin	Route of infection not investigated	*Rhizobium*
Legumes	Infected cells entirely enclosed by diffusion-limiting cell layer and surrounded by vascular bundles; oxygen diffusion via leghemoglobin	Intra/Intercellular	*Rhizobium*

FIGURE 2. External appearance of nodules on roots of *Alnus maritima* (seaside alder) dug on the Eastern Shore of Maryland, USA. Note the formation of multiple nodular lobes.

fected tissues. The mechanics of the control of gas diffusion in actinorhizal nodules varies with, and is likely the driving force for the diversity in, actinorhizal nodule structure. Because nitrogenase, the enzyme involved in symbiotic nitrogen fixation, is sensitive to oxygen concentration, the mechanism of protection of nitrogenase from oxygen is an area of active research in actinorhizal species (Parsons et al., 1987; Rosendahl and Huss-Danell, 1988; Silvester and Harris, 1989; Tjepkema and Murray, 1989; Kleeman et al., 1994; Zeng and Tjepkema, 1994; Alskog and Huss-Danell, 1997; Lundquist, 2000).

Oxygen and Nitrogenase

The relationship between *Frankia* and actinorhizal plants is mutualistic. *Frankia* is capable of fixing atmospheric nitrogen, providing a usable form of nitrogen for its actinorhizal host. In turn, the plant releases enough photosynthate to sustain the microsymbiont. Nitrogen fixation is an energy-intensive process, requiring that large amounts of photosynthate be processed via respiration by *Frankia* to fuel the function of nitrogenase. Respiration requires the presence of oxygen, but oxygen inactivates the nitrogenase enzyme. So, there is a paradoxical relationship between nitrogenase and oxygen; oxygen is both inhibitory to but indirectly needed for the activity of nitrogenase. This conundrum is resolved in nitrogen-fixing organisms by the compartmentalization of nitrogenase such that an oxygen diffusion gradient is established across the compartmental barrier (Silvester, Harris, and Tjepkema, 1990).

The barrier to the diffusion of oxygen can be either intrinsically or extrinsically provided (Gallon, 1981). Extrinsic protection of nitrogenase is furnished by either the host plant or by ambient environmental conditions. An example of host-provided protection is the legume-rhizobial symbiosis, in which the *Rhizobium*-infected cells are enclosed within a diffusion-resistant layer of cortical cells, resulting in a low internal oxygen environment. *Rhizobium* is unable to express nitrogenase when free-living in air. In the majority of actinorhizal species, however, the oxygen-diffusion barrier is intrinsic; the protection of nitrogenase from inactivation by oxygen is provided by the microsymbiont. In this case, a vesicle confers oxygen protection. Vesicles, which are derived from *Frankia*, act as envelopes and form at the tip of hyphae (Figure 3).

FIGURE 3. Light micrograph of *Frankia* vesicles in cortical cells of a nodular lobe from *Alnus maritima* (seaside alder). The red-stained vesicles in this longitudinal section are uniformly spherical in shape and are thought to act as a barrier to oxygen diffusion with the lobe. Note the darkly stained cells, which contain tannins and phenolics. Bar represents 20 μm.

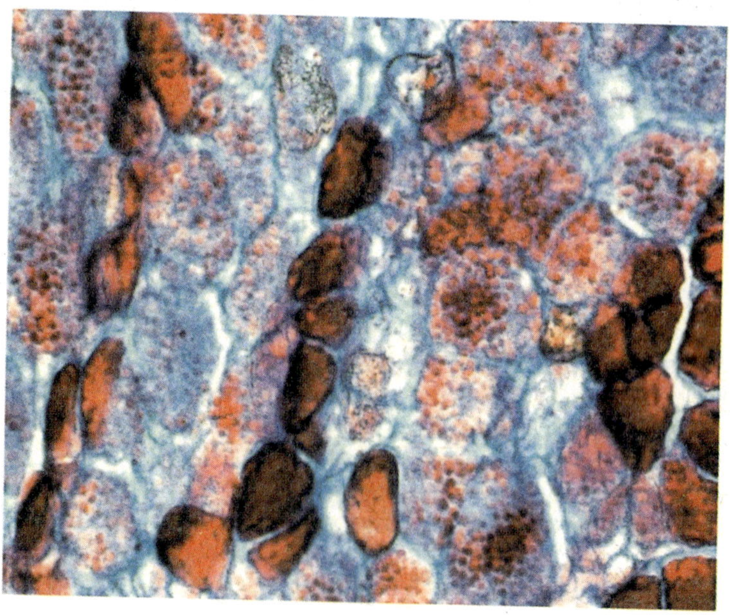

Frankia strains that form hyphae can express the nitrogenase enzyme and fix nitrogen while free-living in air (Callaham, Del Tredici, and Torrey, 1978).

The vesicle is a multi-laminate structure composed of one to many lipid monolayers that function as a barrier to the diffusion of oxygen. Although free-living strains of *Frankia* are capable of forming vesicles, the presence and thickness of the vesicle is dependent upon the oxygen environment in which the strain is grown (Parsons et al., 1987). *Frankia* cells grown in a high-oxygen medium develop thicker vesicles than do *Frankia* cells grown in a low-oxygen medium, and cells grown in an oxygen-free medium do not form a vesicle (Murray, Zhongze, and Torrey, 1985). Within the symbiosis, however, the presence and morphology of the vesicle is determined by the host plant (Huss-Danell, 1997). This was demonstrated using cross-inoculation studies with strains of *Frankia* capable of infecting multiple genera (St-Laurent and Lalonde, 1987; Torrey and Racette, 1989). Further, the strain of *Frankia* within nodules of *Casuarina* does not form a vesicle in symbiosis but does vesiculate when cultured on a high-oxygen medium (Murray, Zhongze, and Torrey, 1985). Heavily suberized cells of the periderm may provide oxygen protection of nitrogenase in the *Frankia* of *Casuarina*.

It has not been determined whether these phenomena result from host-plant regulation of the symbiosis, or whether they are responses of the microsymbiont to ambient conditions within the nodule. Indeed, it is often by effecting changes within the nodule environment that signals between symbiotic partners are executed. Further, changes in plant-growth conditions can trigger such changes in the nodule environment and affect the symbiotic relationship and, therefore, nodule function (Quispel, 1958). Markham (1996) demonstrated that *Alnus rubra* inoculated with crushed-nodule inoculum from parent plants were smaller and showed lower rates of nitrogen fixation when grown at elevations that differed from the parent than plants grown at the same elevation as their parents. This observation suggests coevolution between *Frankia* and *Alnus rubra* and demonstrates a change in the relationship between *Frankia* and *Alnus rubra* triggered by environmental change. For this reason, any attempt to establish an actinorhizal species within a particular environment should involve careful consideration of the source and/or the strain of *Frankia* chosen for inoculation.

Response to Nitrogen Environment–Autoregulation

A well-documented example of the regulation of nitrogen-fixing symbioses in response to changes in the nodule environment was first observed in legumes (MacConnell and Bond, 1957; Parsons et al., 1993) and was later demonstrated in actinorhizal species (Arnone, Kohls, and Baker, 1994; Wall and Huss-Danell, 1997). Auto-regulation is a feedback mechanism in which increasing nitrogen concentrations in plant tissues result in a localized inhibition of nitrogen fixation and in a systemic response that inhibits the formation of new nodules in younger parts of the plant. Auto-regulation is hypothesized to be a way for the plant to regulate nodule numbers and to prevent excessive export of photosynthate. As plant nitrogen concentrations rise, dependence upon the microsymbiont to meet nitrogen needs decreases, and the potential for parasitism by the microsymbiont in meeting its own energy needs increases. The first observable effect of high nitrogen is a decreased rate of nitrogen fixation. This nitrogen-induced inhibition of nitrogen fixation appears to result in altered activities and concentrations of stress-related molecules, such as reactive oxygen species and reactive oxygen-quenching enzymes, and the inhibition triggers the premature senescence of developing nodules (Swaraj, Laura, and Bishnoi, 1993).

This effect can be induced by externally applied nitrogen as well as microsymbiont-fixed atmospheric nitrogen and, in some cases, can be eliminated by correction of phosphate deficiency in the plant (Huss-Danell, 1997). Indeed, earlier studies have demonstrated a stimulatory effect of phosphorus on nodulation (Quispel, 1958; Rhigetti et al., 1982). It is hypothesized that the ratio of nitrogen to phosphorus concentration in the plant is crucial to nodule development (Huss-Danell, 1997).

These observations have implications for the growth of actinorhizal species in controlled environments. For example, as little as 0.7-mM exogenous nitrogen (as NH_4NO_3) significantly inhibited nodulation in *Cowania mexicana* (Righetti, Chard, and Munns, 1982). Similarly, only 1-mM nitrate inhibited nodule number in *Alnus glutinosa*, *Casuarina cunninghamiana*, and *Myrica cerifera* (Kohls and Baker, 1989). Although low-level nitrogen amendment of the soil during germination can prevent the premature death of young seedlings, an excess of nitrogen during later growth and development of actinorhizal species can inhibit nodulation and result in dependence upon applied fertilizers to meet nitrogen needs. It may also promote lush green shoot growth at the expense of root growth and ultimately limit the overall growth potential of the plant (Benoit and Berry, 1990). However, small amounts of nitrogen applied in increasing doses that did not exceed the nitrogen consumption of the plant enhanced both nitrogen fixation and plant growth in *Alnus incana* (Ingestad, 1980). So it appears that the plant can adapt to changing nitrogen conditions as long as the capacity of the plant to use the nitrogen is not exceeded. Further work with the same alder species demonstrated that this species uses fixed atmospheric nitrogen more efficiently than it does applied ammonium (Sellstedt and Huss-Danell, 1986). In this study, NH_4NO_3 was provided to non-symbiotic plants at the same rate that the symbiotic plants accrued fixed nitrogen. The symbiotic plants consistently exhibited greater biomass production and contained more total plant nitrogen than did non-symbiotic plants (133 mg vs. 73 mg total plant nitrogen after 56 d). No difference in biomass production was found between symbiotic and non-symbiotic plants, however, when nitrate was used as the source of applied nitrogen (Sellstedt, 1986).

Nitrogen-Fixing Symbioses and Stress Avoidance

Other abiotic factors that may affect nodulation success include soil moisture (Hopmans et al., 1983; Hennessey, Bair, and McNew, 1985; Hawkins and McDonald, 1993; Batzli and Dawson, 1997; Sayed et al., 1997), temperature (Reddell, Bowen, and Robson, 1985; Hawkins and McDonald, 1993; Sayed et al., 1997), pH (Quispel, 1958; Crannell, Tanaka, and Myrold, 1994), salinity (Hopmans et al., 1983; Reddell, Foster, and Bowen, 1986; Sande and Young, 1992), and levels of other essential minerals like iron (Aronson and Boyer, 1994), molybdenum and cobalt (Huss-Danell, 1997), and calcium (Crannell, Tanaka, and Myrold, 1994). Also important to the plant, however, is whether the capacity to fix nitrogen will protect against stress-induced constraints in growth and development. That is, is there a selective advantage, under stressful environmental conditions, to have the capacity to form nitrogen-fixing relationships?

Under low-nitrogen conditions, it is obviously an advantage to a capable plant to interact with nitrogen-fixing microsymbionts; otherwise, there would have been no evolutionary pressure for the development of such symbioses. This fact forms the basis for the observation that actinorhizal plants are often a dominant understory species in the early stages of plant succession after ecological disturbances and are prevalent on nitrogen-poor soils (Dawson, 1990). The situation is not so clear when nitrogen is not a limiting factor to growth and reproduction.

It is reasonable to hypothesize that a plant that is optimally nitrogen-nourished will be more resistant to other abiotic stresses than a plant that is nitrogen malnourished. Indeed, Reddell, Bowen, and Robson (1985) demonstrated that plants of *Casuarina cunninghamiana* provided with supplemental nitrogen (NH_4NO_3) were less susceptible to chilling (15°C) than were uninoculated plants that were not supplied with supplemental nitrogen. However, the same study showed an increase in susceptibility to both low and high (30°C) temperatures in *Frankia*-inoculated plants as compared with plants grown in nitrogen-amended soil. This increase in susceptibility of inoculated plants was attributed to poor nodulation and inhibition of nitrogen fixation at these temperatures. Interestingly, at moderate temperatures (20 vs. 25°C), differences in root and shoot dry weights and in total nitrogen content between inoculated and nitrogen-supplemented plants were found. At 20°C, total nitrogen content and root/shoot dry weight were higher in plants grown with applied nitrogen, while at 25°C, they were higher in inoculated plants. These results demonstrate a susceptibility of the microsymbiont to environmental factors, but suggest enhanced growth and nitrogen content of inoculated plants over nitrogen-amended plants when grown at microsymbiont-compatible temperatures.

Given what is known about actinorhizal plants, it may be safe to conclude that under low-nitrogen conditions, and without access to an alternative source of nitrogen, plants that form nitrogen-fixing symbioses may be less susceptible to abiotic stress than their non-nitrogen fixing counterparts, provided the stress does not prevent the function of *Frankia*. A more strongly supported case may be made regarding resistance of plants to biotic stress. There is evidence that infection and nodulation by *Frankia* may provide the plant with resistance to pathogens (Baker, Newcomb, and Torrey, 1980). Many researchers have demonstrated the effects of environmental stressors on nodulation and nitrogen fixation, but interactions between *Frankia* and its actinorhizal symbiont are delicate and complex. The tolerance of either organism to a particular stress may conflict with that of the other. In any symbiosis, a balance must be struck between the partners such that one does not exist at the expense of the other. If this should occur, interactions between the two break down, and defense mechanisms are triggered in one or both partners. More work is needed to clarify the interactions between actinorhizal symbionts and how environmental changes might affect the function of such relationships at the ecosystem level.

CONCLUSIONS

Actinorhizal woody plants, because of their nitrogen-fixing capability, have the potential for broad commercial use, both in land-reclamation projects and in sustainable agronomic and horticultural environments, where the capacity to fix atmospheric nitrogen can add significantly to the productivity of poor soils. There is great interest in expanding the nitrogen-fixing habit to non-nitrogen-fixing species to reduce dependence on use of chemical fertilizers to replenish the nitrogen content of depleted soils. Previous efforts to accomplish this goal have focused on agronomically important grasses (Bennett and Ladha, 1992), by using rhizobia as the potential microsymbiont. However, such attempts have been only marginally successful, probably because of the genetic dissimilarity between grasses and legumes. It is postulated that introduction of the nitrogen-fixing habit to novel species may meet with less resistance in more closely related plants that share a genetic predisposition to form root nodule symbioses (Swensen and Mullin, 1997).

In this spirit, an attempt was made to graft the rootstock of the actinorhizal species, *Cowania mexicana,* with the scion of *Fallugia paradoxa,* a non-nitrogen-fixing species of the same family, Rosaceae (Kyle et al., 1986). Reciprocal grafts were also made. This group used an *in vitro* micrografting technique to minimize environmentally introduced stress and seedling-source rootstock as a way to avoid potential incompatibility that could arise in older, more differentiated tissue. Although their results were modest (25% of grafts were successful), and the formation of effective nitrogen-fixing nodules has yet to be demonstrated, this work represents an important step in advancing research on the nodulation process and may have commercial applications for many actinorhizal crop species.

REFERENCES

Allen, O.N. and E.K. Allen. (1981). The Leguminosae. The Univ. of Wisconsin Press, Madison.

Alskog, G. and K. Huss-Danell (1997). Superoxide dismutase, catalase and nitrogenase activities of symbiotic *Frankia (Alnus incana)* in response to different oxygen tensions. *Physiologia Plantarum* 99: 296-292.

Aronson, D.B. and G.L. Boyer (1994). Growth and siderophore formation in six iron-limited strains of *Frankia. Soil Biology and Biochemistry* 26(5): 561-567.

Arnone, J.A., S.J. Kohls, and D.D. Baker (1994). Nitrate effects on nodulation and nitrogenase activity of actinorhizal *Casuarina* studied in split-root experiments. *Soil Biology and Biochemistry* 26(5): 599-606.

Baker, D.D. and C.R. Schwintzer (1990). Introduction. In: eds., C.R. Schwintzer and J.D. Tjepkema. *The Biology of Frankia and Actinorhizal Plants.* San Diego: Academic Press.

Baker, D., W. Newcomb, and J.G. Torrey (1980). Characterization of an ineffective actinorhizal microsymbiont, *Frankia*. sp. *Canadian Journal of Microbiology* 26: 1072-1089.

Balboa, O., G. Avila, and P. Arce (1988). Fine structure of actinorhiza of the Rhamnaceae growing in Chile. I. *Talguenea quinqueneriva* (Gill et Hook). *Protoplasma* 147(2/3): 143-148.

Batzli, J.M. and J.O. Dawson (1997). Physiological and morphological responses of red alder and sitka alder to flooding. *Physiologia Plantarum* 99: 653-663.

Bennett, J. and J.K. Ladha (1992). Introduction: Feasibility of nodulation and nitrogen fixation in rice-potential and prospects. In: eds., G.S. Khush and J. Bennett. *Nodulation and Nitrogen Fixation in Rice*. Manila, Philippines: International Rice Research Institute.

Benoit, L.F. and A.M. Berry (1990). Methods for production and use of actinorhizal plants in forestry, low-maintenance landscapes, and revegetation. In: eds., C.R. Schwintzer and J.D. Tjepkema. *The Biology of Frankia and Actinorhizal Plants*. San Diego: Academic Press.

Benson, D.R. (1982). Isolation of *Frankia* strains from alder actinorhizal root nodules. *Applied Environmental Microbiology* 44: 461-465.

Benson, D.R. and W.B. Silvester (1993). Biology of *Frankia* strains, Actinomycete symbionts of Actinorhizal plants. *Microbiological Reviews* 57: 293-319.

Berg, R.H., B. Langenstein, and W.B. Silvester (1999). Development in the *Datisca-Coriaria* nodule type. *Canadian Journal of Botany* 77: 1334-1350.

Berry, A.M. and L.A. Sunell (1990). The infection process and nodule development. In: eds., C.R. Schwintzer and J.D. Tjepkema. *The Biology of Frankia and Actinorhizal Plants*. San Diego: Academic Press.

Bousquet, J. and M. Lalonde (1990). The Genetics of Actinorhizal Betulaceae. In: eds., C.R. Schwintzer and J.D. Tjepkema. *The Biology of Frankia and Actinorhizal Plants*. San Diego: Academic Press.

Callaham, D., P. Del Tredici, and J.G. Torrey (1978). Isolation and cultivation in vitro of the actinomycete causing root nodulation in *Comptonia*. *Science* 199: 899-902.

Clawson, M.L., M. Caru, and D.R. Benson (1998). Diversity of *Frankia* strains in root nodules of plants from the families Elaeagnaceae and Rhamnaceae. *Applied and Environmental Microbiology* 64(9): 3539-3543.

Crannell, W.K., Y. Tanaka, and D.D. Myrold (1994). Calcium and pH interactions on root nodulation of nursery-grown red alder (*Alnus rubra* Bong.) seedlings by *Frankia*. *Soil Biology and Biochemistry* 26(5): 607-614.

Cronquist, A. (1981). *An Integrated System of Classification of Flowering Plants*. New York: Columbia University Press.

Dawson, J.O. (1990). Interactions among actinorhizal and associated plant species. In: eds., C.R. Schwintzer and J.D. Tjepkema. *The Biology of Frankia and Actinorhizal Plants*. San Diego: Academic Press.

Diem, H.G. and Y.R. Dommergues (1990). Current and potential uses and management of Casuarinaceae in the tropics and subtropics. In: eds., C.R. Schwintzer and J.D. Tjepkema. *The Biology of Frankia and Actinorhizal Plants*. San Diego: Academic Press.

Dirr, M.A. (1998). *Manual of Woody Landscape Plants: Their Identification, Ornamental Characteristics, Propagation and Uses.* 5th ed. Champaign, IL, Stipes Publishing.

Doyle, J.J. (1998). Phylogenetic perspectives on nodulation: Evolving views of plants and symbiotic bacteria. *Trends in Plant Science* 3(12): 473-478.

Foster, C.M., H.T. Horner, and W.R. Graves. (1998). Nodulation response of woody Papilinoid species after inoculation with rhizobia and soils from Hawaii, Asia, and North America. *Plant and Soil* 205:103-111.

Franche, C., L. Laplaze, E. Duhoux, and D. Bogusz (1998). Actinorhizal symbioses: Recent advances in plant molecular and genetic transformation studies. *Critical Reviews in Plant Sciences* 17(1): 1-28.

Gallon, J.R. (1981). The oxygen sensitivity of nitrogenase: A problem for biochemists and microorganisms. *Trends in Biochemical Science* 6: 19-23.

Graves, W.R., M.A. Kroggel, and M.P. Widrlechner. (2002). Photosynthesis and shoot health of five birch and four alder taxa after drought and flooding. *Journal of Environmental Horticulture* 20:36-40.

Hawkins, B.J. and S. McDonald (1993). The influences of temperature and soil water on growth, photosynthesis, and nitrogen fixation of red alder (*Alnus rubra*) seedlings. *Canadian Journal of Forestry Research* 24: 1029-1032.

Hennessey, T.C., L.K. Bair, and R.W. McNew (1985). Variation in response among three *Alnus* spp. clones to progressive water stress. *Plant and Soil* 87: 135-141.

Hopmans, P., L.A. Douglas, P.M. Chalk, and S.G. Delbridge (1983). Effects of soil moisture, mineral nutrition, and salinity on nitrogen fixation (acetylene reduction) by *Allocasuarina verticillata* (Lam.) *Australian Forest Research* 13: 189-196.

Huss-Danell, K. (1997). Actinorhizal symbioses and their N_2 fixation. *Tansley Review* 93: 375-405.

Ingestad, T. (1980). Growth, nutrition, and nitrogen fixation in grey alder at varied rate of nitrogen addition. *Physiologia Plantarum* 50: 353-364.

Kleeman, G., G. Alskog, A.M. Berry, and K. Huss-Danell (1994). Lipid composition and nitrogenase activity of symbiotic *Frankia* (*Alnus incana*) in response to different oxgyen concentrations. *Protoplasma* 183: 107-115.

Kohls, S.J. and D.D. Baker (1989). Effects of substrate nitrate concentration on symbiotic nodule formation in actinorhizal plants. *Plant and Soil* 118: 171-179.

Kyle, N.E., J.L. Jakobek, R.A. Backhaus, J.C. Stutz, and T.L. Righetti (1986). Micrografting between N-fixing and non-N-fixing genera of the Rosaceae. *Botanical Gazette* 147: 243-246.

Lechevalier, M.P. (1994). Taxonomy of the genus *Frankia* (Actinomycetales). *International Journal of Systematic Bacteriology* 44: 1-8.

Lechevalier M.P. and H.A. Lechevalier (1990). Systematics, isolation, and culture of *Frankia*. In: C.R. Schwintzer and J.D. Tjepkema, eds. *The Biology of Frankia and Actinorhizal Plants.* San Diego: Academic Press.

Lepper, M.G. and M. Fleschner (1977). Nitrogen fixation by *Cercocarpus ledifolius* (Rosaceae) in pioneer habitats. *Oecologia* 27: 333-338.

Lundquist, P.-O. (2000). Nitrogenase activity in *Alnus incana* root nodules. Responses to O_2 and short-term N_2 deprivation. *Plant Physiology* 122: 553-561.

MacConnell, J.T. and G. Bond (1957). A comparison of the effect of combined nitrogen on nodulation in non-legumes and legumes. *Plant and Soil* 8: 378-388.

Markham, J.H. (1996). A field test of the degree of coevolution between red alder and *Frankia* populations. PhD Dissertation. University of British Columbia.

Miller, I.M. and D.D. Baker (1985). The initiation, development, and structure of root nodules in *Elaeagnus augustifolia* L. (Elaeagnaceae). *Protoplasma* 128(2/3): 107-119.

Miller, I.M. and D.D. Baker (1986). Nodulation of actinorhizal plants by *Frankia* strains capable of both root hair infection and intercellular penetration. *Protoplasma* 131: 82-91.

Morris, D.E., D.E. Eveleigh, S.C. Riggs, and W.N. Tiffney, Jr. (1974). Nitrogen fixation in the Bayberry (*Myrica pensylvanica*) and its role in coastal succession. *American Journal of Botany* 61(8): 867-870.

Muller, J. (1981). Fossil pollen records of extant angiosperms. *Botanical Reviews* 47: 1-142.

Murray, M.A., Z. Zhongze, and J.G. Torrey (1985). Effect of O_2 on vesicle formation, acetylene reduction, and O_2-uptake kinetics in *Frankia* sp. HFPCcI3 isolated from *Casuarina cunninghamiana*. *Canadian Journal of Microbiology* 31(9): 804-809.

Newcomb, W. and R.M. Heisey (1984). Ultrastructure of actinorhizal root nodules of *Chamaebatia foliolosa* (Rosaceae). *Canadian Journal of Botany* 62(8): 1697-1707.

Normand, P., S. Orso, B. Cournoyer, P. Jeannin, C. Chapelon, J. Dawson, L. Evtushenko, and A.K. Misra (1996). Molecular phylogeny of the genus *Frankia* and related genera and emendation of the family Frankiaceae. *International Journal of Systematic Bacteriology* 46: 1-9.

Parsons, R., W.B. Silvester, S. Harris, W.T.M. Gruijters, and S. Bullivant (1987). *Frankia* vesicles provide inducible and absolute oxygen protection for nitrogenase. *Plant Physiology* 83: 728-731.

Parsons, R., A. Stanforth, J.A. Raven, and J.I. Sprent (1993). Nodule growth and activity may be regulated by a feedback mechanism involving phloem nitrogen. *Plant, Cell and Environment* 16: 125-136.

Paschke, M.W. (1997). Actinorhizal plants in rangelands of the western United States. *Journal of Range Management* 50: 62-72.

Pawlowski, K. and T. Bisseling (1996). Rhizobial and Actinorhizal Symbioses: What are the shared features? *The Plant Cell* 8: 1899-1913.

Quispel, A. (1958). Symbiotic nitrogen fixation in non-leguminous plants. *Acta Botanica Neerlandica* 7: 191-204.

Reddell, P., G.D. Bowen, and A.D. Robson (1985). The effects of soil temperature on plant growth, nodulation and nitrogen fixation in *Casuarina cunninghamiana* Miq. *New Phytologist* 101: 441-450.

Reddell, P., R.C. Foster, and G.D. Bowen (1986). The effects of sodium chloride on growth and nitrogen fixation in *Casuarina obesa* Miq. *New Phytologist* 102: 397-408.

Ribeiro, A., A.D.L. Akkermans, A. van Kammen, T. Bisseling, and K. Pawlowski (1995). A nodule-specific gene encoding a subtilisin-like protease is expressed in early stages of actinorhizal nodule development. *The Plant Cell* 7: 785-794.

Righetti, T.L.,C.H. Chard, and D.N. Munns (1982). Opportunities and approaches for enhancing nitrogen fixation in Purshia, Cowania, and Fallugia browse plants, range management, soil fertility factors, ecotypes. *USDA Forestry Service General Technical Rep INT Intermt Forestry Range Experiment Station Ogden: The Station*.152: 214-224.

Rosendahl, L. and K. Huss-Danell (1988). Effects of elevated oxygen tensions on acetylene reduction in *Alnus incana-Frankia* symbioses. *Physiologia Plantarum* 74: 89-94.

Sande, E. and D.R. Young (1992). Effect of sodium chloride on growth and nitrogenase activity in seedlings of *Myrica cerifera* L. *New Phytologist* 120: 345-350.

Sayed, W.F., C.T. Wheeler, H.H. Zahran, and A.A.M. Shoreit (1997). Effect of temperature and soil moisture on the survival and symbiotic effectiveness of *Frankia* spp. *Biological Fertility of Soils* 25: 349-353.

Schwintzer, C.R., A.M. Berry, and J.D. Tjepkema (1980). Seasonal changes in *Myrica gale* nodules and the *Frankia* sp. endophyte. *Proceedings of the 4th International Symposium on Nitrogen Fixation*, p. 478, Elsevier/North-Holland Biomedical Press, Amsterdam.

Sellstedt, A. (1986). Nitrogen and carbon utilization in *Alnus incana* fixing N_2 or supplied with NO_3^- at the same rate. *Journal of Experimental Botany* 37: 786-797.

Sellstedt, A. and K. Huss-Danell (1986). Biomass production and nitrogen utilization by *Alnus incana* when grown on N_2 or NH_4^+ made available at the same rate. *Planta* 167: 387-394.

St.-Laurent, L. and M. Lalonde (1987). Isolation and characterization of *Frankia* strains isolated from *Myrica gale*. *Canadian Journal of Botany* 65: 1356-1363.

Silvester, W.B. (1977). Dinitrogen fixation by plant associations excluding legumes. In: eds., R.W.F. Hardy and A.H. Gibson. *A Treatise on Dinitrogen Fixation*. New York: Wiley & Sons.

Silvester, W.B. and S.L. Harris (1989). Nodule structure and nitrogenase activity of *Coriaria arborea* in response to varying pO_2. *Plant and Soil* 118: 97-109.

Silvester, W.B., O.O. Balboa, and J.A. Martinez (1985). Nodulation and nitrogen fixation in members of the Rhamnaceae (*Colletia, Retanilla, Talguenea* and *Trevoa*) growing in the Chilean matorral. *Symbiosis* 1: 29-38.

Silvester, W.B., S.L. Harris, and J.D. Tjepkema (1990). Oxygen regulation and hemoglobin. In: eds., C.R. Schwintzer and J.D. Tjepkema. *The Biology of Frankia and Actinorhizal Plants*. San Diego: Academic Press.

Soltis, D.E., P.S. Soltis, D.R. Morgan, S.M. Swenson, B.C. Mullin, J.M. Dowd, and P.G. Martin (1995). Chloroplast gene sequence data suggest a single origin of the predisposition for symbiotic nitrogen fixation in angiosperms. *Proceedings of the National Academy of Sciences USA* 92: 2646-2651.

Sprent, J.I. (1979). The range of nitrogen-fixing organisms. In: *The Biology of Nitrogen-Fixing Organisms*. London: McGraw-Hill.

Sullivan, J.T. and C.W. Ronson (1998). Evolution of rhizobia by acquisition of a 500-kb symbiosis island that integrates into a phe-tRNA gene. *Proceedings of the National Academy of Sciences USA* 3(12): 5145-5149.

Swaraj, K., J.S. Laura, and N.R. Bishnoi (1993). Nitrate induced nodule senescence and changes in activities of enzymes scavenging H_2O_2 in clusterbean (*Cyamopsis tetragonaloba* Taub.) *Journal of Plant Physiology* 141: 202-205.

Swenson, S.M. (1996). The evolution of actinorhizal symbioses: Evidence for multiple origins of the symbiotic association. *American Journal of Botany* 83(11): 1503-1512.

Swenson, S.M. and B.C. Mullin (1997). Phylogenetic relationships among actinorhizal plants. The impact of molecular systematics and implications for the evolution of actinorhizal symbioses. *Physiologia Plantarum* 99: 565-573.

Tjepkema J.D. and J.G. Torrey (1979). Symbiotic nitrogen fixation in actinomycete-nodulated plants. Preface. *Botanical Gazette (Chicago), Supplement* 140: i-ii.

Tjepkema, J.D. and M.A. Murry (1989). Respiration and nitrogenase activity in nodules of *Casuarina cunninghamiana* and cultures of *Frankia* sp. HFP020203: Effects of temperature and partial pressure of O_2. *Plant and Soil* 118: 111-118.

Torrey, J.G. and R.H. Berg (1988). Some morphological features for generic characterization among the Casuarinaceae. *American Journal of Botany* 75: 844-874.

Torrey, J.G. and S. Racette (1989). Specificity among the Casuarinaceae in root nodulation by *Frankia*. *Plant and Soil* 118: 157-164.

Trinick, M.J. (1973). Symbiosis between *Rhizobium* and the non-legume, *Trema aspera*. *Nature* 244: 459-460.

Trinick, M.J. and J. Galbraith (1976). Structure of root nodules formed by *Rhizobium* on the non-legume *Trema cannabina* var. *scabra*. *Archives of Microbiology* 108: 159-166.

Van Dijk, C. and A. Sluimer-Stolk (1990). An ineffective strain type of *Frankia* in the soil of natural stands of *Alnus glutinosa* (L.) Gaertner. *Plant and Soil* 127: 107-121.

Wall, L.G. (2000). The actinorhizal symbiosis. *Journal of Plant Growth Regulation* 19: 167-182.

Wall, L.G. and K. Huss-Danell (1997). Regulation of nodulation in *Alnus incana-Frankia* symbiosis. *Physiologia Plantarum* 99: 594-600.

Wheeler, C.T. and I.M. Miller (1990). Current and potential uses of actinorhizal plants in Europe. In: eds., C.R. Schwintzer and J.D. Tjepkema. *The Biology of Frankia and Actinorhizal Plants*. San Diego: Academic Press.

Zeng, S. and J.D. Tjepkema (1994). The wall of the infected cell may be the major diffusion barrier in nodules of *Myrica gale* L. *Soil Biology and Biochemistry* 26(5): 633-639.

Index

Abscissic acid (ABA)
in bud scales, 182-183
in dormancy induction, 161-164
early research, 160-161
evidence against involvement, 165-167
exogenous application, 163-164,166-167
in other plant systems, 164-165
in signal tranduction, 183
Acer negundo, 169
Acer platanoides, 81
Acer pseudoplatanus, 116,160,169
dormancy induction, 163-164,165
Acer rubrum, 165
Acer saccharum, 165
Acer sp., dormancy induction, 168
Acetabularia acetabulum, 183
Actinidia arguta, 163-164,182
Actinorhizal species, 281-304
Alnus sp., 286
autoregulation, 295-297
background and principles, 282-283
Ceanothus sp., 286
Comptonia peregrina, 288-289
conclusions, 298-299
Elaeagnus sp., 288
Frankia microsymbiont, 289-290
Hippophae rhamnoides, 287
with horticultural merit, 286-290
Myrica sp., 288
nitrogen-fixing symbioses and stress avoidance, 297-298
nodule development and morphology, 290-293
oxygen and nitrogenase, 293-295
phylogenetics and geographic distribution, 283-286
Agricultural University of Norway, 113-156
Allocasuarina, 284
Alnus glutinosa, 92-93,163-164,166
Alnus incana, 81-84
Alnus maritima, 295
Alnus rubra, 295
Alnus sp., 284,286
Amelanchier alnifolia, 181
Anthirrhinum, 126
Apical growth cessation, 81-84
gibberellin and, 96
Apical meristem, growth regulation and, 183-185
Appressed endoplasmic reticulum, 123
Arabidopsis CLV3, 121
Arabidopsis thaliana
ABA in, 164,183
budding in, 101-102
cryptochromes in, 90
floricaula gene expression, 126
gibberellin in, 93
growth regulation in, 185
moisture stress in, 175,178,179
photoperiodic timekeeping in, 86
phytochrome apoproteins, 84-85
phytochromes and, 171
PR-10 proteins in, 266
shoot apical meristem, 120
vegetative storage proteins, 240
ASDA Agricultural Research Service, 5-16

1,3-β-D-Glucan, 128-131
1,3-β-D-Gluconase, 139-140
Betula lutea, 165
Betula papyrifera, 165
Betula pubescens, 81-84,90,92-93,116, 118,169
  chilling in endodormancy, 138
  dormancy induction, 162-163,165, 166-167
  moisture stress and, 175
Betula sp.
  in hormonal research, 160
  phytochrome and, 169-173
Bud scales, 179-182

Caleosins, 141
Camellia japonica, 180-181
Canadian Forest Service, 261-280
Casuarina, 284,295
Casuarina cunninghamiana, 296,297-298
Ceanothus sp., 285,286
Cedrus libani, 163-164
Cell-cell communication
  during development, 118-127
  in dormancy, 113-156. *See also* Endodormancy
  at shoot apical meristem, 118-123
Cercocarpus, 284-285
Ceuthostoma, 284
Chamaebatia, 284-285
Chara corallina model, 185
Chlamydomonas reinhardii, 172
Chlorophyll fluorescence, *Vitis* sp., 220
Cold acclimation
  defined, 6, 17
  dehydrin accumulation in, 29-30
Colletia sp., 285
Common Plant Regulatory Factors (CPRFs), 171
Comptonia peregrina, 288-289
Comptonia sp., 284,291
Concentric configurations of endoplasmic reticulum (CERs), 135-137
Coriaria sp., 285,291
Cornus sericea L., 17-35,168
  cDNA identification and isolation, 19-20
  identification of dehydrin-like gene family, 21-30
  nucleic acid sequence analysis, 20-21
  temperature and dormancy, 174
Cornus sp., phytochrome and, 169-173
Coryllus avellana, 81-84
Cowania, 284-285
Cowania mexicana, 296,299
Critical photoperiod, 77-112
Cryptochromes, 90
Cucurbita pepo, 85
Cupressus arizonica, 93
Cyclin-CDK complexes, 185
Cyclin-dependent kinase, 172
Cytohistological zone, 120-121

Datisca sp., 284,291
Daylength, in *Vitis* sp., 227
Deacclimation, *Vitis* sp., 216,226-227
Deep supercooling. *See* Supercooling
Deficiens gene, 126
Dehydration, 13
Dehydrins, 176
  in *Cornus sericea*, 17-35. *See also* *Cornus sericea* L.
  in *Vaccinium* sp., 37-52
Dessication stress, 17
Differential thermal analysis (DTA), 14
Discaria sp., 285
Dormancy, 113-156. *See also* Endodormancy
Dormin, 160. *See also* Abscissic acid (ABA)
Drought stress, 17,174-179
Dryas, 284-285
Dwarfism, 93

Elaeagnus sp., 285,288
Electrolyte leakage, in *Vitis* sp., 221-222
Endodormancy, 117
  abscissic acid (ABA) in, 161-164, 183
  apical meristem and growth regulation, 183-185
  background and principles, 158-159
  1,3-β-D-glucan in, 129-131
  1,3-β-D-gluconase in, 139-140
  bud scale development, 182-183
  bud scales, 179-182
  cell-cell communication lacking, 127-137
  conclusions, 147
  definition, 114-118
  developmental biology and, 116-118
  diagnosing and controlling, 144-147
  early hormonal research, 159-161
  effects of chilling, 137-138
  GA involvement, 167-168
  hormonal regulation, 157-199
  hormones and translocatable dormancy factor, 168
  moisture stress and, 174-179
  photoperiodic ecotypes and phytochromes, 169-173
  plasmodesmal sphincter models, 131-137
  restoration of communication pathways, 137-144
  spherosomes and, 140-144
  temperature and, 173-174
Endoplasmic reticulum, appressed, 123
Equilibrium freezing, 12-13
Eucalyptus globulus, 93
Extraorgan freezing, 12-13

Fallugia paradoxa, 299
Frankia sp., 281-304. *See also* Actinorhizal species
Fraxinus sp., 182
Freezing
  equilibrium, 12-13
  extraorgan, 12-13

Gibberellin, 90-102
  biosynthesis, 92
  mode of action, 99-100
  photoperiodic regulation, 95-98
  regulation of biosynthesis, 94-95
  shoot regulation and, 90-94
  steps of biosynthesis, 98-99
Glass formation (vitrification), 14-15
Globosa gene, 126
Grapes. *See Vitis*
Gymnostoma sp., 284,291

Heat shock proteins, 28
Helianthus annuus, 164
Hibiscus syriacus, 179
High temperature exotherms (HTEs), 223
Hippophae rhamnoides, 81-84,287
Hippophae sp., 285

Ice nucleation, 7-12
Ice propagation, 7-12
  barriers to, 7-8
Infrared thermography, 10-12
  Vitis sp., 221
Iowa State University, 5-16,37-52

Juniperus scopulorum, 93

Kentrothamnus sp., 285
Knotted-1 gene, 120
KNOTTED1 protein, 127

Late embryogenesis abundant (LEA) proteins, 18-19,266

LEAFY gene, 101
Light distribution, in *Vitis* sp., 227
Lilium speciosum, 164-165
Lipid transfer protein (LPT), 120
Low temperature exotherms (LTEs), 222-224
Lycopersicon escalentum, 12

Magnetic resonance imaging (MRI), 175-176
    Vitis sp., 213-215
Malus domestica L., 8,165
Malus hupehensis, 165
Malus sp.
    bud scales in, 182
    dormancy induction, 168
(University of) Maryland, 237-259
Meristem state, 115-117. *See also* Endodormancy
Mesembryanthemum crystallinum, 179
(University of) Minnesota, 5-16
Molecular genetics, of *Vaccinium* dehydrins, 53-76. *See also* *Vaccinium* sp.
Myrica cerifera, 296
Myrica sp., 284,288,291

Nitrogen cycling, 237-259
    background and principles, 238-239
    bark storage proteins (BSPs), 242-245
    bspA regulation model, 253-254
    nucleoside phosphorylases, 242-245
    phytochrome, 246-250
    Populus sp., 237-259
    secondary messengers, 252-253
    seed storage proteins, 245-246
    source-sink relations, 250-252
    vegetative storage proteins (VSPs), 240-242
Nitrogen fixation, in actinorhizal trees and shrubs, 281-304. *See also* Actinorhizal species
Nuclear magnetic resonance (NMR), 175-176
    Vitis sp., 213-215

Oleosins, 142
Oxidative browning, *Vitis* sp., 220-222

Paclobutrazol, 167
Parasponia sp., 285-286
Pathogenesis-related proteins, 261-280. *See also* PR-10 proteins
Petula pubescens, 101
Pheliinus weirii, 272-273
Photoperiodic ecotypes, 81-84
    phytochrome and, 169-173
Photoperiodism, 77-112
    background and principles, 78-79
    comments and future aspects, 102-105
    gibberellin biosynthesis and, 95-98
    hormonal growth control and, 90-102. *See also* Gibberellin
    photoperiodic timekeeping in trees, 86-90
    phytochrome and, 84-90
    shoot elongation and, 80-84
PHY apoproteins, 84-90
Phytochromes, 84-90
    in nitrogen cycling, 246-250
    in tree species, 85-86
Picea abies, 81-84,85,93,97-98,162, 169,266
Picea engelmannii, 163-164
Picea glauca, 271-272
Picea mariana, 93-94
Picea sitchensis, 93
Pinus contorta, 163-164,166
Pinus elliotti, 85,91
Pinus lambertina Doug., 263-264
Pinus mariana, 91
Pinus monticola D., 263-264
Pinus radiata, 91

Pinus sylvestris, 80,81-84,85,91,94, 185
Plasma membrane intrinsic proteins (PIP), 179
Plasmodesmal sphincters, 131-137
 and concentric configurations of endoplasmic reticulum (CERs), 135-137
 endodormancy sphincter complex, 134-135
 flux-regulating external, 131-133
(University of) Plymouth, 5-16
Populus sp., 81-84,93,172
 endodormancy in, 145
 nitrogen cycling, 239,240-259. *See also* Nitrogen cycling
 photoperiodism in, 88-89
Populus tremula × *alba*, 100
Populus tremula × *tremuloides*, 85-86, 91,93,94,97,98,100-101,170
Populus tremuloides, 15
Populus trichocarpa, 85,183
Populus × *euroamericana*, 241
Potamogeton nodosus, 181-182
PR-10 proteins, 261-280
 antibody preparation, 264
 background and principles, 262-263
 cloning and cDNA characterization of *Pin m III*, 269-270
 cloning and cDNA characterization of sugar pine *Pin II*, 271
 cloning of white spruce PR-10 gene, 271-272
 detection in Douglas fir and white spruce, 272-273
 frost hardiness and, 265-267
 fungal induction, 267-268
 gene promoters, 273-275
 gene structure and homologues of *Pin m III*, 270-271
 seasonal variation, 264-265
 ultrastructural localization of *Pin m III*, 268-269
 in white pine blister rust pathosystem, 263-264
Prunus persica L., 8
Prunus sp., dormancy induction, 168
Pseudomonas syringae, 8
Pseudotsuga menziesii, 85,91,93, 99-100,272-273
Purdue University, 17-35
Purshia sp., 284-285

Quantitative trait locus, in *Vaccinium* (blueberry), 37-52
Quercus robur, 165

Red-osier dogwood. *See Cornus sericea* L.
Retanolla sp., 285
Retinoblastoma gene, 172
Rhizobium, 285-286,291-293,294
Rhododendron sp., 8,181-182
Rhus pseudoacacia, 169
Rhus typhina, 179
Ribes nigram, 10-12
Ribes nigrum, 163-164
Robinia pseudoacacia, 244
ROD (red-osier dehydrins), 22-27
Root pruning, 167-168
Rosa hybrida, 164,185

Salix caprea, 81-84
Salix pentandra, 81-84,90,91,92-93, 96-97,98,100,170,174
Salix sp., phytochrome and, 169-173
Salix viminalis, 162,165,175
Sambucus nigra, 244
Seasonal nitrogen cycling, 237-259. *See also* Nitrogen cycling
Seed storage proteins (SSPs), 245-246
Shepherida sp., 285
Shoot apical meristem
 1,3-β-D-glucan in, 129-131
 cell-cell communication at, 118-123
 extracellular matrix, 121-122
 signal exchange pathways, 121-123

signal exchange via plasmodesmata, 123-127
symplasmic space, 122-123
Sinapis alba, 120-121
Solanum tuberosum, 116,119
Sophoroa japonica, 244
South Dakota State University, 201-235
Spearman-Karber method, 221
Spherosomes, 140-144
Spirogyra sp., 183
Supercooling, 12-14
defined, 7
in *Vitis* sp., 216,222
Syringa vulgaris, 179

Talguena sp., 285
Temperature
in dormancy induction, 173-174
in grapevine freezing tolerance, 225-227
(University of) Tennessee Health Science Center, 37-52
Tilia cordata, 81-84
Trevoa sp., 285
Triphenyl tetrazolium chloride (TTC), 220
Triticum aestivum L. (wheat), 8
(University of) Tromsö (Norway), 77-112

Ulmus glabra, 81-84
University of Maryland, 237-259
University of Minnesota, 5-16
University of Plymouth (UK), 5-16
University of Saskatchewan (Canada), 157-199
University of Tennessee Health Science Center, 37-52
University of Tromsö (Norway), 77-112
University of Wisconsin, 5-16
USDA Agricultural Research Service, 37-52

Vaccinium corymbosum L., 41
Vaccinium darrowi × *V. caesariense*, 41
Vaccinium macrocarpon (cranberry), 8-10
Vaccinium sp. (blueberry)
conclusions, 49-50
materials and methods, 41-43
quantitative trait loci (QTL) mapping, 37-52
results and discussion, 43-49
Vegetative storage proteins (VSPs), 240-245
Vitis amurensis, 202,224-225
Vitis labruscana, 208,209,211, 218-219,222-223
Vitis riparia, 169,176,202,211, 214-216,218,224-225,227, 228-229,229
Vitis rotundifolia, 223,225
Vitis solonis × Othello, 227
Vitis sp., 167,168,201-235. *See also individual species*
bud freezing tolerance, 210-217
bud scales in, 182
bud surface moisture, 216-217
bud water content, 212-213
cane freezing tolerance, 208-210
carbohydrate analysis, 212
controlled freeze testing, 219-224, 221-224
deacclimation, 216
defoliation, 228
dormancy induction, 163
environmental influences, 225-227
evaluating freezing response, 217-224
factors influencing freezing tolerance, 224-229
field analysis, 218-219
freezing event analysis, 220-221
freezing tolerance species chart, 205-207
genetic influences, 224-225
infrared thermography, 221
MRI studies, 213-215

root freezing tolerance, 203-204
rootstocks, 228-229
supercooling, 216
terminal growth in, 179
tissue viability analysis, 219-220
training systems, 228
trunk freezing tolerance, 204-208
vine management practices, 227-229
Vitis vinifera, 168,202,209,211,212, 218-219,222-223,224-225
ABA in, 182-183
moisture stress and, 175

Winter phloem, 128
WIN4 vegetative storage protein, 242
(University of) Wisconsin, 5-16

Y dehydrin motif, 26

Zea mays, 164